Rasim Latifovic
613-947-1816

Geir Evensen
Data Assimilation

Geir Evensen

Data Assimilation
The Ensemble Kalman Filter

With 63 Figures

Springer

PROF. GEIR EVENSEN

Hydro Research Centre, Bergen
PO Box 7190
N 5020 Bergen
Norway

and

Mohn-Sverdrup Center for Global Ocean Studies
and Operational Oceanography
at Nansen Environmental and Remote Sensing Center
Thormølensgt 47
N 5600 Bergen
Norway

e-mail:
Geir.Evensen@hydro.com

Library of Congress Control Number: 2006932964

ISBN-10   3-540-38300-0-X   Springer Berlin Heidelberg New York
ISBN-13   978-3-540-38300-0   Springer Berlin Heidelberg New York

This work is subject to copyright. All rights are reserved, whether the whole or part of the material is concerned, specifically the rights of translation, reprinting, reuse of illustrations, recitation, broadcasting, reproduction on microfilm or in any other way, and storage in data banks. Duplication of this publication or parts thereof is permitted only under the provisions of the German Copyright Law of September 9, 1965, in its current version, and permission for use must always be obtained from Springer-Verlag. Violations are liable to prosecution under the German Copyright Law.

Springer is a part of Springer Science+Business Media
springer.com
© Springer-Verlag Berlin Heidelberg 2007

The use of general descriptive names, registered names, trademarks, etc. in this publication does not imply, even in the absence of a specific statement, that such names are exempt from the relevant protective laws and regulations and therefore free for general use.

Cover design: Erich Kirchner
Typesetting: camera-ready by the author
Production: Christine Adolph
Printing: Krips bv, Meppel
Binding: Stürtz AG, Würzburg

Printed on acid-free paper   30/2133/ca 5 4 3 2 1 0

To Tina and Endre

Preface

The aim of this book is to introduce the formulation and solution of the data assimilation problem. The focus is mainly on methods where the model is allowed to contain errors and where the error statistics evolve through time. So-called strong constraint methods and simple methods where the error statistics are constant in time are only briefly explained, and then as special cases of more general weak constraint formulations.

There is a special focus on the Ensemble Kalman Filter and similar methods. These are methods which have become very popular, both due to their simple implementation and interpretation and their properties with nonlinear models.

The book has been written during several years of work on the development of data assimilation methods and the teaching of data assimilation methods to graduate students. It would not have been completed without the continuous interaction with students and colleagues, and I particularly want to acknowledge the support from Laurent Bertino, Kari Brusdal, François Counillon, Mette Eknes, Vibeke Haugen, Knut Arild Lisæter, Lars Jørgen Natvik, and Jan Arild Skjervheim, with whom I have worked closely for several years. Laurent Bertino and François Counillon also provided much of the material for the chapter on the TOPAZ ocean data assimilation system. Contributions from Laurent Bertino, Theresa Lloyd, Gordon Wilmot, Martin Miles, Jennifer Trittschuh-Vallès, Brice Vallès and Hans Wackernagel, on proof-reading parts of the final version of the book are also much appreciated.

It is hoped that the book will provide a comprehensive presentation of the data assimilation problem and that it will serve as a reference and textbook for students and researchers working with development and application of data assimilation methods.

Bergen, June 2006 *Geir Evensen*

Contents

List of symbols xv

1 Introduction 1

2 Statistical definitions 5
   2.1 Probability density function 5
   2.2 Statistical moments 8
      2.2.1 Expected value 8
      2.2.2 Variance 8
      2.2.3 Covariance 9
   2.3 Working with samples from a distribution 9
      2.3.1 Sample mean 9
      2.3.2 Sample variance 10
      2.3.3 Sample covariance 10
   2.4 Statistics of random fields 10
      2.4.1 Sample mean 10
      2.4.2 Sample variance 10
      2.4.3 Sample covariance 11
      2.4.4 Correlation 11
   2.5 Bias 11
   2.6 Central limit theorem 12

3 Analysis scheme 13
   3.1 Scalar case 13
      3.1.1 State-space formulation 13
      3.1.2 Bayesian formulation 15
   3.2 Extension to spatial dimensions 16
      3.2.1 Basic formulation 16
      3.2.2 Euler–Lagrange equation 17
      3.2.3 Representer solution 19
      3.2.4 Representer matrix 20

		3.2.5	Error estimate	20
		3.2.6	Uniqueness of the solution	22
		3.2.7	Minimization of the penalty function..................	23
		3.2.8	Prior and posterior value of the penalty function	24
	3.3	Discrete form		24

4 Sequential data assimilation 27
4.1 Linear Dynamics 27
4.1.1 Kalman filter for a scalar case...................... 28
4.1.2 Kalman filter for a vector state..................... 29
4.1.3 Kalman filter with a linear advection equation 29
4.2 Nonlinear dynamics.. 32
4.2.1 Extended Kalman filter for the scalar case 32
4.2.2 Extended Kalman filter in matrix form 33
4.2.3 Example using the extended Kalman filter 35
4.2.4 Extended Kalman filter for the mean 36
4.2.5 Discussion 37
4.3 Ensemble Kalman filter 38
4.3.1 Representation of error statistics 38
4.3.2 Prediction of error statistics 39
4.3.3 Analysis scheme.................................... 41
4.3.4 Discussion 43
4.3.5 Example with a QG model 44

5 Variational inverse problems 47
5.1 Simple illustration 47
5.2 Linear inverse problem 50
5.2.1 Model and observations 51
5.2.2 Measurement functional 51
5.2.3 Comment on the measurement equation 51
5.2.4 Statistical hypothesis 52
5.2.5 Weak constraint variational formulation 52
5.2.6 Extremum of the penalty function 53
5.2.7 Euler–Lagrange equations 54
5.2.8 Strong constraint approximation 55
5.2.9 Solution by representer expansions................... 56
5.3 Representer method with an Ekman model 57
5.3.1 Inverse problem 58
5.3.2 Variational formulation 58
5.3.3 Euler–Lagrange equations 59
5.3.4 Representer solution 60
5.3.5 Example experiment 61
5.3.6 Assimilation of real measurements 64
5.4 Comments on the representer method 67

6 Nonlinear variational inverse problems 71
6.1 Extension to nonlinear dynamics 71
6.1.1 Generalized inverse for the Lorenz equations 72
6.1.2 Strong constraint assumption 73
6.1.3 Solution of the weak constraint problem 76
6.1.4 Minimization by the gradient descent method 77
6.1.5 Minimization by genetic algorithms 78
6.2 Example with the Lorenz equations 82
6.2.1 Estimating the model error covariance 82
6.2.2 Time correlation of the model error covariance 83
6.2.3 Inversion experiments 84
6.2.4 Discussion 92

7 Probabilistic formulation 95
7.1 Joint parameter and state estimation 95
7.2 Model equations and measurements 96
7.3 Bayesian formulation 97
7.3.1 Discrete formulation 98
7.3.2 Sequential processing of measurements 99
7.4 Summary 101

8 Generalized Inverse 103
8.1 Generalized inverse formulation 103
8.1.1 Prior density for the poorly known parameters 103
8.1.2 Prior density for the initial conditions 104
8.1.3 Prior density for the boundary conditions 104
8.1.4 Prior density for the measurements 105
8.1.5 Prior density for the model errors 105
8.1.6 Conditional joint density 107
8.2 Solution methods for the generalized inverse problem 108
8.2.1 Generalized inverse for a scalar model 108
8.2.2 Euler–Lagrange equations 109
8.2.3 Iteration in α 111
8.2.4 Strong constraint problem 111
8.3 Parameter estimation in the Ekman flow model 113
8.4 Summary 117

9 Ensemble methods 119
9.1 Introductory remarks 119
9.2 Linear ensemble analysis update 121
9.3 Ensemble representation of error statistics 122
9.4 Ensemble representation for measurements 124
9.5 Ensemble Smoother (ES) 124
9.6 Ensemble Kalman Smoother (EnKS) 126
9.7 Ensemble Kalman Filter (EnKF) 129

		9.7.1 EnKF with linear noise free model 129
		9.7.2 EnKS using EnKF as a prior 130
	9.8	Example with the Lorenz equations 131
		9.8.1 Description of experiments 131
		9.8.2 Assimilation Experiment 132
	9.9	Discussion 137

10 Statistical optimization 139
    10.1 Definition of the minimization problem 139
        10.1.1 Parameters 140
        10.1.2 Model 140
        10.1.3 Measurements 140
        10.1.4 Cost function 141
    10.2 Bayesian formalism 141
    10.3 Solution by ensemble methods 142
        10.3.1 Variance minimizing solution 144
        10.3.2 EnKS solution 144
    10.4 Examples 145
    10.5 Discussion 154

11 Sampling strategies for the EnKF 157
    11.1 Introduction 157
    11.2 Simulation of realizations 158
        11.2.1 Inverse Fourier transform 159
        11.2.2 Definition of Fourier spectrum 159
        11.2.3 Specification of covariance and variance 160
    11.3 Simulating correlated fields 162
    11.4 Improved sampling scheme 163
    11.5 Experiments 167
        11.5.1 Overview of experiments 167
        11.5.2 Impact from ensemble size 170
        11.5.3 Impact of improved sampling for the initial ensemble .. 171
        11.5.4 Improved sampling of measurement perturbations...... 171
        11.5.5 Evolution of ensemble singular spectra 173
        11.5.6 Summary 174

12 Model errors 175
    12.1 Simulation of model errors 175
        12.1.1 Determination of ρ 175
        12.1.2 Physical model 176
        12.1.3 Variance growth due to the stochastic forcing 176
        12.1.4 Updating model noise using measurements 180
    12.2 Scalar model 180
    12.3 Variational inverse problem 181
        12.3.1 Prior statistics 181

	12.3.2 Penalty function 182
	12.3.3 Euler–Lagrange equations 182
	12.3.4 Iteration of parameter 182
	12.3.5 Solution by representer expansions................. 183
	12.3.6 Variance growth due to model errors 184
12.4	Formulation as a stochastic model 185
12.5	Examples... 185
	12.5.1 Case A0 186
	12.5.2 Case A1 186
	12.5.3 Case B 189
	12.5.4 Case C 192
	12.5.5 Discussion 193

13 Square Root Analysis schemes 195
13.1 Square root algorithm for the EnKF analysis 195
    13.1.1 Updating the ensemble mean 196
    13.1.2 Updating the ensemble perturbations 196
    13.1.3 Randomization of the analysis update 197
    13.1.4 Final update equation in the square root algorithms ... 200
13.2 Experiments 201
    13.2.1 Overview of experiments 201
    13.2.2 Impact of the square root analysis algorithm 203

14 Rank issues 207
14.1 Pseudo inverse of $\boldsymbol{C}$ 207
    14.1.1 Pseudo inverse 208
    14.1.2 Interpretation..................................... 209
    14.1.3 Analysis schemes using the pseudo inverse of $\boldsymbol{C}$ 209
    14.1.4 Example 209
14.2 Efficient subspace pseudo inversion 212
    14.2.1 Derivation of the subspace pseudo inverse 212
    14.2.2 Analysis schemes based on the subspace pseudo inverse 216
    14.2.3 An interpretation of the subspace pseudo inversion 217
14.3 Subspace inversion using a low-rank $\boldsymbol{C}_{\epsilon\epsilon}$ 218
    14.3.1 Derivation of the pseudo inverse 218
    14.3.2 Analysis schemes using a low-rank $\boldsymbol{C}_{\epsilon\epsilon}$ 219
14.4 Implementation of the analysis schemes 220
14.5 Rank issues related to the use of a low-rank $\boldsymbol{C}_{\epsilon\epsilon}$ 221
14.6 Experiments with $m \gg N$ 224
14.7 Summary... 229

15 An ocean prediction system 231
15.1 Introduction 231
15.2 System configuration and EnKF implementation 232
15.3 Nested regional models................................... 235
15.4 Summary.. 236

16 Estimation in an oil reservoir simulator 239
16.1 Introduction 239
16.2 Experiment 241
   16.2.1 Parameterization 242
   16.2.2 State vector 243
16.3 Results... 245
16.4 Summary.. 248

A Other EnKF issues 249
A.1 Local analysis... 249
A.2 Nonlinear measurements in the EnKF 251
A.3 Assimilation of non-synoptic measurements 253
A.4 Time difference data 254
A.5 Ensemble Optimal Interpolation (EnOI) 255
A.6 Chronology of ensemble assimilation developments 255
   A.6.1 Applications of the EnKF 255
   A.6.2 Other ensemble based filters 264
   A.6.3 Ensemble smoothers 264
   A.6.4 Ensemble methods for parameter estimation 264
   A.6.5 Nonlinear filters and smoothers 265

References... 267

Index 277

List of symbols

a	De-correlation lengths in variogram models (11.2–11.4); Error in initial condition for scalar models (5.5), (5.22), (8.22) and (12.22)
$A(z)$	Vertical diffusion coefficient in Ekman model, Sect. 5.3
$A_0(z)$	First guess of vertical diffusion coefficient in Ekman model, Sect. 5.3
$\boldsymbol{a}$	Error in initial condition for vector models (5.70), (6.8–6.10), (7.2)
$\boldsymbol{A}_i$	Ensemble matrix at time t_i, Chap. 9
$\boldsymbol{A}$	Ensemble matrix, Chap. 9
$\boldsymbol{b}$	Vector of coefficients solved for in the analysis scheme (3.38)
$\boldsymbol{b}(\boldsymbol{x},t)$	Error in boundary condition, Chap. 7
$\boldsymbol{b}_0$	Stochastic error in upper condition of Ekman model, Sect. 5.3
$\boldsymbol{b}_H$	Stochastic error in lower condition of Ekman model, Sect. 5.3
c	Constant in Fourier spectrum (11.10) and (11.14); constant multiplier used when simulating model errors (12.20)
c_i	Multiplier used when simulating model errors (12.54)
c_d	Wind-drag coefficient in Ekman model, Sect. 5.3
c_{d_0}	First guess value of wind-drag coefficient in Ekman model, Sect. 5.3
c_{rep}	Constant multiplier used when modelling model errors in representer method, Chap. 12
$C_{\psi\psi}$	Covariance of a scalar state variable ψ
$C_{c_d c_d}$	Covariance of error in wind-drag c_{d_0}
$C_{AA}(z_1,z_2)$	Covariance of error in vertical diffusion $A_0(z)$
$C_{\psi\psi}(\boldsymbol{x}_1,\boldsymbol{x}_2)$	Covariance of scalar field $\psi(\boldsymbol{x})$ (2.25)

List of symbols

$C_{\psi\psi}$	Covariance of a discrete ψ (sometimes short for $C_{\psi\psi}(\boldsymbol{x}_1,\boldsymbol{x}_2)$)
$C_{\psi\psi}(\boldsymbol{x}_1,\boldsymbol{x}_2)$	Covariance of a vector of scalar state variables $\boldsymbol{\psi}(\boldsymbol{x})$
$C_{\epsilon\epsilon}$	Used for variance of ϵ in Chap. 3
C_{aa}	Scalar initial error covariance
C_{qq}	Model error covariance
$\boldsymbol{C}$	Matrix to be inverted in the ensemble analysis schemes, Chap. 9
$\boldsymbol{C}_{\epsilon\epsilon}$	Covariance of measurement errors $\boldsymbol{\epsilon}$
$\boldsymbol{C}_{\epsilon\epsilon}^{\mathrm{e}}$	Low-rank representation of measurement error covariance
$\boldsymbol{C}_{aa}$	Initial error covariance
$\boldsymbol{C}_{qq}$	Model error covariance
d	Measurement
$\boldsymbol{d}$	Vector of measurements
$\boldsymbol{D}$	Perturbed measurements, Chap. 9
$\boldsymbol{D}_j$	Perturbed measurements at data time j, Chap. 9
$\boldsymbol{E}$	Measurement perturbations, Chap. 9
$f(\boldsymbol{x})$	Arbitrary function, e.g. in (3.55)
$f(\psi)$	Probability density, e.g. $f(\psi)$ or $f(\boldsymbol{\psi})$ where $\boldsymbol{\psi}$ is a vector or a vector of fields
F	Distribution function (2.1)
$g(\boldsymbol{x})$	Arbitrary function, e.g. in (3.55) and (3.59)
G	Model operator for a scalar state; linear (4.1) or nonlinear (4.14)
$\boldsymbol{G}$	Model operator for a vector state; linear (4.11) or nonlinear (4.21), (9.1) and (7.1)
$h()$	Arbitrary function used at different occasions
H	Depth of bottom boundary in Ekman model, Sect. 5.3
$\boldsymbol{h}$	Innovation vector (3.51); spatial distance vector (11.1)
$i(j)$	Time index corresponding to measurement j, Fig. 7.1
$\boldsymbol{I}$	Identity matrix
J	Number of measurement times, Fig. 7.1
$\boldsymbol{k}$	Vertical unit vector $(0,0,1)$ in Ekman model Sect. 5.3; wave number $\boldsymbol{k} = (\kappa,\lambda)$, Chap. 11
k_{h}	Permeability, Chap. 16
$\boldsymbol{K}$	Kalman gain matrix (3.85)
m_j	Total number of measurements at measurement time j, Fig. 7.1
m	Sometimes used as abbreviation for m_j
$m(\psi)$	Nonlinear measurement functional in the Appendix

M	Total number of measurements over the assimilation interval
$\boldsymbol{M}$	Measurement matrix for a discrete state vector (3.76); measurement matrix operator (10.20)
n	Dimension of state vector $n = n_\psi + n_\alpha$, Chap. 9 and 10
n_α	Number of parameters, Chaps. 9 and 10
n_ψ	Dimension of model state, Chaps. 7–10
n_x	Gridsize in x-direction, Chap. 11
n_y	Gridsize in y-direction, Chap. 11
N	Sample or ensemble size
p	Error of first guess of a scalar or scalar field, Chap. 3; matrix rank, Chap. 14; probability (6.24)
p_A	Error of first guess vertical diffusion coefficient in Ekman model, Sect. 5.3
p_{c_d}	Error of first guess wind drag coefficient in Ekman model, Sect. 5.3
P	Reservoir pressure, Chap. 16
q	Stochastic error of scalar model used in Kalman filter formulations
$\boldsymbol{q}(i)$	Discrete model error at time t_i, (6.16)
$\boldsymbol{q}$	Stochastic error of vector model used in Kalman filter formulations
$\boldsymbol{Q}$	Ensemble of model noise, Sect. 11.4
r	De-correlation length in Fourier space (11.10)
r_1	De-correlation length in principal direction in Fourier space (11.11)
r_2	De-correlation length orthogonal to principal direction in Fourier space (11.11)
r_x	De-correlation length in principal direction in physical space (11.23)
r_y	De-correlation length orthogonal to principal direction in physical space (11.23)
$\boldsymbol{r}(\boldsymbol{x},t)$	Vector of representer functions (3.39) and (5.48)
$\boldsymbol{r}$	Matrix of representers (3.80)
$\boldsymbol{R}$	Representer matrix (3.63)
R_s	Gas in a fluid state at reservoir conditions, Chap. 16
R_v	Oil in a gas state at reservoir conditions, Chap. 16
S_w	Water saturation, Chap. 16
S_g	Gas saturation, Chap. 16

xviii List of symbols

S_o	Oil saturation, Chap. 16
$s(\boldsymbol{x}, t)$	Vector of adjoints of representer functions (5.49)
$\boldsymbol{S}_j$	Measurement of ensemble perturbations at data time j, Chap. 9
$\boldsymbol{S}$	Measurement of ensemble perturbations, Chap. 9
t	Time variable
T	Final time of assimilation period for some examples
u	Dependent variable (5.99)
$\boldsymbol{u}(z)$	Horizontal velocity vector in Ekman model, Sect. 5.3
$\boldsymbol{u}_0(z)$	Initial condition for velocity vector in Ekman model, Sect. 5.3
$\boldsymbol{U}$	Left singular vectors from the singular value decomposition, Sect. 11.4 and (14.68)
$\boldsymbol{U}_0$	Left singular vectors from the singular value decomposition (14.19)
$\boldsymbol{U}_1$	Left singular vectors from the singular value decomposition (14.52)
$\boldsymbol{v}$	Dummy vector (5.101)
$\boldsymbol{V}$	Right singular vectors from the singular value decomposition, Sect. 11.4 and (14.68)
$\boldsymbol{V}_0$	Right singular vectors from the singular value decomposition (14.19)
$\boldsymbol{V}_1$	Right singular vectors from the singular value decomposition (14.52)
w_k	Random realization with mean equal to zero and variance equal to one (11.33)
W_{aa}	Inverse of scalar initial error covariance
$\boldsymbol{W}$	Matrix (14.63) and (14.64)
$\boldsymbol{W}_{\psi\psi}(\boldsymbol{x}_1, \boldsymbol{x}_2)$	Functional inverse of $\boldsymbol{C}_{\psi\psi}(\boldsymbol{x}_1, \boldsymbol{x}_2)$, e.g. (3.27)
$\boldsymbol{W}_{aa}(\boldsymbol{x}_1, \boldsymbol{x}_2)$	Functional inverse of initial error covariance
$\boldsymbol{W}_{aa}$	Inverse of initial error covariance
$\boldsymbol{W}_{\eta\eta}$	Smoothing weight (6.19)
$\boldsymbol{W}_{\epsilon\epsilon}$	Matrix inverse of the covariance $\boldsymbol{C}_{\epsilon\epsilon}$
$\boldsymbol{x}$	Independent spatial variable
x_n	x-position in grid $x_n = n\Delta x$, Chap 11
$\boldsymbol{X}_0$	Matrix (14.26) and (14.51)
$\boldsymbol{X}_1$	Matrix (14.30) and (14.55)
$\boldsymbol{X}_2$	Matrix (14.34) and (14.59)
x, y, z	Dependent variables in Lorenz equations (6.5–6.6)

$\boldsymbol{x}$	Dependent variable $\boldsymbol{x}^{\mathrm{T}} = (x, y, z)$ in Lorenz equations
$\boldsymbol{x}_0$	Initial condition $\boldsymbol{x}_0^{\mathrm{T}} = (x_0, y_0, z_0)$ in Lorenz equations
y_m	y-position in grid $y_m = m\Delta y$, Chap 11
$\boldsymbol{Y}$	Matrix (14.65)
$\boldsymbol{Z}$	Matrix of eigenvectors from eigenvalue decomposition
$\boldsymbol{Z}_1$	Matrix of eigenvectors from eigenvalue decomposition (14.27)
$\boldsymbol{Z}_p$	Matrix of p first eigenvectors from eigenvalue decomposition (14.15)
$\mathcal{B}$	Penalty function in measurement space, e.g. $\mathcal{B}[\boldsymbol{b}]$ in (3.66)
$\mathcal{D}$	Model domain
$\partial\mathcal{D}$	Boundary of model domain
$\mathcal{H}$	Hamiltonian, used in hybrid Monte Carlo algorithm (6.25)
$\boldsymbol{\mathcal{H}}$	Hessian operator (second derivative of model operator)
$\mathcal{J}$	Penalty function, e.g. $\mathcal{J}[\boldsymbol{\psi}]$
$\mathcal{M}$	Scalar measurement functional (3.24)
$\boldsymbol{\mathcal{M}}$	Vector of measurement functionals
$\mathcal{N}$	Normal distribution
$\boldsymbol{\mathcal{P}}$	Matrix to be inverted in representer method (3.50)
α_1, α_2	Coefficients used in Chap. 3
α_{ij}	Coefficient used in (16.1
$\boldsymbol{\alpha}(\boldsymbol{x})$	Poorly known model parameters to be estimated, Chap. 7
$\boldsymbol{\alpha}'(\boldsymbol{x})$	Errors in model parameters, Chaps. 7 and 10
β	Coefficient in Lorenz equations (6.7); constant (also β_{ini} and β_{mes}), Sect. 11.4
$\delta\psi$	Variation of ψ
ϵ	Real measurement error, Chap. 3
$\boldsymbol{\epsilon}_{\mathcal{M}}$	Representation errors in measurement operator, Sect. 5.2.3
$\boldsymbol{\epsilon}_d$	Actual measurement errors, Sect. 5.2.3
$\boldsymbol{\epsilon}$	Random or real measurement errors, Sect. 5.2.3
η	Smoothing operators used in gradient method (6.19)
γ	Constant used in smoothing norm analysis (6.32); step length (6.22)
$\gamma(\boldsymbol{h})$	Variogram (11.1)
$\kappa_2(\boldsymbol{A})$	Condition number, Chap. 11
κ_l	Wave number in x direction, Chap. 11
λ_p	Wave number in y direction, Chap. 11
λ	Eigenvalue (13.8); scalar adjoint variable (5.37)

xx   List of symbols

$\boldsymbol{\lambda}$	Vector adjoint variable
$\boldsymbol{\Lambda}$	Diagonal matrix of eigenvalues from eigenvalue decomposition
$\boldsymbol{\Lambda}_1$	Diagonal matrix of eigenvalues from eigenvalue decomposition (14.26)
$\boldsymbol{\Lambda}_p$	Diagonal matrix of eigenvalues from eigenvalue decomposition (14.14)
μ	Sample mean (2.20)
$\mu(\boldsymbol{x})$	Sample mean (2.23)
ω	Frequency variable (6.33)
ω_i	Unit variance noise process (8.10)
$\boldsymbol{\Omega}$	Error covariance of ω_i noise process (8.12)
ϕ	Scalar variable, Chap. 2
$\phi(\boldsymbol{x})$	Porosity in Chap. 16
$\phi_{l,p}$	Uniform random number (11.10)
Φ	Random scalar variable, Chap. 2
π	3.1415927
$\boldsymbol{\pi}$	Momentum variable used in hybrid Monte Carlo algorithm (6.25)
ψ	Scalar state variable (has covariance $C_{\psi\psi}$)
$\psi(\boldsymbol{x})$	Scalar state variable field (has error covariance $\boldsymbol{C}_{\psi\psi}(\boldsymbol{x}_1, \boldsymbol{x}_2)$)
$\widehat{\psi}(\boldsymbol{k})$	Fourier transform of $\psi(\boldsymbol{x})$, Chap. 11
$\boldsymbol{\psi}$	Vector state variable, e.g. from a discretized $\psi(\boldsymbol{x})$ (has error covariance $\boldsymbol{C}_{\psi\psi}$)
$\boldsymbol{\psi}(\boldsymbol{x})$	Vector of scalar state variables (has error covariance $\boldsymbol{C}_{\psi\psi}(\boldsymbol{x}_1, \boldsymbol{x}_2)$)
Ψ	Random scalar variable, Chap. 2
Ψ_0	Best guess initial condition for dynamical scalar models, may be function of $\boldsymbol{x}$
$\boldsymbol{\psi}(\boldsymbol{x})$	Vector of fields, sometimes written just $\boldsymbol{\psi}$
$\boldsymbol{\psi}_0$	Estimate of initial condition $\boldsymbol{\Psi}_0$, Chap. 7
$\boldsymbol{\psi}_b$	Estimate of boundary condition $\boldsymbol{\Psi}_b$, Chap. 7
$\boldsymbol{\Psi}$	Combined state vector, Chap. 10
$\boldsymbol{\Psi}_0$	Best guess initial condition
$\boldsymbol{\Psi}_b$	Best guess boundary condition, Chap. 7
ρ	Correlation parameter (11.33); coefficient in Lorenz equations (6.6)
$\boldsymbol{\Sigma}$	Matrix of singular values from the singular value decomposition, Sect. 11.4 and (14.68)

$\boldsymbol{\Sigma}_0$	Matrix of singular values from the singular value decomposition (14.19)
$\boldsymbol{\Sigma}_1$	Matrix of singular values from the singular value decomposition (14.52)
σ	Standard deviation defined in Chap. 2; used as coefficient in Lorenz equations (6.5); singular values, Sect. 11.4, Chap. 13 and Chap. 14
τ	De-correlation time (12.1)
θ	Pseudo temperature variable used in simulated annealing algorithm; rotation of principal direction (11.11)
Θ	Random rotation used in SQRT analysis scheme, Chap. 13
ξ	Random number used in Metropolis algorithm
$\boldsymbol{\xi}$	Coordinate running over boundary of model domain
$\mathbf{1}_N$	$N \times N$ matrix with all elements equal to 1
$\delta()$	Dirac delta function (3.24)
$\boldsymbol{\delta}_{\psi_i}$	Vector used to extract a component of the state vector, Chap. 7
$E[]$	Expectation operator
$\mathcal{O}()$	Order of magnitude function
$\Re$	Space of real numbers, e.g. $\Re^{n \times m}$ for a real $n \times m$ matrix

1
Introduction

Does the solution of a dynamical model with conditions have any statistical meaning or scientific purpose?

A model consists of a number of mathematical equations which are defined to represent the interaction between various variables through certain physical processes. In many cases the model excludes several processes or scales which are believed to have less importance for the applications at hand. Even if the model is a perfect representation of reality, its solution will not describe reality unless we have perfect knowledge about the initial and boundary conditions which are often difficult to prescribe with high accuracy.

From a single model integration we obtain a solution or realization without knowledge about its uncertainty. In fact, the model solution is just one out of infinitively many equally likely realizations. Thus, we should really consider the time evolution of the probability density function (pdf) for the model state. With knowledge of the pdf for the model state we can extract information about the most likely estimate of the model state as well as its uncertainty.

In many applications we have an approximate dynamical model with uncertain estimates of initial and boundary conditions. In addition we may have measurements of the model solution collected at different space and time locations. The computation of the pdf of the model solution conditioned on the measured observations defines the data assimilation or inverse problem considered in the following chapters.

The accurate representation of the full pdf becomes extremely expensive for high dimensional simulation models. Thus, data assimilation and inverse methods must normally represent the pdf using statistical moments or an ensemble of model states and then search for estimators such as the mean and maximum likelihood with the associated covariance representing uncertainty.

There is now a large class of different data assimilation and inverse methods which for practical and computational efficiency implement different statistical and conceptual approximations. The different methods have different properties which may depend on the dynamical system to which they are applied. Some methods will work well with linear dynamics but be completely useless for nonlinear dynamics. Other methods may handle nonlinearity well but computational requirements limit their use to low dimensional dynamical systems.

Parameter estimation in dynamical models is a field of research which has developed side by side with the developments in data assimilation. Traditionally one searches for a set of parameters in the model which results in a model solution that is consistent with a set of measurements. The methods used are in many cases strongly related to traditional data assimilation methods. Still there has been limited communication between the two communities. Statements like "one should not fiddle with model parameters but focus on the estimation of the state" has been followed by statements similar to "state estimation does not provide any scientific knowledge, what matters is to identify the parameters". So who should we trust?

This book aims to explain the fundamental data assimilation and inverse problem and the derivation and properties of the various methods which can be used to solve it. It may serve as a text book for students who take an introductory course in data assimilation and inverse methods, but is also intended as a reference book on the interpretation and implementation of advanced ensemble methods. The book has been organized with fairly basic discussions of traditional sequential and variational assimilation methods in the first chapters. This is followed by a more elaborate discussion of the fully nonlinear combined state and parameter estimation problem while the final part of the book is giving an extensive discussion on the practical implementation of ensemble methods.

Note also that much of the code used in the ensemble Kalman filter experiments is available from the EnKF home page:

$$\text{http://enkf.nersc.no,}$$

together with other information which is useful for the implementation of the EnKF.

The outline of the book is the following:

Chap. 2 summarizes basic statistical notation. This is just meant to be a quick reference and it does not give a complete introduction to the subject.

In Chap. 3 we consider the time independent inverse problem; i.e. given a first guess of a variable or model state and a set of measurements, what is the best estimate of the state given the prior estimate and the measurements. A linear unbiased variance minimizing analysis scheme is derived and shown to be the optimal solution as long as the prior error statistics are Gaussian.

In Chap. 4 we introduce the time evolution of the model state through a dynamical model and show how this problem can be solved using the Kalman Filter (KF), the Extended Kalman Filter (EKF) and the Ensemble Kalman Filter (EnKF). The methods rely on the analysis scheme derived in the previous chapter, and differ in the representation of error statistics and how this evolve in time. Simple examples are used to illustrate the properties of the methods and we indicate issues related to the use of the methods with nonlinear dynamics.

Chap. 5 introduces the variational inverse problem. It discusses the implications of using the model as a weak or strong constraint but focus on the solution of the weak constraint problem. The Euler–Lagrange equations are derived and it is shown how they can be solved using the representer method.

In Chap. 6 the nonlinear variational inverse problem is considered. This may alternatively be solved using substitution methods like gradient descent. The different methods are used in simple examples which illustrate the properties of the nonlinear variational inverse problem.

Then in Chap. 7 we reformulate the data assimilation or inverse problem as a combined state and parameter estimation problem using Bayesian statistics. A fundamental result from this chapter is that if measurements at different times are independent, they can be processed sequentially in time. Thus, the Bayesian problem becomes a sequence of Bayesian subproblems. This result is exploited when deriving sequential data assimilation algorithms for the nonlinear assimilation problem in the following chapters.

In Chap. 8 the generalized inverse formulation is derived from the Bayesian formulation, and it is shown that the solution becomes the maximum likelihood estimator of the joint conditional pdf. Further, Euler–Lagrange equations for the generalized inverse are derived and they include the parameter estimation case which is solved in a simple illustration.

The ensemble methods are rederived in Chap. 9 starting from the Bayesian formulation. This leads to the Ensemble Smoother (ES) and the Ensemble Kalman Smoother (EnKS) as ensemble methods for solving the generalized inverse problem. The ensemble Kalman filter (EnKF) is then derived as a special case of the EnKS where information is only carried forward in time. Finally, the ensemble methods are examined in an example with the chaotic Lorenz equations.

In Chap. 10 a simple but nonlinear optimization problem is considered. It is shown that it can be solved using a statistical minimization method based on the EnKS. This example illustrates the impact of non-Gaussian statistics in the ensemble analysis update.

Chap. 11 discusses is some detail issues related to the sampling of ensemble realizations. It presents a simple methodology for generating random realizations of smooth pseudo-random fields with anisotropic covariance structure. An improved sampling scheme is presented which can be used to generate an ensemble with better rank properties, and experiments are presented which demonstrate the impact of ensemble size and the improved sampling algorithm.

Chap. 12 discusses the use of model errors and in particular the case of time correlated model errors. It includes a simple example illustrating how model errors, as well as model bias and model parameters, can be estimated using the EnKF and EnKS.

Recently developed square root schemes, which avoid the perturbation of measurements, are discussed in Chap. 13. The derivation of the square root schemes is discussed and it is shown that an additional randomization of the ensemble updates is still required. The square root schemes are evaluated and compared with results from the original EnKF scheme.

In Chap. 14 we discuss how it is possible to consistently compute the inversion in the different analyses schemes when the number of measurements is much larger than the number of ensemble members. This discussion leads to the final form of the EnKF analysis scheme where different pseudo-inversion schemes can be used in combination with either the traditional analysis update or the square root analysis. In particular the development of a sub-space inversion has lead to a very efficient algorithm which is useful even with very large data sets.

An operational ocean prediction system, which is based on the EnKF, is presented in Chap. 15. The purpose is to illustrate what is really possible today using state of the art ocean circulation models together with advanced data assimilation schemes.

Another application, based on a reservoir simulation model, is given in Chap. 16, where both the model state and model parameters are estimated using the EnKF.

Finally, in the Appendix, some special issues related to the practical implementation of the EnKF, as well as ensemble methods in general, are given. This involves the use of a local analysis update, the use of nonlinear and non-synoptic measurements and the use of so-called time difference data. In addition a chronological listing of previous developments related to the EnKF and other ensemble methods is included.

2
Statistical definitions

Basic statistical definitions which will be used in the following chapters are explained. The following is only meant to be a quick reference on statistical notation and definitions and more elaborate textbooks can be used for a comprehensive introduction to the subject.

2.1 Probability density function

Given a continuous random variable Ψ, we can associate a *distribution function* $F(\psi)$. This is also named the cumulative density function or probability distribution function, and it describes the probability that a realization of Ψ takes a value less than or equal to ψ. We can relate it to a continuous probability density function $f(\psi)$, through

$$F(\psi) = \int_{-\infty}^{\psi} f(\psi')d\psi', \qquad (2.1)$$

thus $f(\psi)$, when it exists, is just the derivative of the distribution function

$$f(\psi) = \frac{\partial F(\psi)}{\partial \psi}. \qquad (2.2)$$

The probability density function (pdf) gives the probability that a random variable Ψ will take a particular value ψ. If a probability distribution has density $f(\psi)$, then the infinitesimal interval $(\psi, \psi + d\psi)$ has probability $f(\psi)d\psi$.

The pdf must satisfy the conditions

$$f(\psi) \geq 0 \quad \text{for all } \psi, \qquad (2.3)$$

which states that the probability for Ψ to take a value ψ, must be positive or zero, and

$$\int_{-\infty}^{\infty} f(\psi)d\psi = 1, \tag{2.4}$$

that is, the probability of finding Ψ in the space of real numbers $\Re^1$, is equal to one.

Further, given $f(\psi)$, the probability that ψ takes a value in the interval $[\psi_a, \psi_b]$ is

$$\Pr(\Psi \in [\psi_a, \psi_b]) = \int_{\psi_a}^{\psi_b} f(\psi)d\psi. \tag{2.5}$$

The most common and useful distribution is the one called the *normal or Gaussian distribution*. It is defined by its *mean* and *variance* and has a bell shaped or Gaussian form. It represents a family of distributions of the same general form, characterized by their mean μ, and the variance σ^2. The *standard normal distribution* is a normal distribution with a mean of zero and a variance of one. The normal distribution has the pdf

$$f(\psi) = \frac{1}{\sigma\sqrt{2\pi}} \exp\left(-\frac{(\psi - \mu)^2}{2\sigma^2}\right). \tag{2.6}$$

A convenient aspect of a normal population distribution is that the following empirical "rule of thumb" can be applied to the data: $\mu \pm \sigma$ spans approximately 68% of the realizations, $\mu \pm 2\sigma$ spans approximately 95% of the realizations, and $\mu \pm 3\sigma$ spans about 99% of the realizations.

The *joint pdf* describes the probability of two events together. Given two random variables Ψ and Φ we can define the joint pdf $f(\psi, \phi)$.

The *conditional pdf* describes the probability of some event Ψ, assuming the event Φ. The conditional pdf is denoted $f(\psi|\phi)$ which is read as the pdf for Ψ given Φ. It is often called the *posterior pdf*.

The *marginal pdf* is the pdf of one event, ignoring any information about the other event. It is obtained by integrating the joint pdf over the ignored event; e.g. the marginal pdf for Ψ is $f(\psi) = \int_{-\infty}^{\infty} f(\psi, \phi)d\phi$.

We also have that

$$f(\psi|\phi) = \frac{f(\psi, \phi)}{f(\phi)}, \tag{2.7}$$

or equivalently

$$f(\psi, \phi) = f(\psi|\phi)f(\phi) = f(\phi|\psi)f(\psi). \tag{2.8}$$

The variables Ψ and Φ are said to be independent if $f(\psi, \phi) = f(\psi)f(\phi)$.

From 2.8 we can write

$$f(\psi|\phi) = \frac{f(\psi)f(\phi|\psi)}{f(\phi)}. \tag{2.9}$$

This is Bayes' theorem which is a general result in probability theory giving the conditional probability distribution of a random variable Ψ given Φ in terms of the conditional probability distribution of variable Φ given Ψ, often named the *likelihood*, and the marginal probability distribution of Ψ alone. In the context of Bayesian probability theory, the marginal probability distribution of Ψ alone is usually called the *prior* probability distribution or simply the prior. The conditional distribution of Ψ given the "data" Φ is called the *posterior* probability distribution or just the posterior. This is a general result and will be used extensively in the following chapters.

In this book we will in several occasions refer to and use Bayesian statistics to derive and explain data assimilation methods and their properties. In particular we will use a probability density function $f(\boldsymbol{\psi})$, for the event $\boldsymbol{\psi} \in \Re^n$. This is again related to the distribution function $F(\boldsymbol{\psi})$, of the random variable $\boldsymbol{\Psi} \in \Re^n$, through the equation

$$F(\psi_1, \ldots, \psi_n) = \int_{-\infty}^{\psi_1} \cdots \int_{-\infty}^{\psi_n} f(\psi_1', \ldots, \psi_n') d\psi_1' \ldots d\psi_n', \qquad (2.10)$$

and the pdf is again defined as the derivative of the distribution function.

The pdf is a positive function of dimension n and it has the property that

$$\int_{-\infty}^{\infty} \cdots \int_{-\infty}^{\infty} f(\psi_1, \ldots, \psi_n) d\psi_1 \ldots d\psi_n = 1. \qquad (2.11)$$

Thus, the probability that $\boldsymbol{\psi}$ is located somewhere in $\Re^n$ is one. For each value of $\boldsymbol{\psi}$, $f(\boldsymbol{\psi})$ gives the probability for this particular state. The pdf $f(\boldsymbol{\psi})$ is also named the joint pdf for $(\psi_1, \ldots, \psi_n)$.

This joint pdf can be factorized into

$$f(\psi_1, \ldots, \psi_n) = f(\psi_1) f(\psi_2|\psi_1) f(\psi_3|\psi_1, \psi_2) \cdots f(\psi_n|\psi_1, \ldots, \psi_{n-1}). \qquad (2.12)$$

Here $f(\psi_2|\psi_1)$ is the likelihood of ψ_2 given ψ_1, and if $n = 2$ we get just $f(\psi_1, \psi_2) = f(\psi_1) f(\psi_2|\psi_1)$, which is interpreted as the probability of ψ_1 times the likelihood of ψ_2 given ψ_1.

If the events, $(\psi_1, \ldots, \psi_n)$ are independent we can write

$$f(\psi_1, \ldots, \psi_n) = f(\psi_1) f(\psi_2) \cdots f(\psi_n). \qquad (2.13)$$

We will make frequent use of the pdf of a model state $\boldsymbol{\psi}$, and the likelihood function for a vector of measurements $\boldsymbol{d}$, of the state which is written as $f(\boldsymbol{d}|\boldsymbol{\psi})$. The joint pdf of the state and the measurements can be written

$$f(\boldsymbol{\psi}, \boldsymbol{d}) = f(\boldsymbol{\psi}) f(\boldsymbol{d}|\boldsymbol{\psi}) = f(\boldsymbol{d}) f(\boldsymbol{\psi}|\boldsymbol{d}), \qquad (2.14)$$

and we must have

$$f(\boldsymbol{\psi}|\boldsymbol{d}) = \frac{f(\boldsymbol{\psi}) f(\boldsymbol{d}|\boldsymbol{\psi})}{f(\boldsymbol{d})}, \qquad (2.15)$$

where the denominator is just the integral of the numerator, which normalizes the numerator such that the expression integrates to one. This is Bayes' theorem, and in this context it states that the pdf of the model state given a set of measurements is proportional to the pdf of the model state times the likelihood function for the measurements.

2.2 Statistical moments

The probability density function $f(\psi)$, contains a huge amount of information, especially for high dimensional systems, and actually much more information than is normally needed. Instead of working with the full density it is often convenient to define statistical moments of the density. These are defined from the general expression of the expected value of a function $h(\Psi)$,

$$E[h(\Psi)] = \int_{-\infty}^{\infty} h(\psi) f(\psi) d\psi. \tag{2.16}$$

2.2.1 Expected value

The expected value of a random variable Ψ with distribution $f(\psi)$, is defined as

$$\mu = E[\Psi] = \int_{-\infty}^{\infty} \psi f(\psi) d\psi. \tag{2.17}$$

The expected value (or expectation) of a random variable represents the average one "expects" if an infinite number of samples are drawn from the distribution. Note that the value itself may not be expected in the general sense, it may be unlikely or even impossible, dependent on the shape of $f(\psi)$.

2.2.2 Variance

If Ψ is a random variable, the variance is given by

$$\begin{aligned} \sigma^2 = E\left[(\Psi - E[\Psi])^2\right] &= \int_{-\infty}^{\infty} (\psi - E[\Psi])^2 f(\psi) d\psi \\ &= E[\Psi^2] - E[\Psi]^2. \end{aligned} \tag{2.18}$$

That is, it is the expected value of the square of the deviation of Ψ from its own mean. In other words, it is the average of the square of the distance of each data point from the mean. It is thus the mean squared deviation. The second line in 2.18 is often used for the practical computation of the variance. It is just the second moment minus the square of the first moment.

An inconvenience is that the variance has a unit which is the square of the data unit. For this reason it is common to use the square root of the variance which is named the *standard deviation*, denoted σ. It can also easily be shown that the variance does not depend on the mean, thus the variance of $\Psi + b$ is the same as the variance of Ψ. On the other hand the variance of $a\Psi$ is $a^2\sigma^2$.

2.2.3 Covariance

Given two random variables Ψ and Φ and their respective probability density functions $f(\psi)$ and $f(\phi)$, from which we can define the joint probability $f(\psi,\phi) = f(\psi|\phi)f(\phi) = f(\phi|\psi)f(\psi)$, their covariance is defined as

$$
\begin{aligned}
E\Big[\big(\Psi - E[\Psi]\big)\big(\Phi - E[\Phi]\big)\Big] & \\
&= \iint_{-\infty}^{\infty} (\psi - E[\Psi])(\phi - E[\Phi])f(\psi,\phi)d\psi d\phi \quad (2.19)\\
&= \iint_{-\infty}^{\infty} \psi\phi f(\psi,\phi)d\psi d\phi - E[\Psi]E[\Phi].
\end{aligned}
$$

Note that the same conditions (2.3) and (2.4) also apply for $f(\psi,\phi)$. In the case when the random variables Ψ and Φ are independent, $f(\psi,\phi) = f(\psi)f(\phi)$ and the covariance becomes zero.

2.3 Working with samples from a distribution

Clearly when the dimension of a probability function increases to more than about 3–4 it becomes very impractical, if not impossible, to evaluate the integrals by numerical integration on a regular grid. Suppose the dimension is 10 and we need 10 grid points in each direction to have a proper representation of the density. A grid with 10^{10} nodes would then have to be stored which would require 40 Giga bytes of storage and 10^{10} additions would be needed to calculate the integral.

Fortunately there is an alternative to the direct numerical integration which often works very well even for high dimensional systems. The approach is called the Markov Chain Monte Carlo (MCMC) methods, (see e.g. *Robert and Casella*, 2004), and assumes that we have available a large number N, of realizations from the distribution $f(\psi)$.

2.3.1 Sample mean

Having a sample of independent realizations from the distribution $f(\psi)$, i.e. ψ_i, for $i = 1, N$, then the sample mean $\overline{\psi}$, is given by

$$\mu = E[\psi] \simeq \overline{\psi} = \frac{1}{N}\sum_{i=1}^{N}\psi_i. \quad (2.20)$$

The "expected value" terminology is meant to connote that $E[\Psi]$ is, in some sense, the "best guess" as to the possible outcome of Ψ, or said in another way; the expected value is the value we expect to obtain if infinitely many data are present, and the sample mean of these is computed. This is a reason why $E[\Psi]$ is often called the mean of Ψ.

2.3.2 Sample variance

The variance can be calculated from the formula

$$\sigma^2 = E\left[\left(\Psi - E[\Psi]\right)^2\right]$$
$$\simeq \overline{\left(\psi - \overline{\psi}\right)^2} = \frac{1}{N-1} \sum_{i=1}^{N} \left(\psi_i - \overline{\psi}\right)^2, \qquad (2.21)$$

where the denominator $N-1$ is used instead of N to ensure that the formula (2.21) becomes an unbiased estimator for the variance.

2.3.3 Sample covariance

The covariance can be calculated from the formula

$$\mathrm{Cov}(\psi, \phi) = E\left[\left(\Psi - E[\Psi]\right)\left(\Phi - E[\Phi]\right)\right]$$
$$\simeq \overline{\left(\psi - \overline{\psi}\right)\left(\phi - \overline{\phi}\right)} = \frac{1}{N-1} \sum_{i=1}^{N} \left(\psi_i - \overline{\psi}\right)\left(\phi_i - \overline{\phi}\right). \qquad (2.22)$$

2.4 Statistics of random fields

Of special interest for us will be the statistics of so-called random fields $\Psi(\boldsymbol{x})$ where Ψ is now a function of $\boldsymbol{x} = (x, y, z, \ldots)$.

2.4.1 Sample mean

Having an ensemble of independent samples from the distribution $f(\psi(\boldsymbol{x}))$, i.e. $\psi_i(\boldsymbol{x})$, for $i = 1, N$, then the sample mean is given by

$$\mu(\boldsymbol{x}) \simeq \overline{\psi(\boldsymbol{x})} = \frac{1}{N} \sum_{i=1}^{N} \psi_i(\boldsymbol{x}). \qquad (2.23)$$

2.4.2 Sample variance

The sample variance of an ensemble of independent samples from the distribution $f(\psi(\boldsymbol{x}))$, is given as

$$\sigma^2(\boldsymbol{x}) \simeq \overline{\left(\psi(\boldsymbol{x}) - \overline{\psi(\boldsymbol{x})}\right)^2} = \frac{1}{N-1} \sum_{i=1}^{N} \left(\psi_i(\boldsymbol{x}) - \overline{\psi(\boldsymbol{x})}\right)^2. \qquad (2.24)$$

2.4.3 Sample covariance

The covariance between two different locations x_1 and x_2 for the random fields are given by

$$C_{\psi\psi}(x_1, x_2) \simeq \overline{\bigl(\psi(x_1) - \overline{\psi(x_1)}\bigr)\bigl(\psi(x_2) - \overline{\psi(x_2)}\bigr)}$$
$$= \frac{1}{N-1} \sum_{j=1}^{N} \bigl(\psi_j(x_1) - \overline{\psi(x_1)}\bigr)\bigl(\psi_j(x_2) - \overline{\psi(x_2)}\bigr). \tag{2.25}$$

Note that if $x_1 = x_2$, then (2.25) reduces to the definition of variance.

The covariance of Ψ between the two locations x_1 and x_2 defines how values of Ψ, at different locations, are "varying together" or "covarying". For example, if the random fields Ψ are smooth we will expect that neighboring points are correlated or covarying. The covariance can therefore be a measure of smoothness.

2.4.4 Correlation

The correlation between the random variables $\Psi(x_1)$ and $\Psi(x_2)$ is defined by

$$\operatorname{Cor}\bigl(\psi(x_1), \psi(x_2)\bigr) = \frac{C(x_1, x_2)}{\sigma(x_1)\sigma(x_2)}. \tag{2.26}$$

Thus, the correlation is just a normalized covariance.

2.5 Bias

One meaning is involved in what is called a biased sample; if some elements are more likely to be chosen in the sample than others, and those have a higher/lower value of the quantity being estimated, the outcome will be higher/lower than the true value.

Another kind of bias in statistics does not involve biased samples, but rather the use of a statistics whose average value differs from the value of the quantity being estimated. Suppose we are trying to estimate the true value ψ^t of a parameter ψ using an estimator $\hat{\psi}$ (that is, some function of the observed data). Then the bias of $\hat{\psi}$ is defined to be

$$E\bigl[\hat{\psi}\bigr] - \psi^t. \tag{2.27}$$

In words, this would be "the expected value of the estimator $\hat{\psi}$ minus the true value ψ^t". This may be rewritten as

$$E\bigl[\hat{\psi} - \psi^t\bigr], \tag{2.28}$$

which would read "the expected value of the difference between the estimator and the true value".

An example of a biased estimator of variance is

$$\sigma_{\text{biased}}^2 = \frac{1}{N} \sum_{i=1}^{N} (\psi_i - \overline{\psi})^2, \qquad (2.29)$$

which differs from the formula (2.21) by the division by N rather than $N-1$. The proof that this is a biased estimator of the variance is left as an exercise.

2.6 Central limit theorem

The central limit theorem can be used to say something about the convergence of the moments of a sample with increasing sample size.

Assume that we draw a number of samples of the random variable Ψ, each with sample size N. We then have the following:

- The sample mean $\mu(\psi)$ from (2.23), computed from the different samples is normally distributed, independent of the distribution for Ψ.
- The standard deviation of $\mu(\psi)$ as computed from the different samples tends towards $\sigma(\Psi)/\sqrt{N}$.

Thus, if we compute the sample mean from a given sample, we can expect that the error in the computed sample mean is normally distributed and given by $\sigma(\Psi)/\sqrt{N}$. Importantly, the error decreases proportional to $1/\sqrt{N}$.

The amazing and counter-intuitive property of the central limit theorem is that no matter what the shape of the original distribution, the sampling distribution of the mean approaches a normal distribution. Furthermore, for most distributions, a normal distribution is approached very quickly as N increases.

3
Analysis scheme

This chapter discusses the problem of how to combine a model prediction of a state variable at a given time with a set of measurements available at this particular time. It is assumed that error statistics of the model prediction as well as the measurements are known and characterized by the respective error covariances. Based on this information the so-called analysis scheme used in linear data assimilation methods is presented in some detail. First the theory is derived for the scalar case and then it is extended to the case with a spatial dimension. An extensive analysis of the properties of the analysis scheme is given and this introduces notation and concepts which are also valid for the time dependent problems treated in the following chapters.

3.1 Scalar case

We start by deriving the optimal linear and unbiased estimator for a scalar state variable combined with a single measurement.

3.1.1 State-space formulation

Given two different estimates of the true state ψ^{t} (e.g. a temperature at a particular location and time):

$$\psi^{\text{f}} = \psi^{\text{t}} + p^{\text{f}}, \tag{3.1}$$
$$d = \psi^{\text{t}} + \epsilon, \tag{3.2}$$

where ψ^{f} may be a model forecast or a first-guess estimate and d is a measurement of ψ^{t}. The term p^{f} denotes the unknown error in the forecast and ϵ is the unknown measurement error. The problem is now, to find an improved analyzed estimate ψ^{a} of ψ^{t}. Thus, additional information about the error terms must be supplied and we make the following assumptions:

$$\overline{p^{\mathrm{f}}} = 0, \qquad\qquad \overline{(p^{\mathrm{f}})^2} = C^{\mathrm{f}}_{\psi\psi},$$
$$\overline{\epsilon} = 0, \qquad\qquad \overline{(\epsilon)^2} = C_{\epsilon\epsilon}, \qquad (3.3)$$
$$\overline{\epsilon p^{\mathrm{f}}} = 0.$$

Here the overbar denotes ensemble averaging or expected value.

We now seek a linear estimator
$$\psi^{\mathrm{a}} = \psi^{\mathrm{t}} + p^{\mathrm{a}} = \alpha_1 \psi^{\mathrm{f}} + \alpha_2 d, \qquad (3.4)$$

where we define
$$\overline{p^{\mathrm{a}}} = 0, \qquad \overline{(p^{\mathrm{a}})^2} = C^{\mathrm{a}}_{\psi\psi}. \qquad (3.5)$$

The definition (3.5) means that we assume that the error p^{a}, in the analyzed estimate is unbiased. Thus, the analyzed estimate itself becomes an unbiased estimate of the true state ψ^{t}, i.e. $\overline{\psi^{\mathrm{a}}} = \psi^{\mathrm{t}}$.

Inserting the estimates (3.1) and (3.2) in (3.4) we get
$$\psi^{\mathrm{t}} + p^{\mathrm{a}} = \alpha_1(\psi^{\mathrm{t}} + p^{\mathrm{f}}) + \alpha_2(\psi^{\mathrm{t}} + \epsilon). \qquad (3.6)$$

The expectation of this equation is
$$\psi^{\mathrm{t}} = \alpha_1 \psi^{\mathrm{t}} + \alpha_2 \psi^{\mathrm{t}} = (\alpha_1 + \alpha_2)\psi^{\mathrm{t}}. \qquad (3.7)$$

Thus, we must have
$$\alpha_1 + \alpha_2 = 1, \qquad \text{or} \qquad \alpha_1 = 1 - \alpha_2, \qquad (3.8)$$

and a linear unbiased estimator for ψ^{t} is given as
$$\begin{aligned} \psi^{\mathrm{a}} &= (1 - \alpha_2)\psi^{\mathrm{f}} + \alpha_2 d \\ &= \psi^{\mathrm{f}} + \alpha_2(d - \psi^{\mathrm{f}}). \end{aligned} \qquad (3.9)$$

Using (3.1), (3.2) and (3.4) in this equation gives an expression for the error in the analysis
$$p^{\mathrm{a}} = p^{\mathrm{f}} + \alpha_2(\epsilon - p^{\mathrm{f}}). \qquad (3.10)$$

The error variance is then using (3.3)
$$\begin{aligned} \overline{(p^{\mathrm{a}})^2} = C^{\mathrm{a}}_{\psi\psi} &= \overline{(p^{\mathrm{f}} + \alpha_2(\epsilon - p^{\mathrm{f}}))^2} \\ &= \overline{(p^{\mathrm{f}})^2 + 2\alpha_2 p^{\mathrm{f}}(\epsilon - p^{\mathrm{f}}) + \alpha_2^2 \epsilon^2 - 2\epsilon p^{\mathrm{f}} + (p^{\mathrm{f}})^2} \\ &= C^{\mathrm{f}}_{\psi\psi} - 2\alpha_2 C^{\mathrm{f}}_{\psi\psi} + \alpha_2^2(C_{\epsilon\epsilon} + C^{\mathrm{f}}_{\psi\psi}), \end{aligned} \qquad (3.11)$$

and the minimum variance is defined by
$$\frac{dC^{\mathrm{a}}_{\psi\psi}}{d\alpha_2} = -2C^{\mathrm{f}}_{\psi\psi} + 2\alpha_2(C_{\epsilon\epsilon} + C^{\mathrm{f}}_{\psi\psi}) = 0. \qquad (3.12)$$

Solving for α_2 gives

$$\alpha_2 = \frac{C^{\mathrm{f}}_{\psi\psi}}{C_{\epsilon\epsilon} + C^{\mathrm{f}}_{\psi\psi}}, \qquad (3.13)$$

and the analyzed estimate becomes

$$\psi^{\mathrm{a}} = \psi^{\mathrm{f}} + \frac{C^{\mathrm{f}}_{\psi\psi}}{C_{\epsilon\epsilon} + C^{\mathrm{f}}_{\psi\psi}} \left(d - \psi^{\mathrm{f}}\right). \qquad (3.14)$$

Further, the error variance of the analyzed estimate is now from (3.11) and (3.13)

$$\begin{aligned}
C^{\mathrm{a}}_{\psi\psi} &= C^{\mathrm{f}}_{\psi\psi} - 2\frac{C^{\mathrm{f}}_{\psi\psi}}{C_{\epsilon\epsilon} + C^{\mathrm{f}}_{\psi\psi}} C^{\mathrm{f}}_{\psi\psi} + \left(\frac{C^{\mathrm{f}}_{\psi\psi}}{C_{\epsilon\epsilon} + C^{\mathrm{f}}_{\psi\psi}}\right)^2 \left(C_{\epsilon\epsilon} + C^{\mathrm{f}}_{\psi\psi}\right) \\
&= C^{\mathrm{f}}_{\psi\psi} - \frac{(C^{\mathrm{f}}_{\psi\psi})^2}{C_{\epsilon\epsilon} + C^{\mathrm{f}}_{\psi\psi}} = C^{\mathrm{f}}_{\psi\psi} \left(1 - \frac{C^{\mathrm{f}}_{\psi\psi}}{C_{\epsilon\epsilon} + C^{\mathrm{f}}_{\psi\psi}}\right).
\end{aligned} \qquad (3.15)$$

3.1.2 Bayesian formulation

Given a probability density function $f(\psi)$ for the first-guess estimate ψ^{f}, and a likelihood function $f(d|\psi)$ for the measurement d; then, from Chap. 2 we have Bayes' theorem

$$f(\psi|d) \propto f(\psi) f(d|\psi). \qquad (3.16)$$

Thus, the posterior density for ψ given the measurement d, is proportional to the product of the prior density for ψ times the likelihood function for the measurement d.

Again consider the two estimates (3.1) and (3.2) of the true state ψ^{t}. In the case with Gaussian statistics we can define the prior and likelihood as

$$f(\psi) \propto \exp\left(-\frac{1}{2}\left(\psi - \psi^{\mathrm{f}}\right)\left(C^{\mathrm{f}}_{\psi\psi}\right)^{-1}\left(\psi - \psi^{\mathrm{f}}\right)\right) \qquad (3.17)$$

and

$$f(d|\psi) \propto \exp\left(-\frac{1}{2}(\psi - d) C_{\epsilon\epsilon}^{-1}(\psi - d)\right). \qquad (3.18)$$

Thus, the posterior density can be written as

$$f(\psi|d) \propto \exp\left(-\frac{1}{2}\mathcal{J}[\psi]\right), \qquad (3.19)$$

where

$$\mathcal{J}[\psi] = \left(\psi - \psi^{\mathrm{f}}\right)\left(C^{\mathrm{f}}_{\psi\psi}\right)^{-1}\left(\psi - \psi^{\mathrm{f}}\right) + (\psi - d) C_{\epsilon\epsilon}^{-1}(\psi - d). \qquad (3.20)$$

The least squares solution ψ^{a}, that gives a minimum for $\mathcal{J}$, also gives a maximum of $f(\psi|d)$, i.e. it is the maximum likelihood estimate. This will always be true as long as all the error terms are normally distributed.

The minimum value of $\mathcal{J}$ is found from

$$\frac{d\mathcal{J}}{d\psi} = 2\left(\psi - \psi^{\mathrm{f}}\right)\left(C^{\mathrm{f}}_{\psi\psi}\right)^{-1} + 2\left(\psi - d\right) C^{-1}_{\epsilon\epsilon} = 0. \tag{3.21}$$

Solving for ψ gives again the result ψ^{a} in (3.14), thus, the minimum variance estimate is also the maximum likelihood estimate in the case with Gaussian priors.

3.2 Extension to spatial dimensions

Now we extend the discussion to involve a variable $\psi^{\mathrm{f}}(\boldsymbol{x})$, with a spatial dimension which may be one or larger, e.g. $\boldsymbol{x} = (x, y, z)$ for a three dimensional space. In the following discussion we adopt the notation used by *Bennett* (1992) who gave a similar derivation for the time dependent problem.

3.2.1 Basic formulation

Assume now a multidimensional variable (e.g. a temperature field), and a vector of measurements $\boldsymbol{d} \in \Re^M$, which is related to the true state through the measurement functional $\boldsymbol{\mathcal{M}} \in \Re^M$, with M being the number of measurements:

$$\psi^{\mathrm{f}}(\boldsymbol{x}) = \psi^{\mathrm{t}}(\boldsymbol{x}) + p^{\mathrm{f}}(\boldsymbol{x}), \tag{3.22}$$

$$\boldsymbol{d} = \boldsymbol{\mathcal{M}}[\psi^{\mathrm{t}}(\boldsymbol{x})] + \boldsymbol{\epsilon}. \tag{3.23}$$

The term $p^{\mathrm{f}}(\boldsymbol{x})$ is the error in the first-guess field $\psi^{\mathrm{f}}(\boldsymbol{x})$, relative to the truth $\psi^{\mathrm{t}}(\boldsymbol{x})$. Further, we have defined the vector of measurement errors $\boldsymbol{\epsilon} \in \Re^M$. The measurement errors may be a composite of errors introduced when measuring the variable and additional representation errors introduced when constructing the measurement functional. This will be discussed in more detail in the following chapters.

As an example of a measurement functional, a direct measurement would be represented by a functional of the form

$$\mathcal{M}_i[\psi(\boldsymbol{x})] = \int_\mathcal{D} \psi(\boldsymbol{x})\delta(\boldsymbol{x} - \boldsymbol{x}_i)d\boldsymbol{x} = \psi(\boldsymbol{x}_i), \tag{3.24}$$

where $\boldsymbol{x}_i$ is the measurement location, $\delta(\boldsymbol{x}-\boldsymbol{x}_i)$ is the Dirac delta function, and the subscript i denotes the component i of the measurement functional. Note that in some of the following equations we will use a subscript on the vector form of the measurement functional, e.g. $\boldsymbol{\mathcal{M}}_{(3)}[\delta\psi(\boldsymbol{x}_3)]$ which just denote that the integration is performed on the dummy variable $\boldsymbol{x}_3$ rather than $\boldsymbol{x}$ as is used in (3.24).

3.2 Extension to spatial dimensions

The actual values of the errors $p^f(\boldsymbol{x})$ and $\boldsymbol{\epsilon}$ are not known. Thus, to make progress, a statistical hypothesis must be used, and we make the following assumptions:

$$\overline{p^f(\boldsymbol{x})} = 0, \qquad \overline{p^f(\boldsymbol{x}_1)p^f(\boldsymbol{x}_2)} = C^f_{\psi\psi}(\boldsymbol{x}_1, \boldsymbol{x}_2),$$
$$\overline{\boldsymbol{\epsilon}} = \boldsymbol{0}, \qquad \overline{\boldsymbol{\epsilon}\boldsymbol{\epsilon}^T} = \boldsymbol{C}_{\epsilon\epsilon}, \qquad (3.25)$$
$$\overline{p^f(\boldsymbol{x})\boldsymbol{\epsilon}} = \boldsymbol{0}.$$

Thus, the means of the errors in the first-guess and the measurements are zero, and there are no cross correlations between these error terms. Further, we have knowledge of the forecast or first-guess error covariance between two points in space $C^f_{\psi\psi}(\boldsymbol{x}_1, \boldsymbol{x}_2)$, and the observation error covariance matrix $\boldsymbol{C}_{\epsilon\epsilon} \in \Re^{M \times M}$. Note that the error covariance differs from the sample covariance as defined in (2.22) by referring to the true (unknown) state rather than the sample average.

We are now defining a variational functional

$$\mathcal{J}[\psi] = \iint_{\mathcal{D}} \bigl(\psi^f(\boldsymbol{x}_1) - \psi(\boldsymbol{x}_1)\bigr) W^f_{\psi\psi}(\boldsymbol{x}_1, \boldsymbol{x}_2) \bigl(\psi^f(\boldsymbol{x}_2) - \psi(\boldsymbol{x}_2)\bigr) d\boldsymbol{x}_1 d\boldsymbol{x}_2 \qquad (3.26)$$
$$+ \bigl(\boldsymbol{d} - \mathcal{M}_{(3)}[\psi_3]\bigr)^T \boldsymbol{W}_{\epsilon\epsilon} \bigl(\boldsymbol{d} - \mathcal{M}_{(4)}[\psi_4]\bigr),$$

where $W^f_{\psi\psi}(\boldsymbol{x}_1, \boldsymbol{x}_2)$ is defined as a functional inverse of $C^f_{\psi\psi}(\boldsymbol{x}_1, \boldsymbol{x}_2)$ from

$$\int_{\mathcal{D}} C^f_{\psi\psi}(\boldsymbol{x}_1, \boldsymbol{x}_2) W^f_{\psi\psi}(\boldsymbol{x}_2, \boldsymbol{x}_3) d\boldsymbol{x}_2 = \delta(\boldsymbol{x}_1 - \boldsymbol{x}_3), \qquad (3.27)$$

and $\boldsymbol{W}_{\epsilon\epsilon}$ is the inverse of the measurement error covariance matrix $\boldsymbol{C}_{\epsilon\epsilon}$. Here we have used subscripts on the measurement operator and its argument, e.g. $\mathcal{M}_{(3)}[\psi_3]$ indicating that the dummy variable for the integration is $\boldsymbol{x}_3$. This has no implications in this expression but it will be useful in the following derivation.

The variational functional (3.26) measures, in a weighted sense, the distance between an estimate $\psi(\boldsymbol{x})$ and the forecast or first-guess $\psi^f(\boldsymbol{x})$, plus the distance between the estimate and the observations $\boldsymbol{d}$. The field $\psi(\boldsymbol{x})$ which minimizes (3.26) is named $\psi^a(\boldsymbol{x})$. The use of inverses of the error covariances as weights, ensures that the variance minimizing estimate becomes equal to the maximum likelihood estimate in the case with Gaussian error statistics.

3.2.2 Euler–Lagrange equation

To minimize the variational functional, (3.26), we can calculate the variational derivative of $\mathcal{J}[\psi]$ and require that it approaches zero when the arbitrary perturbation $\delta\psi(\boldsymbol{x})$ goes to zero. Thus, we have

$$\delta\mathcal{J} = \mathcal{J}[\psi + \delta\psi] - \mathcal{J}[\psi] = \mathcal{O}(\delta\psi^2). \qquad (3.28)$$

3 Analysis scheme

Evaluating (3.28) gives

$$\delta \mathcal{J} = -2 \iint_\mathcal{D} \delta\psi(\boldsymbol{x}_1) W^{\text{f}}_{\psi\psi}(\boldsymbol{x}_1, \boldsymbol{x}_2)(\psi^{\text{f}}(\boldsymbol{x}_2) - \psi(\boldsymbol{x}_2)) d\boldsymbol{x}_1 d\boldsymbol{x}_2$$
$$- 2\boldsymbol{\mathcal{M}}_{(3)}[\delta\psi(\boldsymbol{x}_3)]^{\text{T}} \boldsymbol{W}_{\epsilon\epsilon}(\boldsymbol{d} - \boldsymbol{\mathcal{M}}_{(4)}[\psi(\boldsymbol{x}_4)])$$
$$+ \mathcal{O}(\delta\psi^2) = \mathcal{O}(\delta\psi^2). \quad (3.29)$$

Thus, to have an extrema of $\mathcal{J}$ we must have

$$\iint_\mathcal{D} \delta\psi(\boldsymbol{x}_1) W^{\text{f}}_{\psi\psi}(\boldsymbol{x}_1, \boldsymbol{x}_2)(\psi^{\text{f}}(\boldsymbol{x}_2) - \psi^{\text{a}}(\boldsymbol{x}_2)) d\boldsymbol{x}_1 d\boldsymbol{x}_2$$
$$+ \boldsymbol{\mathcal{M}}_{(3)}[\delta\psi(\boldsymbol{x}_3)]^{\text{T}} \boldsymbol{W}_{\epsilon\epsilon}(\boldsymbol{d} - \boldsymbol{\mathcal{M}}_{(4)}[\psi^{\text{a}}(\boldsymbol{x}_4)]) = 0. \quad (3.30)$$

To proceed we need to get the second term in under the integral and both terms need to be proportional to $\delta\psi$. We will now show that

$$\boldsymbol{\mathcal{M}}_{(3)}[\delta\psi(\boldsymbol{x}_3)]^{\text{T}} = \int_\mathcal{D} \delta\psi(\boldsymbol{x}_1) \boldsymbol{\mathcal{M}}^{\text{T}}_{(3)}[\delta(\boldsymbol{x}_1 - \boldsymbol{x}_3)] d\boldsymbol{x}_1. \quad (3.31)$$

We start by writing out the measurement of a Dirac delta function, $\delta(\boldsymbol{x}_1 - \boldsymbol{x}_3)$, as

$$\mathcal{M}_{i(3)}[\delta(\boldsymbol{x}_1 - \boldsymbol{x}_3)] = \int_\mathcal{D} \delta(\boldsymbol{x}_1 - \boldsymbol{x}_3) \delta(\boldsymbol{x}_3 - \boldsymbol{x}_i) d\boldsymbol{x}_3 = \delta(\boldsymbol{x}_1 - \boldsymbol{x}_i), \quad (3.32)$$

for $i = 1, \ldots, M$ where M is the number of measurements. The subscript (3) on $\mathcal{M}_i$ defines the variable the functional is operating on, thus, the integration variable is $\boldsymbol{x}_3$. Multiplying this equation with $\delta\psi(\boldsymbol{x}_1)$ and integrating in $\boldsymbol{x}_1$ now gives

$$\int_\mathcal{D} \delta\psi(\boldsymbol{x}_1) \mathcal{M}_{i(3)}[\delta(\boldsymbol{x}_1 - \boldsymbol{x}_3)] d\boldsymbol{x}_1 = \int_\mathcal{D} \delta\psi(\boldsymbol{x}_1) \delta(\boldsymbol{x}_1 - \boldsymbol{x}_i) d\boldsymbol{x}_1$$
$$= \mathcal{M}_{i(1)}[\delta\psi(\boldsymbol{x}_1)] \quad (3.33)$$
$$= \mathcal{M}_{i(3)}[\delta\psi(\boldsymbol{x}_3)].$$

where in the last line, we changed the dummy variable for the integration to $\boldsymbol{x}_3$. Thus, we have obtained (3.31).

We also have that

$$\int_\mathcal{D} C^{\text{f}}_{\psi\psi}(\boldsymbol{x}_1, \boldsymbol{x}_2) \mathcal{M}^{\text{T}}_{i(3)}[\delta(\boldsymbol{x}_2 - \boldsymbol{x}_3)] d\boldsymbol{x}_2 = C^{\text{f}}_{\psi\psi}(\boldsymbol{x}_1, \boldsymbol{x}_i)$$
$$= \mathcal{M}_{i(2)}[C^{\text{f}}_{\psi\psi}(\boldsymbol{x}_1, \boldsymbol{x}_2)]. \quad (3.34)$$

Note that the second term of (3.30), i.e. the measurement term, is constant in the integration with respect to $\boldsymbol{x}_2$. Equations (3.32–3.34) are verified for $i = 1, \ldots, M$, and their results can be generalized and substituted into (3.30) which then leads to

3.2 Extension to spatial dimensions

$$\iint_D \delta\psi(\boldsymbol{x}_1)\Big(W^{\mathrm{f}}_{\psi\psi}(\boldsymbol{x}_1,\boldsymbol{x}_2)\big(\psi^{\mathrm{f}}(\boldsymbol{x}_2)-\psi^{\mathrm{a}}(\boldsymbol{x}_2)\big) \\ +\boldsymbol{\mathcal{M}}^{\mathrm{T}}_{(3)}[\delta(\boldsymbol{x}_1-\boldsymbol{x}_3)]\boldsymbol{W}_{\epsilon\epsilon}\big(\boldsymbol{d}-\boldsymbol{\mathcal{M}}_{(4)}[\psi^{\mathrm{a}}(\boldsymbol{x}_4)]\big)\Big)d\boldsymbol{x}_1 d\boldsymbol{x}_2 = 0, \tag{3.35}$$

or since this must be true for all $\delta\psi$ we must have

$$W^{\mathrm{f}}_{\psi\psi}(\boldsymbol{x}_1,\boldsymbol{x}_2)\big(\psi^{\mathrm{f}}(\boldsymbol{x}_2)-\psi^{\mathrm{a}}(\boldsymbol{x}_2)\big) \\ +\boldsymbol{\mathcal{M}}^{\mathrm{T}}_{(3)}[\delta(\boldsymbol{x}_1-\boldsymbol{x}_3)]\boldsymbol{W}_{\epsilon\epsilon}\big(\boldsymbol{d}-\boldsymbol{\mathcal{M}}_{(4)}[\psi^{\mathrm{a}}(\boldsymbol{x}_4)]\big) = 0. \tag{3.36}$$

This is the Euler–Lagrange equation for the variational problem, of which the solution ψ^{a} must be a minimum of $\mathcal{J}$.

Now multiply (3.36) with $C^{\mathrm{f}}_{\psi\psi}(\boldsymbol{x},\boldsymbol{x}_1)$ and integrate with respect to $\boldsymbol{x}_1$. Using the definition (3.27) and the identity (3.34) we get the Euler–Lagrange equation of the form

$$\psi^{\mathrm{a}}(\boldsymbol{x}) - \psi^{\mathrm{f}}(\boldsymbol{x}) = \boldsymbol{\mathcal{M}}^{\mathrm{T}}_{(3)}[C^{\mathrm{f}}_{\psi\psi}(\boldsymbol{x},\boldsymbol{x}_3)]\boldsymbol{W}_{\epsilon\epsilon}\big(\boldsymbol{d}-\boldsymbol{\mathcal{M}}_{(4)}[\psi^{\mathrm{a}}_4]\big). \tag{3.37}$$

3.2.3 Representer solution

A problem with the Euler–Lagrange equation (3.37) is that ψ^{a} is contained on both sides of the equality sign. To resolve this we first define the vector $\boldsymbol{b} \in \Re^M$ as

$$\boldsymbol{b} = \boldsymbol{W}_{\epsilon\epsilon}\big(\boldsymbol{d}-\boldsymbol{\mathcal{M}}_{(4)}[\psi^{\mathrm{a}}_4]\big), \tag{3.38}$$

and then seek a solution of the form

$$\psi^{\mathrm{a}}(\boldsymbol{x}) = \psi^{\mathrm{f}}(\boldsymbol{x}) + \boldsymbol{b}^{\mathrm{T}}\boldsymbol{r}(\boldsymbol{x}), \tag{3.39}$$

where we have introduced the vector of representers $\boldsymbol{r}(\boldsymbol{x}) \in \Re^M$.

Inserting this into (3.37) gives

$$\psi^{\mathrm{f}}(\boldsymbol{x}) - \psi^{\mathrm{f}}(\boldsymbol{x}) + \boldsymbol{b}^{\mathrm{T}}\boldsymbol{r}(\boldsymbol{x}) = \boldsymbol{\mathcal{M}}^{\mathrm{T}}_{(3)}[C^{\mathrm{f}}_{\psi\psi}(\boldsymbol{x},\boldsymbol{x}_3)]\boldsymbol{b}, \tag{3.40}$$

Thus, we get the influence functions or representers $\boldsymbol{r}(\boldsymbol{x})$ defined as

$$\boldsymbol{r}(\boldsymbol{x}) = \boldsymbol{\mathcal{M}}_{(3)}[C^{\mathrm{f}}_{\psi\psi}(\boldsymbol{x},\boldsymbol{x}_3)]. \tag{3.41}$$

Now using (3.39) in (3.38) gives

$$\begin{aligned}\boldsymbol{b} &= \boldsymbol{W}_{\epsilon\epsilon}\big(\boldsymbol{d}-\boldsymbol{\mathcal{M}}_{(4)}[\psi^{\mathrm{f}}_4 + \boldsymbol{b}^{\mathrm{T}}\boldsymbol{r}_4]\big) \\ &= \boldsymbol{W}_{\epsilon\epsilon}\big(\boldsymbol{d}-\boldsymbol{\mathcal{M}}_{(4)}[\psi^{\mathrm{f}}_4]\big) - \boldsymbol{W}_{\epsilon\epsilon}\boldsymbol{\mathcal{M}}_{(4)}[\boldsymbol{b}^{\mathrm{T}}\boldsymbol{r}_4] \\ &= \boldsymbol{W}_{\epsilon\epsilon}\big(\boldsymbol{d}-\boldsymbol{\mathcal{M}}_{(4)}[\psi^{\mathrm{f}}_4]\big) - \boldsymbol{W}_{\epsilon\epsilon}\boldsymbol{b}^{\mathrm{T}}\boldsymbol{\mathcal{M}}_{(4)}[\boldsymbol{r}_4],\end{aligned} \tag{3.42}$$

because of the linearity of $\boldsymbol{\mathcal{M}}$. Rearranging gives

$$\boldsymbol{b} + \boldsymbol{W}_{\epsilon\epsilon}\boldsymbol{b}^{\mathrm{T}}\boldsymbol{\mathcal{M}}_{(4)}[\boldsymbol{r}_4]) = \boldsymbol{W}_{\epsilon\epsilon}\big(\boldsymbol{d}-\boldsymbol{\mathcal{M}}_{(4)}[\psi^{\mathrm{f}}_4]\big), \tag{3.43}$$

and, multiplying from the left with $C_{\epsilon\epsilon}$, we obtain

$$C_{\epsilon\epsilon}b + b^{\mathrm{T}}\mathcal{M}_{(4)}[r_4] = d - \mathcal{M}_{(4)}[\psi_4^{\mathrm{f}}], \tag{3.44}$$

or

$$(\mathcal{M}_{(4)}^{\mathrm{T}}[r_4] + C_{\epsilon\epsilon})b = d - \mathcal{M}_{(4)}[\psi_4^{\mathrm{f}}], \tag{3.45}$$

which is a linear system of equations for b. Rewriting by using (3.41) the equation becomes

$$\left(\mathcal{M}_{(3)}\mathcal{M}_{(4)}^{\mathrm{T}}[C_{\psi\psi}^{\mathrm{f}}(x_3,x_4)] + C_{\epsilon\epsilon}\right)b = d - \mathcal{M}_{(4)}[\psi^{\mathrm{f}}(x_4)]. \tag{3.46}$$

A solution can now be found from the equations (3.39), (3.41) and (3.45).

3.2.4 Representer matrix

Note that with direct measurements as given in (3.24), we have

$$\mathcal{M}_{i(3)}\mathcal{M}_{j(4)}^{\mathrm{T}}[C_{\psi\psi}^{\mathrm{f}}(x_3,x_4)] = C_{\psi\psi}^{\mathrm{f}}(x_i,x_j). \tag{3.47}$$

The matrix $C_{\psi\psi}^{\mathrm{f}}(x_i,x_j)$ is often called the representer matrix and with direct measurements it describes the covariances of the first-guess between the two locations x_i and x_j.

3.2.5 Error estimate

It is possible to derive an error estimate for the analysis (3.39). The simplest is to use the procedure as derived by *Bennett* (1992) for the time dependent problem. From the definition of the error covariance in (3.25) we can write

$$C_{\psi\psi}^{\mathrm{a}}(x_1,x_2) = \overline{(\psi^{\mathrm{t}}(x_1) - \psi^{\mathrm{a}}(x_1))(\psi^{\mathrm{t}}(x_2) - \psi^{\mathrm{a}}(x_2))}, \tag{3.48}$$

and insert the equation for the analysis to get

$$\begin{aligned}C_{\psi\psi}^{\mathrm{a}}(x_1,x_2) &= \overline{(\psi_1^{\mathrm{t}} - \psi_1^{\mathrm{f}} - b^{\mathrm{T}}r_1)(\psi_2^{\mathrm{t}} - \psi_2^{\mathrm{f}} - b^{\mathrm{T}}r_2)} \\ &= \overline{(\psi_1^{\mathrm{t}} - \psi_1^{\mathrm{f}})(\psi_2^{\mathrm{t}} - \psi_2^{\mathrm{f}})} - 2\overline{(\psi_1^{\mathrm{t}} - \psi_1^{\mathrm{f}})b^{\mathrm{T}}r_2} + \overline{r_1^{\mathrm{T}}bb^{\mathrm{T}}r_2}.\end{aligned} \tag{3.49}$$

We have used that b is a function of ψ and the representers r, are functions of the covariance matrix and then $\overline{\psi}$. Further, we used the property $(AB)^{\mathrm{T}} = B^{\mathrm{T}}A^{\mathrm{T}}$ for matrices A and B, and that the covariance is symmetrical in x_1 and x_2.

The first term is just $C_{\psi\psi}^{\mathrm{f}}$ while the two other terms will be treated next and we now define for convenience

$$\mathcal{P} = \mathcal{M}_{(3)}\mathcal{M}_{(4)}^{\mathrm{T}}[C_{\psi\psi}^{\mathrm{f}}(x_3,x_4)] + C_{\epsilon\epsilon}, \tag{3.50}$$

and the residual or innovation

$$h = d - \mathcal{M}_{(4)}[\psi_4^{\text{f}}]. \tag{3.51}$$

Using (3.41), (3.50) and (3.51) in (3.45) gives $b = \mathcal{P}^{-1}h$. Furthermore, by using (3.23), (3.25), (3.41) and (3.45), in addition to the two definitions above, the second term in (3.49) becomes

$$\begin{aligned}
&\overline{-2(\psi_1^{\text{t}} - \psi_1^{\text{f}})b^{\text{T}}r_2} \\
&= -2\overline{(\psi_1^{\text{t}} - \psi_1^{\text{f}})(\mathcal{P}^{-1}h)^{\text{T}}}r_2 \\
&= -2\overline{(\psi_1^{\text{t}} - \psi_1^{\text{f}})\left(\mathcal{P}^{-1}(d - \mathcal{M}_{(4)}[\psi_4^{\text{f}}])\right)^{\text{T}}}r_2 \\
&= -2\overline{(\psi_1^{\text{t}} - \psi_1^{\text{f}})\left(\mathcal{P}^{-1}(\mathcal{M}_{(4)}[\psi_4^{\text{t}}] + \epsilon - \mathcal{M}_{(4)}[\psi_4^{\text{f}}])\right)^{\text{T}}}r_2 \\
&= -2\overline{(\psi_1^{\text{t}} - \psi_1^{\text{f}})\mathcal{M}_{(4)}^{\text{T}}[\psi_4^{\text{t}} - \psi_4^{\text{f}}]}\mathcal{P}^{-1}r_2 + 0 \\
&= -2\mathcal{M}_{(4)}^{\text{T}}[\overline{(\psi_1^{\text{t}} - \psi_1^{\text{f}})(\psi_4^{\text{t}} - \psi_4^{\text{f}})}]\mathcal{P}^{-1}r_2 \\
&= -2\mathcal{M}_{(4)}^{\text{T}}[C_{\psi\psi}^{\text{f}}(x_1, x_4)]\mathcal{P}^{-1}r_2 \\
&= -2r_1^{\text{T}}\mathcal{P}^{-1}r_2.
\end{aligned} \tag{3.52}$$

Here we have also used that $\bar{\epsilon} = 0$ from (3.25), and that $\mathcal{P}$ is a symmetrical function of the covariance and can be moved outside the averaging.

Further, using $(\mathcal{P}^{-1}h)^{\text{T}} = h^{\text{T}}\mathcal{P}^{-1}$, the last term becomes

$$\begin{aligned}
&r_1^{\text{T}}\overline{bb^{\text{T}}}r_2 \\
&= r_1^{\text{T}}\mathcal{P}^{-1}\overline{hh^{\text{T}}}\mathcal{P}^{-1}r_2 \\
&= r_1^{\text{T}}\mathcal{P}^{-1}\overline{(d - \mathcal{M}_{(1)}[\psi_1^{\text{f}}])(d - \mathcal{M}_{(2)}[\psi_2^{\text{f}}])^{\text{T}}}\mathcal{P}^{-1}r_2 \\
&= r_1^{\text{T}}\mathcal{P}^{-1}\overline{(\mathcal{M}_{(1)}[\psi_1^{\text{t}}] + \epsilon - \mathcal{M}_{(1)}[\psi_1^{\text{f}}])(\mathcal{M}_{(2)}[\psi_2^{\text{t}}] + \epsilon - \mathcal{M}_{(1)}[\psi_2^{\text{f}}])^{\text{T}}}\mathcal{P}^{-1}r_2 \\
&= r_1^{\text{T}}\mathcal{P}^{-1}\overline{(\mathcal{M}_{(1)}[\psi_1^{\text{t}} - \psi_1^{\text{f}}] + \epsilon)(\mathcal{M}_{(2)}[\psi_2^{\text{t}} - \psi_2^{\text{f}}] + \epsilon)^{\text{T}}}\mathcal{P}^{-1}r_2 \\
&= r_1^{\text{T}}\mathcal{P}^{-1}(\mathcal{M}_{(1)}\mathcal{M}_{(2)}^{\text{T}}[\overline{(\psi_1^{\text{t}} - \psi_1^{\text{f}})(\psi_2^{\text{t}} - \psi_2^{\text{f}})}] + \overline{\epsilon\epsilon^{\text{T}}})\mathcal{P}^{-1}r_2 \\
&= r_1^{\text{T}}\mathcal{P}^{-1}\mathcal{P}\mathcal{P}^{-1}r_2 \\
&= r_1^{\text{T}}\mathcal{P}^{-1}r_2.
\end{aligned} \tag{3.53}$$

Thus, an error estimate is given as

$$C_{\psi\psi}^{\text{a}}(x_1, x_2) = C_{\psi\psi}^{\text{f}}(x_1, x_2) \\
- r^{\text{T}}(x_1)\left(\mathcal{M}_{(3)}\mathcal{M}_{(4)}^{\text{T}}[C_{\psi\psi}^{\text{f}}(x_3, x_4)] + C_{\epsilon\epsilon}\right)^{-1}r(x_2). \tag{3.54}$$

where the definition for $\mathcal{P}$ has been used.

3.2.6 Uniqueness of the solution

By expressing the solution as in (3.39) not all arbitrary functions can be represented. To show that the solution (3.39) is the unique variance minimizing linear solution we proceed with the following argumentation using a geometrical formulation, identical to the formulation used for the time dependent problem by *Bennett* (1992). First define the inner product

$$< f(\boldsymbol{x}_1), g(\boldsymbol{x}_2) > = \iint_{\mathcal{D}} f(\boldsymbol{x}_1) W^{\mathrm{f}}_{\psi\psi}(\boldsymbol{x}_1, \boldsymbol{x}_2) g(\boldsymbol{x}_2) d\boldsymbol{x}_1 d\boldsymbol{x}_2. \qquad (3.55)$$

Note that

$$< C^{\mathrm{f}}_{\psi\psi}(\boldsymbol{x}_3, \boldsymbol{x}_1), \psi(\boldsymbol{x}_2) >$$
$$= \iint_{\mathcal{D}} C^{\mathrm{f}}_{\psi\psi}(\boldsymbol{x}_3, \boldsymbol{x}_1) W^{\mathrm{f}}_{\psi\psi}(\boldsymbol{x}_1, \boldsymbol{x}_2) \psi(\boldsymbol{x}_2) d\boldsymbol{x}_1 d\boldsymbol{x}_2 \qquad (3.56)$$
$$= \psi(\boldsymbol{x}_3),$$

thus, $C^{\mathrm{f}}_{\psi\psi}(\boldsymbol{x}_3, \boldsymbol{x}_1)$ is a "reproducing kernel" for the inner product (3.55) and the expression (3.56) is true for every field ψ in any point $\boldsymbol{x}$.

Recalling the definition of the representer (3.41) we get

$$< \boldsymbol{r}(\boldsymbol{x}_1), \psi(\boldsymbol{x}_2) > = < \mathcal{M}_{(1)}[C^{\mathrm{f}}_{\psi\psi}(\boldsymbol{x}_3, \boldsymbol{x}_1)], \psi(\boldsymbol{x}_2) >$$
$$= \mathcal{M}_{(1)}[< C^{\mathrm{f}}_{\psi\psi}(\boldsymbol{x}_3, \boldsymbol{x}_1), \psi(\boldsymbol{x}_2) >] \qquad (3.57)$$
$$= \mathcal{M}_{(1)}[\psi(\boldsymbol{x}_1)]$$

Thus, the measurement of a field $\psi(\boldsymbol{x})$ is equivalent to projecting the field onto the representer using the inner product (3.55).

The penalty function (3.26) can now be written entirely in terms of inner products as

$$\mathcal{J}[\psi] = < \psi^{\mathrm{f}} - \psi, \psi^{\mathrm{f}} - \psi > + (\boldsymbol{d} - < \psi, \boldsymbol{r} >)^{\mathrm{T}} \boldsymbol{W}_{\epsilon\epsilon}(\boldsymbol{d} - < \psi, \boldsymbol{r} >). \qquad (3.58)$$

Assume now that the minimizing solution is expressed as

$$\psi^{\mathrm{a}}(\boldsymbol{x}) = \psi^{\mathrm{f}}(\boldsymbol{x}) + \boldsymbol{b}^{\mathrm{T}} \boldsymbol{r}(\boldsymbol{x}) + g(\boldsymbol{x}), \qquad (3.59)$$

where $g(\boldsymbol{x})$ is an arbitrary function orthogonal to the representers, i.e.

$$< g, \boldsymbol{r} > = \boldsymbol{0}. \qquad (3.60)$$

Because of this identity the field g may be regarded as unobservable. Substituting (3.59) into (3.58) gives

$$\mathcal{J}[\psi^{\mathrm{a}}] = < \boldsymbol{r}^{\mathrm{T}}\boldsymbol{b} + g, \boldsymbol{r}^{\mathrm{T}}\boldsymbol{b} + g >$$
$$+ (\boldsymbol{d} - < \psi^{\mathrm{a}}, \boldsymbol{r} >)^{\mathrm{T}} \boldsymbol{W}_{\epsilon\epsilon}(\boldsymbol{d} - < \psi^{\mathrm{a}}, \boldsymbol{r} >)$$
$$= \boldsymbol{b}^{\mathrm{T}} < \boldsymbol{r}, \boldsymbol{r}^{\mathrm{T}} > \boldsymbol{b} + \boldsymbol{b}^{\mathrm{T}} < \boldsymbol{r}, g > + < g, \boldsymbol{r}^{\mathrm{T}} > \boldsymbol{b} + < g, g > \qquad (3.61)$$
$$+ (\boldsymbol{d} - < \psi^{\mathrm{f}}, \boldsymbol{r} > -\boldsymbol{b}^{\mathrm{T}} < \boldsymbol{r}, \boldsymbol{r}^{\mathrm{T}} > - < g, \boldsymbol{r} >)^{\mathrm{T}}$$
$$\times \boldsymbol{W}_{\epsilon\epsilon}(\boldsymbol{d} - < \psi^{\mathrm{f}}, \boldsymbol{r} > -\boldsymbol{b}^{\mathrm{T}} < \boldsymbol{r}, \boldsymbol{r}^{\mathrm{T}} > - < g, \boldsymbol{r} >).$$

Defining the residual
$$h = d - <\psi^f, r>, \quad (3.62)$$
and using the definition of the representer matrix,
$$R = <r_3, r_4^T> = \mathcal{M}_{(3)}[r_3^T], \quad (3.63)$$
and (3.41) and (3.47), we get the penalty function of the form
$$\mathcal{J}[\psi^a] = b^T R b + <g, g> + (h - Rb)^T W_{\epsilon\epsilon}(h - Rb). \quad (3.64)$$

The original penalty function (3.26) has now been reduced to a compact form where the disposable parameters are b and $g(x)$. If ψ minimizes $\mathcal{J}$ then clearly $<g, g> = 0$ and thus
$$g(x) \equiv 0. \quad (3.65)$$
The unobservable field g must be discarded, reducing $\mathcal{J}$ from the infinite dimensional quadratic form (3.26) to the finite dimensional quadratic form
$$\mathcal{B}[b] = b^T R b + (h - Rb)^T W_{\epsilon\epsilon}(h - Rb), \quad (3.66)$$
where $\mathcal{B}[b] = \mathcal{J}[\psi^a]$.

3.2.7 Minimization of the penalty function

The minimizing solution for b can again be found by setting the variational derivative of (3.66) with respect to b equal to zero,
$$\mathcal{B}[b + \delta b] - \mathcal{B}[b] = 2\delta b^T R b + 2\delta b^T R W_{\epsilon\epsilon}(Rb - h) + \mathcal{O}(\delta b^2) = \mathcal{O}(\delta b^2), \quad (3.67)$$
which gives
$$\delta b^T \left(Rb + RW_{\epsilon\epsilon}(Rb - h)\right) = 0, \quad (3.68)$$
or
$$Rb + RW_{\epsilon\epsilon}(Rb - h) = 0, \quad (3.69)$$
since δb is arbitrary. This equation can be written as
$$R(b + W_{\epsilon\epsilon}Rb - W_{\epsilon\epsilon}h) = 0, \quad (3.70)$$
which leads to the standard linear system of equations
$$(R + C_{\epsilon\epsilon})b = h, \quad (3.71)$$
or
$$b = \mathcal{P}^{-1}h, \quad (3.72)$$
as the solution for b. Note that we have used that $R = \mathcal{M}_{(i)}[r_i]$ for all i.

3.2.8 Prior and posterior value of the penalty function

Inserting the first-guess value ψ^{f}, into the penalty function (3.58) gives

$$\mathcal{J}[\psi^{\mathrm{f}}] = (\boldsymbol{d} - <\psi^{\mathrm{f}}, \boldsymbol{r}>)^{\mathrm{T}} \boldsymbol{W}_{\epsilon\epsilon}(\boldsymbol{d} - <\psi^{\mathrm{f}}, \boldsymbol{r}>) = \boldsymbol{h}^{\mathrm{T}} \boldsymbol{W}_{\epsilon\epsilon} \boldsymbol{h}. \tag{3.73}$$

This is known as the prior value of the penalty function.

Similarly by inserting the minimizing solution (3.72) into the penalty function (3.66) we get the following,

$$\begin{aligned}
\mathcal{J}[\boldsymbol{\mathcal{P}}^{-1}\boldsymbol{h}] &= (\boldsymbol{\mathcal{P}}^{-1}\boldsymbol{h})^{\mathrm{T}} \boldsymbol{R}(\boldsymbol{\mathcal{P}}^{-1}\boldsymbol{h}) + (\boldsymbol{h} - \boldsymbol{R}\boldsymbol{\mathcal{P}}^{-1}\boldsymbol{h})^{\mathrm{T}} \boldsymbol{W}_{\epsilon\epsilon}(\boldsymbol{h} - \boldsymbol{R}\boldsymbol{\mathcal{P}}^{-1}\boldsymbol{h}) \\
&= \boldsymbol{h}^{\mathrm{T}} \boldsymbol{\mathcal{P}}^{-1} \boldsymbol{R} \boldsymbol{\mathcal{P}}^{-1} \boldsymbol{h} + \boldsymbol{h}^{\mathrm{T}} (\boldsymbol{R}\boldsymbol{\mathcal{P}}^{-1} - \boldsymbol{I}) \boldsymbol{W}_{\epsilon\epsilon} (\boldsymbol{R}\boldsymbol{\mathcal{P}}^{-1} - \boldsymbol{I}) \boldsymbol{h} \\
&= \boldsymbol{h}^{\mathrm{T}} \left\{ \boldsymbol{\mathcal{P}}^{-1} \boldsymbol{R} \boldsymbol{\mathcal{P}}^{-1} + (\boldsymbol{R}\boldsymbol{\mathcal{P}}^{-1} - \boldsymbol{I}) \boldsymbol{W}_{\epsilon\epsilon} (\boldsymbol{R}\boldsymbol{\mathcal{P}}^{-1} - \boldsymbol{I}) \right\} \boldsymbol{h} \\
&= \boldsymbol{h}^{\mathrm{T}} \left\{ \boldsymbol{\mathcal{P}}^{-1} \boldsymbol{R} \boldsymbol{\mathcal{P}}^{-1} + \boldsymbol{\mathcal{P}}^{-1} (\boldsymbol{R} - \boldsymbol{\mathcal{P}}) \boldsymbol{W}_{\epsilon\epsilon} (\boldsymbol{R} - \boldsymbol{\mathcal{P}}) \boldsymbol{\mathcal{P}}^{-1} \right\} \boldsymbol{h} \\
&= \boldsymbol{h}^{\mathrm{T}} \boldsymbol{\mathcal{P}}^{-1} \left\{ \boldsymbol{R} + (\boldsymbol{R} - \boldsymbol{\mathcal{P}}) \boldsymbol{W}_{\epsilon\epsilon} (\boldsymbol{R} - \boldsymbol{\mathcal{P}}) \right\} \boldsymbol{\mathcal{P}}^{-1} \boldsymbol{h} \\
&= \boldsymbol{h}^{\mathrm{T}} \boldsymbol{\mathcal{P}}^{-1} \left\{ \boldsymbol{R} + \boldsymbol{C}_{\epsilon\epsilon} \right\} \boldsymbol{\mathcal{P}}^{-1} \boldsymbol{h} \\
&= \boldsymbol{h}^{\mathrm{T}} \boldsymbol{\mathcal{P}}^{-1} \boldsymbol{\mathcal{P}} \boldsymbol{\mathcal{P}}^{-1} \boldsymbol{h} \\
&= \boldsymbol{h}^{\mathrm{T}} \boldsymbol{\mathcal{P}}^{-1} \boldsymbol{h} \\
&= \boldsymbol{h}^{\mathrm{T}} \boldsymbol{b},
\end{aligned} \tag{3.74}$$

as long as $\boldsymbol{b}$ is given from (3.72). This is known as the posterior value of the penalty function.

It is explained by *Bennett* (2002, section 2.3) that the reduced penalty function is a $\mathcal{X}_M^2$ variable. Thus, we have a mean to test the validity of our statistical assumptions, by checking if the value of reduced penalty function is a Gaussian variable with mean equal to M and variance equal to $2M$. This could be done rigorously by repeated minimizations of the penalty function using different data sets.

3.3 Discrete form

When discretized on a numerical grid, (3.22–3.23) are written as

$$\psi^{\mathrm{f}} = \psi^{\mathrm{t}} + \boldsymbol{p}^{\mathrm{f}}, \tag{3.75}$$

$$\boldsymbol{d} = \boldsymbol{M}\psi^{\mathrm{t}} + \boldsymbol{\epsilon}, \tag{3.76}$$

where $\boldsymbol{M}$, now called the measurement matrix, is the discrete representation of $\mathcal{M}$.

The statistical null hypothesis $\mathcal{H}_0$ is then

$$\begin{aligned}
\overline{\boldsymbol{p}^{\mathrm{f}}} &= \boldsymbol{0}, & \overline{\boldsymbol{p}^{\mathrm{f}}(\boldsymbol{p}^{\mathrm{f}})^{\mathrm{T}}} &= \boldsymbol{C}_{\psi\psi}^{\mathrm{f}}, \\
\overline{\boldsymbol{\epsilon}} &= \boldsymbol{0}, & \overline{\boldsymbol{\epsilon}\boldsymbol{\epsilon}^{\mathrm{T}}} &= \boldsymbol{C}_{\epsilon\epsilon}, \\
\overline{\boldsymbol{p}^{\mathrm{f}}\boldsymbol{\epsilon}^{\mathrm{T}}} &= \boldsymbol{0}.
\end{aligned} \tag{3.77}$$

3.3 Discrete form

By using the same statistical procedure as in Sect. 3.1, or alternatively by minimizing the variational functional

$$J[\psi^a] = (\psi^f - \psi^a)^T (C^f_{\psi\psi})^{-1}(\psi^f - \psi^a) + (d - M\psi^a)^T W_{\epsilon\epsilon}(d - M\psi^a), \quad (3.78)$$

with respect to ψ^a, one get,

$$\psi^a = \psi^f + r^T b, \quad (3.79)$$

where the influence functions (e.g. error covariance functions for direct measurements) are given as

$$r = M C^f_{\psi\psi}, \quad (3.80)$$

i.e. "measurements" of the error covariance matrix $C^f_{\psi\psi}$. Thus, r is a matrix where each row contains a representer for a particular measurement. The coefficients b are determined from the system of linear equations

$$(M C^f_{\psi\psi} M^T + C_{\epsilon\epsilon}) b = d - M\psi^f. \quad (3.81)$$

In addition the error estimate (3.54) becomes

$$C^a_{\psi\psi} = C^f_{\psi\psi} - r^T \left(M C^f_{\psi\psi} M^T + C_{\epsilon\epsilon} \right)^{-1} r. \quad (3.82)$$

Thus, the inverse estimate ψ^a, is given by the first-guess ψ^f, plus a linear combination of influence functions $r^T b$, one for each of the measurements. The coefficients b are clearly small if the first-guess is close to the data, and large if the residual between the data and the first-guess is large.

Note that a more common way of writing the previous equations is the following:

$$\psi^a = \psi^f + K(d - M\psi^f), \quad (3.83)$$
$$C^a_{\psi\psi} = (I - KM) C^f_{\psi\psi}, \quad (3.84)$$
$$K = C^f_{\psi\psi} M^T (M C^f_{\psi\psi} M^T + C_{\epsilon\epsilon})^{-1}, \quad (3.85)$$

where the matrix K is often called the Kalman gain. This can be derived directly from (3.79)–(3.82) by rearranging terms, and it is the standard way of writing the analysis equations for the Kalman filter to be discussed in Chap. 4. The numerical evaluation of these equations, however, is simpler and more efficient using the form (3.79)–(3.82)

4
Sequential data assimilation

In the previous chapter we considered a time independent problem and computed the best conditional estimate given a prior estimate and measurements of the state.

For time dependent problems, sequential data assimilation methods use the analysis scheme from the previous chapter to sequentially update the model state. Such methods have proven useful for many applications in meteorology and oceanography, including operational weather prediction systems where new observations are sequentially assimilated into the model when they become available.

If a model forecast $\boldsymbol{\psi}^\mathrm{f}$, and the forecast error covariance $\boldsymbol{C}^\mathrm{f}_{\psi\psi}$, are known at a time t_k, where we have available measurements $\boldsymbol{d}$, with a measurement error covariance matrix $\boldsymbol{C}_{\epsilon\epsilon}$, it is possible to calculate an improved analysis $\boldsymbol{\psi}^\mathrm{a}$, with its analyzed error covariance $\boldsymbol{C}^\mathrm{a}_{\psi\psi}$. A major issue is then how to estimate or predict the error covariance $\boldsymbol{C}^\mathrm{f}_{\psi\psi}$ for the model forecast at the time t_k.

This chapter will briefly outline the Kalman Filter (KF) originally proposed by *Kalman* (1960), which introduces an equation for the time evolution of the error covariance matrix. Further, the problems associated with the use of the KF with nonlinear dynamical models will be illustrated. Finally a basic introduction is given to the Ensemble Kalman Filter (EnKF) proposed by *Evensen* (1994a).

4.1 Linear Dynamics

For linear dynamics the optimal sequential assimilation method is the Kalman filter. In the Kalman filter an additional equation for the second-order statistical moment is integrated forward in time to predict error statistics for the model forecast. The error statistics are then used to calculate a variance minimizing estimate whenever measurements are available.

4.1.1 Kalman filter for a scalar case

Assume now that a discrete dynamical model for the true state of a scalar ψ can be written as

$$\psi^{\mathrm{t}}(t_k) = G\psi^{\mathrm{t}}(t_{k-1}) + q(t_{k-1}), \tag{4.1}$$

$$\psi^{\mathrm{t}}(t_0) = \Psi_0 + a, \tag{4.2}$$

where G is a linear model operator, q is the model error over one time step and Ψ_0 is an initial condition with error a.

The model error is normally not known and a numerical model will therefore evolve according to

$$\psi^{\mathrm{f}}(t_k) = G\psi^{\mathrm{a}}(t_{k-1}) \tag{4.3}$$

$$\psi^{\mathrm{a}}(t_0) = \Psi_0. \tag{4.4}$$

That is, given a best estimate ψ^{a}, for ψ at time t_{k-1}, a forecast ψ^{f}, is calculated at time t_k, using the approximate equation (4.3).

Now subtract (4.3) from (4.1) to get

$$\psi_k^{\mathrm{t}} - \psi_k^{\mathrm{f}} = G(\psi_{k-1}^{\mathrm{t}} - \psi_{k-1}^{\mathrm{a}}) + q_{k-1}. \tag{4.5}$$

where we have defined $\psi_k = \psi(t_k)$ and $q_k = q(t_k)$. The error covariance matrix for the forecast at time t_k is

$$\begin{aligned} C_{\psi\psi}^{\mathrm{f}}(t_k) &= \overline{(\psi_k^{\mathrm{t}} - \psi_k^{\mathrm{f}})^2} \\ &= \overline{G^2(\psi_{k-1}^{\mathrm{t}} - \psi_{k-1}^{\mathrm{a}})^2} + \overline{q_{k-1}^2} + \overline{2G(\psi_{k-1}^{\mathrm{t}} - \psi_{k-1}^{\mathrm{a}})q_{k-1}} \\ &= G^2 C_{\psi\psi}^{\mathrm{a}}(t_{k-1}) + C_{qq}(t_{k-1}). \end{aligned} \tag{4.6}$$

We have defined the error covariance for the model state

$$C_{\psi\psi}^{\mathrm{a}}(t_{k-1}) = \overline{(\psi_{k-1}^{\mathrm{t}} - \psi_{k-1}^{\mathrm{a}})^2}, \tag{4.7}$$

the model error covariance

$$C_{qq}(t_{k-1}) = \overline{q_{k-1}^2}, \tag{4.8}$$

and the initial error covariance

$$C_{\psi\psi}(t_0) = C_{aa} = \overline{a^2}. \tag{4.9}$$

It is also assumed that there are no correlations between the error in the state, $\psi_{k-1}^{\mathrm{t}} - \psi_{k-1}^{\mathrm{a}}$, the model error q_{k-1}, and the initial error a.

Thus, we have a consistent set of dynamical equations for the model evolution (4.3 and 4.4), and the error (co)variance evolution (4.6), (4.8) and (4.9). At times when there are measurements available, an analyzed estimate can be calculated using the equations (3.14) and (3.15), and when there are no measurements available we just set $\psi^{\mathrm{a}} = \psi^{\mathrm{f}}$ and $C_{\psi\psi}^{\mathrm{a}} = C_{\psi\psi}^{\mathrm{f}}$, and continue the integration. These equations define the Kalman filter for a linear scalar model, and thus constitute the optimal sequential data assimilation method for this model given that the priors are all Gaussian and unbiased.

4.1.2 Kalman filter for a vector state

If the true state $\psi^{\mathrm{t}}(\boldsymbol{x})$ is discretized on a numerical grid, it can be represented by the state vector $\boldsymbol{\psi}^{\mathrm{t}}$. It is assumed that the true state evolves according to a dynamical model

$$\boldsymbol{\psi}_k^{\mathrm{t}} = \boldsymbol{G}\boldsymbol{\psi}_{k-1}^{\mathrm{t}} + \boldsymbol{q}_{k-1}, \qquad (4.10)$$

where $\boldsymbol{G}$ is a linear model operator (matrix) and $\boldsymbol{q}$ is the unknown model error over one time step. In this case a numerical model will evolve according to

$$\boldsymbol{\psi}_k^{\mathrm{f}} = \boldsymbol{G}\boldsymbol{\psi}_{k-1}^{\mathrm{a}}. \qquad (4.11)$$

That is, given the best possible estimate for $\boldsymbol{\psi}$ at time t_{k-1}, a forecast is calculated at time t_k, using the approximate equation (4.11).

The error covariance equation is derived using a similar procedure as was used for (4.6), and becomes

$$\boldsymbol{C}_{\psi\psi}^{\mathrm{f}}(t_k) = \boldsymbol{G}\boldsymbol{C}_{\psi\psi}^{\mathrm{a}}(t_{k-1})\boldsymbol{G}^{\mathrm{T}} + \boldsymbol{C}_{qq}(t_{k-1}). \qquad (4.12)$$

Thus, the standard Kalman filter consists of the dynamical equations (4.11) and (4.12) together with the analysis equations (3.83–3.85) or alternatively (3.79–3.82).

4.1.3 Kalman filter with a linear advection equation

Here we illustrate the properties of the KF when used with a one-dimensional linear advection model on a periodic domain of length 1000 m. The model has a constant advection speed, $u = 1$ m/s, the grid spacing is $\Delta x = 1$ m and the time step is $\Delta t = 1$ s.

Given an initial condition, the solution of this model is exactly known, and this allows us to run realistic experiments with zero model errors to examine the impact of the dynamical evolution of the error covariance.

The true initial state is sampled from a distribution $\mathcal{N}$, with mean equal to zero, variance equal to one, and a spatial de-correlation length of 20 m.

The first guess solution is generated by drawing another sample from $\mathcal{N}$ and adding this to the true state. The initial ensemble is then generated by adding samples drawn from $\mathcal{N}$ to the first guess solution. Thus, the initial state is assumed to have an error variance equal to one.

Four measurements of the true solution, distributed regularly in the model domain, are assimilated every 5th time step. The measurements are contaminated by errors of variance equal to 0.01, and we have assumed uncorrelated measurement errors.

The length of the integration is 300 s, which is 50 s longer than the time needed for the solution to advect from one measurement to the next (i.e. 250 s).

In Fig. 4.1 an example is shown where the model errors have been set to zero. The plots illustrate the convergence of the estimated solution at various

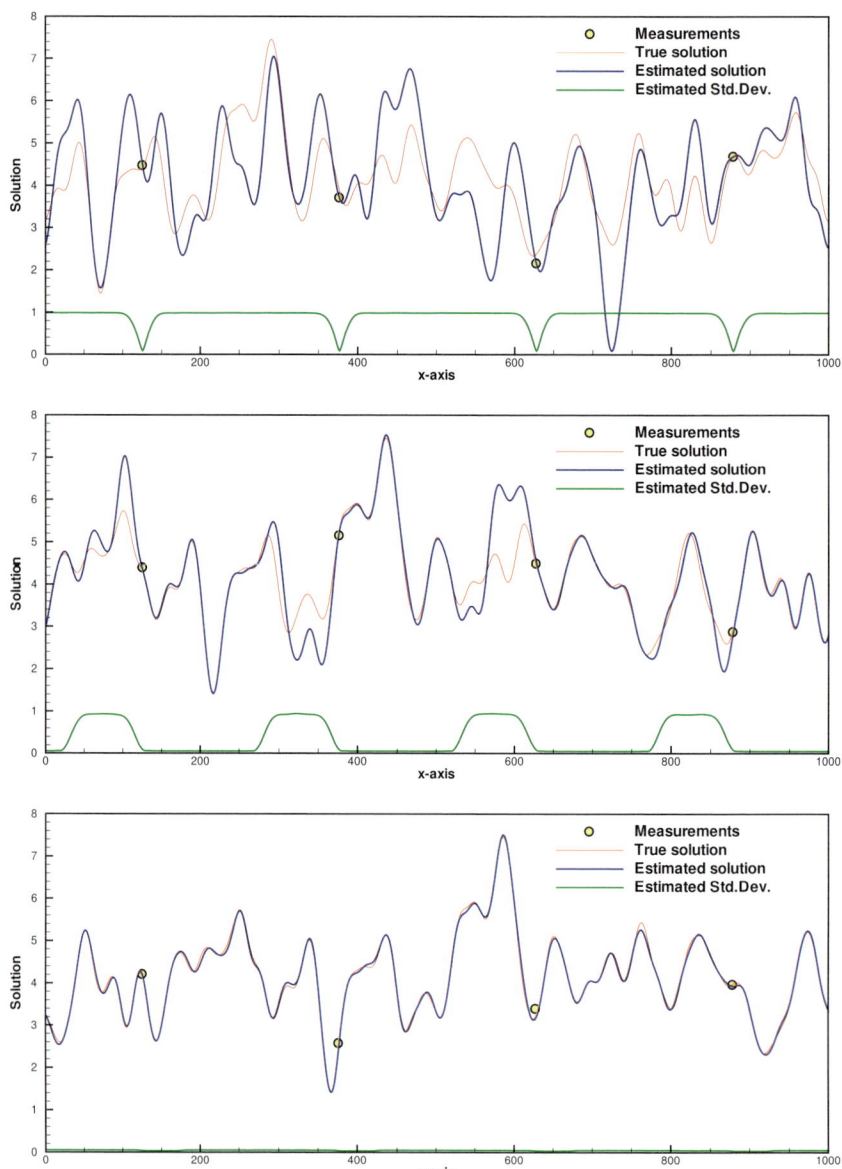

Fig. 4.1. Kalman filter experiment: reference solution, measurements, estimate and standard deviation at three different times $t = 5$ *(top)*, $t = 150$ *(middle)*, and $t = 300$ *(bottom)*

4.1 Linear Dynamics 31

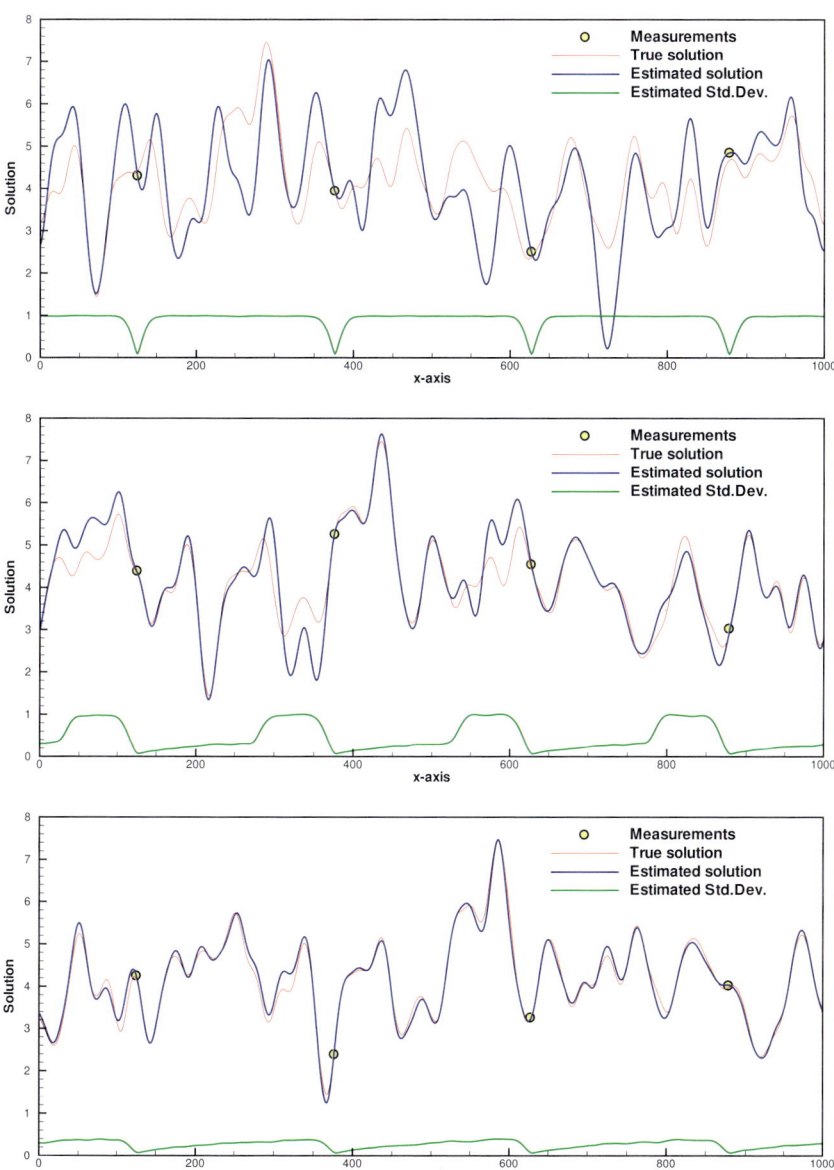

Fig. 4.2. Kalman filter experiment when system noise is included: reference solution, measurements, estimate and standard deviation at three different times $t = 5$ *(top)*, $t = 150$ *(middle)*, and $t = 300$ *(bottom)*

times during the experiment, and show how information from measurements is propagated with the advection speed and how the error variance is reduced every time measurements are assimilated.

The first plot shows the result of the first update with the four measurements. Near the measurement locations, the estimated solution is clearly consistent with the true solution and the measurements, and the error variance is reduced accordingly. The second plot is taken at $t = 150$ s, i.e. after 30 updates with measurements. Now the information from the measurements has propagated to the right with the advection speed. This is seen both from direct comparison of the estimate with the true solution, and from the estimated variance. The final plot is taken at $t = 300$ s and the estimate is now in good agreement with the true solution throughout the model domain. Note also a further reduction of the error variance to the right of the measurements. This is caused by the further introduction of information from the measurements to the already accurate estimate. In this case the estimated errors will converge towards zero since we experience a further accumulation of information and error reduction every 250 s of integration.

The impact of model errors is illustrated in Fig. 4.2. Here we note a linear increase in error variance to the right of the measurements. This is caused by the addition of model errors every time step. It is also clear that the estimated solution deteriorates far from the measurement in the advection direction. The converged error variance is larger than in the previous case. It turns out that for linear models with regular measurements at fixed locations and stationary error statistics, the error variance converges to an estimate where the increase of error variance from model errors balances the reduction from the updates with measurements.

In fact these examples were actually run using the Ensemble Kalman Filter discussed below, but for a linear model the EnKF will converge exactly to the KF with increasing ensemble size.

4.2 Nonlinear dynamics

For nonlinear dynamics the extended Kalman filter (EKF) may be applied, in which an approximate linearized equation is used for the prediction of error statistics.

4.2.1 Extended Kalman filter for the scalar case

Assume now that we have a nonlinear scalar model

$$\psi_k^t = G(\psi_{k-1}^t) + q_{k-1}, \tag{4.13}$$

where $G(\psi)$ is a nonlinear model operator and q is again the unknown model error over one time step. A numerical model will evolve according to the approximate equation

$$\psi_k^{\mathrm{f}} = G(\psi_{k-1}^{\mathrm{a}}). \tag{4.14}$$

Subtracting (4.14) from (4.13) gives

$$\psi_k^{\mathrm{t}} - \psi_k^{\mathrm{f}} = G(\psi_{k-1}^{\mathrm{t}}) - G(\psi_{k-1}^{\mathrm{a}}) + q_{k-1}. \tag{4.15}$$

Now use the Taylor expansion

$$\begin{aligned} G(\psi_{k-1}^{\mathrm{t}}) &= G(\psi_{k-1}^{\mathrm{a}}) + G'(\psi_{k-1}^{\mathrm{a}})(\psi_{k-1}^{\mathrm{t}} - \psi_{k-1}^{\mathrm{a}}) \\ &+ \frac{1}{2} G''(\psi_{k-1}^{\mathrm{a}})(\psi_{k-1}^{\mathrm{t}} - \psi_{k-1}^{\mathrm{a}})^2 + \cdots, \end{aligned} \tag{4.16}$$

in (4.15) to get

$$\begin{aligned} \psi_k^{\mathrm{t}} - \psi_k^{\mathrm{f}} &= G'(\psi_{k-1}^{\mathrm{a}})(\psi_{k-1}^{\mathrm{t}} - \psi_{k-1}^{\mathrm{a}}) \\ &+ \frac{1}{2} G''(\psi_{k-1}^{\mathrm{a}})(\psi_{k-1}^{\mathrm{t}} - \psi_{k-1}^{\mathrm{a}})^2 + \cdots + q_{k-1}. \end{aligned} \tag{4.17}$$

By squaring and taking the expected value an equation for the evolution of the error variance $C_{\psi\psi}^{\mathrm{f}}(t_k)$ becomes

$$\begin{aligned} C_{\psi\psi}^{\mathrm{f}}(t_k) &= \overline{(\psi_k^{\mathrm{t}} - \psi_k^{\mathrm{f}})^2} \\ &= \overline{(\psi_{k-1}^{\mathrm{t}} - \psi_{k-1}^{\mathrm{a}})^2}(G'(\psi_{k-1}^{\mathrm{a}}))^2 \\ &+ \overline{(\psi_{k-1}^{\mathrm{t}} - \psi_{k-1}^{\mathrm{a}})^3}G'(\psi_{k-1}^{\mathrm{a}})G''(\psi_{k-1}^{\mathrm{a}}) \\ &+ \frac{1}{4}\overline{(\psi_{k-1}^{\mathrm{t}} - \psi_{k-1}^{\mathrm{a}})^4}(G''(\psi_{k-1}^{\mathrm{a}}))^2 + \cdots + C_{qq}(t_{k-1}). \end{aligned} \tag{4.18}$$

This equation can be closed by discarding moments of third and higher order, which results in an approximate equation for the error variance,

$$C_{\psi\psi}^{\mathrm{f}}(t_k) \simeq C_{\psi\psi}^{\mathrm{a}}(t_{k-1})(G'(\psi_{k-1}^{\mathrm{a}}))^2 + C_{qq}(t_{k-1}). \tag{4.19}$$

Together with the equations for the analyzed estimate and error variance, (3.14) and (3.15), the dynamical equations (4.14) and (4.19) constitute the extended Kalman filter (EKF) in the case with a scalar state variable.

It is clear that we now have an approximate equation for the error covariance evolution, due to the linearization and closure assumption used. Thus, the usefulness of the EKF will depend on the properties of the model dynamics.

The EKF can also be formulated for measurements which are related to the state variables by a nonlinear operator (see *Gelb*, 1974).

4.2.2 Extended Kalman filter in matrix form

The derivation of the EKF in matrix form is based on the same principles as for the scalar case and can be found in a number of books on control theory (see e.g. *Jazwinski*, 1970, *Gelb*, 1974). Again we assume a nonlinear model, but now the true state vector at time t_k is calculated from

$$\boldsymbol{\psi}_k^{\text{t}} = \boldsymbol{G}(\boldsymbol{\psi}_{k-1}^{\text{t}}) + \boldsymbol{q}_{k-1}, \tag{4.20}$$

and a forecast is calculated from the approximate equation

$$\boldsymbol{\psi}_k^{\text{f}} = \boldsymbol{G}(\boldsymbol{\psi}_{k-1}^{\text{a}}), \tag{4.21}$$

where the model is now dependent of both time and space. The error statistics are then described by the error covariance matrix $\boldsymbol{C}_{\psi\psi}^{\text{f}}(t_k)$ which evolves according to the equation

$$\begin{aligned}
\boldsymbol{C}_{\psi\psi}^{\text{f}}(t_k) &= \boldsymbol{G}'_{k-1}\boldsymbol{C}_{\psi\psi}^{\text{a}}(t_{k-1})\boldsymbol{G}'^{\text{T}}_{k-1} + \boldsymbol{C}_{qq}(t_{k-1}) \\
&+ \boldsymbol{G}'_{k-1}\boldsymbol{\Theta}_{\psi\psi\psi}(t_{k-1})\boldsymbol{\mathcal{H}}^{\text{T}}_{k-1} + \frac{1}{4}\boldsymbol{\mathcal{H}}_{k-1}\boldsymbol{\Gamma}_{\psi\psi\psi\psi}(t_{k-1})\boldsymbol{\mathcal{H}}^{\text{T}}_{k-1} \\
&+ \frac{1}{3}\boldsymbol{G}'_{k-1}\boldsymbol{\Gamma}_{\psi\psi\psi\psi}(t_{k-1})\boldsymbol{\mathcal{T}}^{\text{T}}_{k-1} \\
&+ \frac{1}{4}\boldsymbol{\mathcal{H}}_{k-1}\boldsymbol{C}_{\psi\psi}^{\text{a}}(t_{k-1})\boldsymbol{C}_{\psi\psi}^{\text{aT}}(t_{k-1})\boldsymbol{\mathcal{H}}^{\text{T}}_{k-1} \\
&+ \frac{1}{6}\boldsymbol{\mathcal{H}}_{k-1}\boldsymbol{C}_{\psi\psi}^{\text{a}}(t_{k-1})\boldsymbol{\Theta}^{\text{T}}_{\psi\psi\psi}(t_{k-1})\boldsymbol{\mathcal{T}}^{\text{T}}_{k-1} \\
&+ \frac{1}{36}\boldsymbol{\mathcal{T}}_{k-1}\boldsymbol{\Theta}_{\psi\psi\psi}(t_{k-1})\boldsymbol{\Theta}^{\text{T}}_{\psi\psi\psi}(t_{k-1})\boldsymbol{\mathcal{T}}^{\text{T}}_{k-1} + \cdots,
\end{aligned} \tag{4.22}$$

where $\boldsymbol{C}_{qq}(t_{k-1})$ is the model error covariance matrix, $\boldsymbol{G}'_{k-1}$ is the Jacobi matrix or tangent linear operator,

$$\boldsymbol{G}'_{k-1} = \left.\frac{\partial \boldsymbol{G}(\boldsymbol{\psi})}{\partial \boldsymbol{\psi}}\right|_{\boldsymbol{\psi}_{k-1}}, \tag{4.23}$$

$\boldsymbol{\Theta}_{\psi\psi\psi}$ is the third order statistical moment, $\boldsymbol{\Gamma}_{\psi\psi\psi\psi}$ is the fourth order statistical moment, $\boldsymbol{\mathcal{H}}$ is the Hessian, consisting of second order derivatives of the nonlinear model operator, and $\boldsymbol{\mathcal{T}}$ is an operator containing the third order derivatives of the model operator.

The EKF is based on the assumption that the contribution from all the higher order terms in (4.22), are negligible. By discarding these terms we are left with the approximate error covariance equation

$$\boldsymbol{C}_{\psi\psi}^{\text{f}}(t_k) \simeq \boldsymbol{G}'_{k-1}\boldsymbol{C}_{\psi\psi}^{\text{a}}(t_{k-1})\boldsymbol{G}'^{\text{T}}_{k-1} + \boldsymbol{C}_{qq}(t_{k-1}). \tag{4.24}$$

The analogy between the vector and scalar cases is obvious.

A discussion of the case where higher order approximations for the error variance evolution is used, is given by *Miller* (1994).

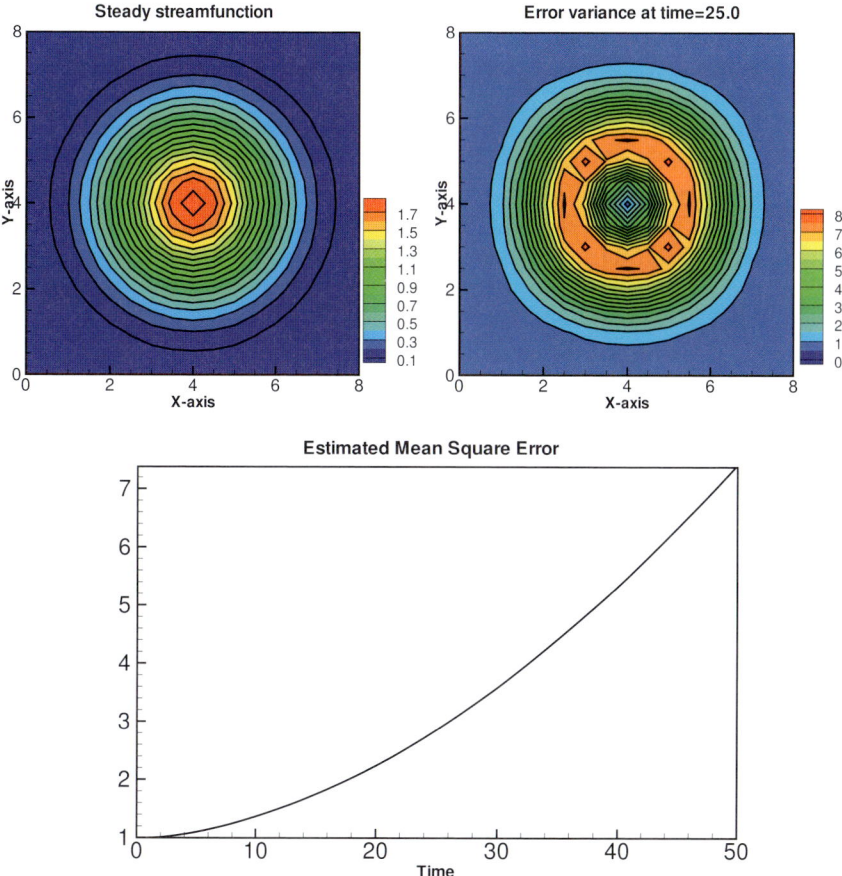

Fig. 4.3. Example of an EKF experiment from *Evensen* (1992). The upper left plot shows the stream function defining the velocity field of a stationary eddy. The upper right plot shows the resulting error variance in the model domain after integration from $t = 0$ till $t = 25$, note the large errors at locations where velocities are high. The lower plot illustrates the exponential time evolution of the estimated variance averaged over the model domain

4.2.3 Example using the extended Kalman filter

Evensen (1992) provided the first application of the EKF with a nonlinear ocean circulation model. The model was a multi-layer quasi-geostrophic model which represents well the mesoscale ocean variability. It solves a conservation equation for potential vorticity.

In *Evensen* (1992) properties of the EKF with this particular model were examined. It was found that the linear evolution equation for the error covariance matrix leads to an unbounded linear instability. This was demonstrated in an experiment using a steady background flow defined by an eddy standing on a flat bottom and with no beta effect (see left plot in Fig. 4.3). Thus, vorticity is just advected along the stream lines with a velocity defined by the stream function.

The particular stream function results in a velocity shear and thus supports standard sheared flow instability. Thus, if we add a perturbation and advect it using the linearized equations the perturbation will grow exponentially. This is exactly what is observed in the upper right and lower plots of Fig. 4.3. We started out with an initial variance equal to one in all of the model domain and observe a strong error variance growth at locations of large velocity and velocity shear in the eddy. The estimated mean square errors, which equals the trace of $\boldsymbol{C}_{\psi\psi}$ divided by the number of grid points, indicate the exponential error variance growth.

This linear instability is not realistic. In the real world we would expect the instability to saturate at a certain climatological amplitude. As an example, in the atmosphere it is always possible to define a maximum and minimum pressure which is never exceeded, and the same applies for the eddy field in the ocean. A variance estimate which indicates unphysical amplitudes of the variability cannot be accepted, and this is in fact what the EKF may provide.

The main result from this work was the finding of an apparent closure problem in the error covariance evolution equation. The EKF applies a closure scheme where third- and higher-order moments in the error covariance evolution equation are discarded. This results in an unbounded error variance growth or linear instability in the error covariance equation in some dynamical models. If an exact error covariance evolution equation could be used all linear instabilities will saturate due to nonlinear effects. This saturation is missing in the EKF, as was later confirmed by *Miller et al.* (1994), *Gauthier et al.* (1993) and *Bouttier* (1994).

In particular *Miller et al.* (1994) gave a comprehensive discussion on the application of the EKF with the chaotic Lorenz model. The too simplified closure resulted in an estimated solution which was only acceptable in a fairly short time interval, and thereafter unreliable. This was explained by a poor prediction of error covariances $\boldsymbol{C}^{\mathrm{f}}_{\psi\psi}$, resulting in insufficient gain $\boldsymbol{K}$, because of a decaying mode which reflects the stability of the attractor.

A generalization of the EKF, where third and fourth order moments and evolution equations for these were included, was also examined by *Miller et al.* (1994) and it was shown that this more sophisticated closure scheme provided a better evolution of error statistics which also resulted in sufficient gain to keep the estimate on track.

4.2.4 Extended Kalman filter for the mean

The previous derivation is the most commonly used for the EKF. A weakness of the formulation is that the so-called central forecast is used as the estimate. The central forecast is a single model realization initialized with the best estimate of the initial state. For nonlinear dynamics the central forecast is not equal to the expected value, and it is not clear how it shall be interpreted.

A different approach is to derive a model for the evolution of the first moment or mean. This is done by expanding $G(\psi)$ around $\overline{\psi}$ to get

$$G(\psi) = G(\overline{\psi}) + G'(\overline{\psi})(\psi - \overline{\psi}) + \frac{1}{2}G''(\overline{\psi})(\psi - \overline{\psi})^2 + \cdots . \quad (4.25)$$

Inserting this in (4.13) and taking the expectation or ensemble average results in the equation

$$\overline{\psi_k} = G(\overline{\psi_{k-1}}) + \frac{1}{2}G''(\overline{\psi_{k-1}})C_{\psi\psi}(t_{k-1}) + \cdots . \quad (4.26)$$

In the vector case this equation becomes

$$\overline{\boldsymbol{\psi}_k} = \boldsymbol{G}(\overline{\boldsymbol{\psi}_{k-1}}) + \frac{1}{2}\mathcal{H}_{k-1}\boldsymbol{C}_{\psi\psi}(t_{k-1}) + \cdots . \quad (4.27)$$

One may argue that for a statistical estimator it makes more sense to work with the mean than a central forecast, after all, the central forecast does not have any statistical interpretation. This can be illustrated by running an atmospheric model without assimilation updates. The central forecast then becomes just one realization out of infinitively many possible realizations and it is not clear how one may relate this to the climatological error covariance estimate. On the other hand the equation for the mean will provide an estimate which converges towards the climatological mean and the covariance estimate thus describes the error variance of the climatological mean. Until now, all applications of the EKF for data assimilation in ocean and atmospheric models have used an equation for the central forecast. However, the interpretation using the equation for the mean will later on support the development of the Ensemble Kalman Filter.

4.2.5 Discussion

There are two major drawbacks of the Kalman filter for data assimilation in high dimensional and nonlinear dynamical models.

The first is related to storage and computational issues. If the dynamical model has n unknowns in the state vector, then the error covariance matrix $\boldsymbol{C}_{\psi\psi}$ has n^2 unknowns. Furthermore, the evolution of the error covariance matrix in time requires the cost of $2n$ model integrations. Thus, clearly, the KF and EKF in the present form, can only be practically used with fairly low-dimensional dynamical models.

The second issue is related to the use of the EKF with nonlinear dynamical models, which requires a linearization when deriving the error covariance evolution equation. This linearization leads to a poor error covariance evolution and for some models unstable error covariance growth. This may be resolved using higher order closure schemes. Unfortunately, such an approach is not practical for a high dimensional model, since the fourth order moment requires storage of n^4 elements. In general one may conclude that a more consistent closure is needed in the error covariance equation.

4.3 Ensemble Kalman filter

Another sequential data assimilation method which has received a lot of attention is named the Ensemble Kalman Filter (EnKF). The method was originally proposed as a stochastic or Monte Carlo alternative to the deterministic EKF by *Evensen* (1994a). The EnKF was designed to resolve the two major problems related to the use of the EKF with nonlinear dynamics in large state spaces, i.e. the use of an approximate closure scheme and the huge computational requirements associated with the storage and forward integration of the error covariance matrix.

The EnKF has gained popularity because of its simple conceptual formulation and relative ease of implementation, e.g. it requires no derivation of a tangent linear operator or adjoint equations and no integrations backward in time. Furthermore, the computational requirements are affordable and comparable to other popular sophisticated assimilation methods such as the representer method by *Bennett* (1992), *Bennett et al.* (1993), *Bennett and Chua* (1994), *Bennett et al.* (1996) and the 4DVAR method which has been much studied by the meteorological community (see e.g. *Talagrand and Courtier*, 1987, *Courtier and Talagrand*, 1987, *Courtier et al.*, 1994, *Courtier*, 1997).

We will adapt a three stage presentation starting with the representation of error statistics using an ensemble of model states, then an alternative to the traditional error covariance equation is proposed for the prediction of error statistics, and finally a consistent analysis scheme is presented.

4.3.1 Representation of error statistics

The error covariance matrices for the predicted and the analyzed estimate, $\boldsymbol{C}^{\mathrm{f}}_{\psi\psi}$ and $\boldsymbol{C}^{\mathrm{a}}_{\psi\psi}$, are in the Kalman filter defined in terms of the true state as

$$\boldsymbol{C}^{\mathrm{f}}_{\psi\psi} = \overline{(\boldsymbol{\psi}^{\mathrm{f}} - \boldsymbol{\psi}^{\mathrm{t}})(\boldsymbol{\psi}^{\mathrm{f}} - \boldsymbol{\psi}^{\mathrm{t}})^{\mathrm{T}}}, \tag{4.28}$$

$$\boldsymbol{C}^{\mathrm{a}}_{\psi\psi} = \overline{(\boldsymbol{\psi}^{\mathrm{a}} - \boldsymbol{\psi}^{\mathrm{t}})(\boldsymbol{\psi}^{\mathrm{a}} - \boldsymbol{\psi}^{\mathrm{t}})^{\mathrm{T}}}, \tag{4.29}$$

where the ensemble averaging defined by the overline converges to the expectation value in the case of an infinite ensemble size. However, the true state is not known, and we therefore define the ensemble covariance matrices around the ensemble mean $\overline{\boldsymbol{\psi}}$,

$$\left(\boldsymbol{C}^{\mathrm{e}}_{\psi\psi}\right)^{\mathrm{f}} = \overline{\left(\boldsymbol{\psi}^{\mathrm{f}} - \overline{\boldsymbol{\psi}^{\mathrm{f}}}\right)\left(\boldsymbol{\psi}^{\mathrm{f}} - \overline{\boldsymbol{\psi}^{\mathrm{f}}}\right)^{\mathrm{T}}}, \tag{4.30}$$

$$\left(\boldsymbol{C}^{\mathrm{e}}_{\psi\psi}\right)^{\mathrm{a}} = \overline{\left(\boldsymbol{\psi}^{\mathrm{a}} - \overline{\boldsymbol{\psi}^{\mathrm{a}}}\right)\left(\boldsymbol{\psi}^{\mathrm{a}} - \overline{\boldsymbol{\psi}^{\mathrm{a}}}\right)^{\mathrm{T}}}, \tag{4.31}$$

where now the overline denote an average over the ensemble. Thus, we can use an interpretation where the ensemble mean is the best estimate and the spreading of the ensemble around the mean is a natural definition of the error in the ensemble mean.

Since the error covariances as defined in (4.30) and (4.31) are defined as ensemble averages, there will clearly exist an infinite number of ensembles with an error covariance equal to $\boldsymbol{C}^{\mathrm{e}}_{\psi\psi}$. Thus, instead of storing a full covariance matrix, we can represent the same error statistics using an appropriate ensemble of model states. Given an error covariance matrix, an ensemble of finite size will provide an approximation to the error covariance matrix. However, when the size of the ensemble N increases, the errors in the Monte Carlo sampling will decrease proportional to $1/\sqrt{N}$.

Suppose now that we have N model states in the ensemble, each of dimension n. Each of these model states can be represented as a single point in an n-dimensional state space. All the ensemble members together will constitute a cloud of points in the state space. Such a cloud of points can, in the limit when N goes to infinity, be described using a probability density function

$$f(\boldsymbol{\psi}) = \frac{dN}{N}, \tag{4.32}$$

where dN is the number of points in a small unit volume and N is the total number of points. With knowledge about either $f(\boldsymbol{\psi})$ or the ensemble representing $f(\boldsymbol{\psi})$ we can calculate whichever statistical moments (such as mean, covariances etc.) we want whenever they are needed.

The conclusion so far is that the information contained by a full probability density function can be exactly represented by an infinite ensemble of model states.

4.3.2 Prediction of error statistics

In *Evensen* (1994a) it was shown that a Monte Carlo method can be used to solve an equation for the time evolution of the probability density of the model state, as an alternative to using the approximate error covariance equation in the EKF.

For a nonlinear model where we appreciate that the model is not perfect and contains model errors, we can write it as a stochastic differential equation as

$$d\boldsymbol{\psi} = \boldsymbol{G}(\boldsymbol{\psi})dt + \boldsymbol{h}(\boldsymbol{\psi})d\boldsymbol{q}. \tag{4.33}$$

This equation states that an increment in time will yield an increment in $\boldsymbol{\psi}$, which in addition, is influenced by a random contribution from the stochastic forcing term $\boldsymbol{h}(\boldsymbol{\psi})\,d\boldsymbol{q}$, representing the model errors. The $d\boldsymbol{q}$ term describes a vector Brownian motion process with covariance $\boldsymbol{C}_{qq}dt$. The nonlinear model operator $\boldsymbol{G}$ is not an explicit function of the random variable $d\boldsymbol{q}$ so the Ito interpretation of the stochastic differential equation is used instead of the Stratonovich interpretation (see *Jazwinski*, 1970).

When additive Gaussian model errors forming a Markov process are used one can derive the Fokker-Planck equation (also named Kolmogorov's equation) which describes the time evolution of the probability density $f(\boldsymbol{\psi})$ of the model state,

$$\frac{\partial f}{\partial t} + \sum_i \frac{\partial (g_i f)}{\partial \psi_i} = \frac{1}{2} \sum_{i,j} \frac{\partial^2 f(\boldsymbol{h}\boldsymbol{C}_{qq}\boldsymbol{h}^T)_{ij}}{\partial \psi_i \partial \psi_j}, \qquad (4.34)$$

where g_i is the component number i of the model operator $\boldsymbol{G}$ and $\boldsymbol{h}\boldsymbol{C}_{qq}\boldsymbol{h}^{\mathrm{T}}$ is the covariance matrix for the model errors.

This equation does not apply any important approximations and can be considered as the fundamental equation for the time evolution of error statistics. A detailed derivation is given in *Jazwinski* (1970). The equation describes the change of the probability density in a local "volume" which is dependent on the divergence term describing a probability flux into the local "volume" (impact of the dynamical equation) and the diffusion term which tends to flatten the probability density due to the effect of stochastic model errors. If (4.34) could be solved for the probability density function, it would be possible to calculate statistical moments like the mean and the error covariance for the model forecast to be used in the analysis scheme.

A linear model for a Gauss-Markov process in which the initial condition is assumed to be taken from a normal distribution will have a probability density which is completely characterized by its mean and covariance for all times. One can then derive exact equations for the evolution of the mean and the covariance as a simpler alternative than solving the full Kolmogorov's equation. Such moments of Kolmogorov's equation, including the error covariance (4.12), are easy to derive, and several methods are illustrated by *Jazwinski* (1970, examples 4.19–4.21). This is actually what is done in the KF.

For a nonlinear model, the mean and covariance matrix will not in general characterize the time evolution of $f(\boldsymbol{\psi})$. They do, however, determine the mean path and the dispersion about that path, and it is possible to solve approximate equations for the moments, which is the procedure characterizing the EKF.

The EnKF applies a so-called Markov Chain Monte Carlo (MCMC) method to solve (4.34). The probability density is then represented by a large ensemble of model states as discussed in the previous section. By integrating these model states forward in time according to the model dynamics, as described by the stochastic differential (4.33), this ensemble prediction is equivalent to solving the Fokker Planck equation using a MCMC method.

Different dynamical models can have stochastic terms embedded within the nonlinear model operator and the derivation of the associated Fokker Planck equation may become very complex. Fortunately, the Fokker Planck equation is not needed, since it is sufficient to know that it exists and the MCMC method solves it.

4.3.3 Analysis scheme

The KF analysis scheme uses the definitions of $C^{\mathrm{f}}_{\psi\psi}$ and $C^{\mathrm{a}}_{\psi\psi}$ as given by (4.28) and (4.29). We will now give a derivation of the analysis scheme using the ensemble covariances as defined by (4.30) and (4.31).

As was shown by *Burgers et al.* (1998) it is essential that the observations are treated as random variables having a distribution with mean equal to the first guess observations and covariance equal to $C_{\epsilon\epsilon}$. Thus, we start by defining an ensemble of observations

$$d_j = d + \epsilon_j, \qquad (4.35)$$

where j counts from 1 to the number of ensemble members N. It is ensured that the simulated random measurement errors have mean equal to zero. Next we define the ensemble covariance matrix of the measurement errors as

$$C^{\mathrm{e}}_{\epsilon\epsilon} = \overline{\epsilon\epsilon^{\mathrm{T}}}, \qquad (4.36)$$

and, of course, in the limit of an infinite ensemble size this matrix will converge towards the prescribed error covariance matrix $C_{\epsilon\epsilon}$ used in the standard Kalman filter.

The following discussion is valid both using an exactly prescribed $C_{\epsilon\epsilon}$ and an ensemble representation $C^{\mathrm{e}}_{\epsilon\epsilon}$ of $C_{\epsilon\epsilon}$. The use of $C^{\mathrm{e}}_{\epsilon\epsilon}$ introduces an additional approximation which sometimes is convenient when implementing the analysis scheme. This approximation can be justified since normally the true observation error covariance matrix is poorly known and the errors introduced by the ensemble representation can be made less than the uncertainty in the true $C_{\epsilon\epsilon}$ by choosing a large enough ensemble size. Further, the use of an ensemble representation for $C_{\epsilon\epsilon}$, has less impact than the use of an ensemble representation for $C^{\mathrm{f}}_{\psi\psi}$. Further, $C_{\epsilon\epsilon}$ only appears in the computation of the coefficients for the influence functions $C^{\mathrm{f}}_{\psi\psi} M^{\mathrm{T}}$ while $C^{\mathrm{f}}_{\psi\psi}$ appears both in the computation of the coefficients and it determines the influence functions. Note, however that there are specific issues related to the rank of $C^{\mathrm{e}}_{\epsilon\epsilon}$ when the number of measurements becomes large as is discussed in Chap. 14.

The analysis step in the EnKF consists of the following updates performed on each of the model state ensemble members

$$\psi^{\mathrm{a}}_j = \psi^{\mathrm{f}}_j + \left(C^{\mathrm{e}}_{\psi\psi}\right)^{\mathrm{f}} M^{\mathrm{T}} \left(M \left(C^{\mathrm{e}}_{\psi\psi}\right)^{\mathrm{f}} M^{\mathrm{T}} + C^{\mathrm{e}}_{\epsilon\epsilon}\right)^{-1} \left(d_j - M\psi^{\mathrm{f}}_j\right). \qquad (4.37)$$

With a finite ensemble size, this equation will be an approximation. Further, if the number of measurements is larger than the number of ensemble members, the matrices $M\left(C^{\mathrm{e}}_{\psi\psi}\right)^{\mathrm{f}} M^{\mathrm{T}}$ and $C^{\mathrm{e}}_{\epsilon\epsilon}$ will be singular, and a pseudo inversion must be used (see Chap. 14).

Equation (4.37) implies that

$$\overline{\psi^{\mathrm{a}}} = \overline{\psi^{\mathrm{f}}} + \left(C^{\mathrm{e}}_{\psi\psi}\right)^{\mathrm{f}} M^{\mathrm{T}} \left(M\left(C^{\mathrm{e}}_{\psi\psi}\right)^{\mathrm{f}} M^{\mathrm{T}} + C^{\mathrm{e}}_{\epsilon\epsilon}\right)^{-1} \left(\overline{d} - M\overline{\psi^{\mathrm{f}}}\right), \qquad (4.38)$$

where $\overline{d} = d$ is the first guess vector of measurements. Thus, the relation between the analyzed and predicted ensemble mean is identical to the relation between the analyzed and predicted state in the standard Kalman filter, apart from the use of $\left(C^{\mathrm{e}}_{\psi\psi}\right)^{\mathrm{f,a}}$ and $C^{\mathrm{e}}_{\epsilon\epsilon}$ instead of $C^{\mathrm{f,a}}_{\psi\psi}$ and $C_{\epsilon\epsilon}$. Note that the introduction of an ensemble of observations does not make any difference for the update of the ensemble mean since this does not affect (4.38).

If the mean $\overline{\psi^{\mathrm{a}}}$ is considered to be the best estimate, then it is an arbitrary choice whether one updates the mean using the first guess observations d, or if one updates each of the ensemble members using the perturbed observations (4.35). However, it will now be shown that by updating each of the ensemble members using the perturbed observations one also creates a new ensemble with the correct error statistics for the analysis. The updated ensemble can then be integrated forward in time till the next observation time.

We now derive the analyzed error covariance estimate resulting from the analysis scheme given above, but using the standard Kalman filter form for the analysis equations. First, (4.37) and (4.38) are used to obtain

$$\psi^{\mathrm{a}}_j - \overline{\psi^{\mathrm{a}}} = (I - K_{\mathrm{e}} M)\left(\psi^{\mathrm{f}}_j - \overline{\psi^{\mathrm{f}}}\right) + K_{\mathrm{e}}\left(d_j - \overline{d}\right), \qquad (4.39)$$

where we have used the definition of the Kalman gain,

$$K_{\mathrm{e}} = \left(C^{\mathrm{e}}_{\psi\psi}\right)^{\mathrm{f}} M^{\mathrm{T}} \left(M\left(C^{\mathrm{e}}_{\psi\psi}\right)^{\mathrm{f}} M^{\mathrm{T}} + C^{\mathrm{e}}_{\epsilon\epsilon}\right)^{-1}. \qquad (4.40)$$

The derivation is then as follows,

$$\begin{aligned}
\left(C^{\mathrm{e}}_{\psi\psi}\right)^{\mathrm{a}} &= \overline{\left(\psi^{\mathrm{a}} - \overline{\psi^{\mathrm{a}}}\right)\left(\psi^{\mathrm{a}} - \overline{\psi^{\mathrm{a}}}\right)^{\mathrm{T}}} \\
&= \overline{\left((I - K_{\mathrm{e}} M)\left(\psi^{\mathrm{f}} - \overline{\psi^{\mathrm{f}}}\right) + K_{\mathrm{e}}\left(d - \overline{d}\right)\right)(\cdots)^{\mathrm{T}}} \\
&= (I - K_{\mathrm{e}} M)\overline{\left(\psi^{\mathrm{f}} - \overline{\psi^{\mathrm{f}}}\right)\left(\psi^{\mathrm{f}} - \overline{\psi^{\mathrm{f}}}\right)^{\mathrm{T}}}(I - K_{\mathrm{e}} M)^{\mathrm{T}} \\
&\quad + K_{\mathrm{e}} \overline{\left(d - \overline{d}\right)\left(d - \overline{d}\right)^{\mathrm{T}}} K_{\mathrm{e}}^{\mathrm{T}} \qquad (4.41)\\
&= (I - K_{\mathrm{e}} M)\left(C^{\mathrm{e}}_{\psi\psi}\right)^{\mathrm{f}}\left(I - M^{\mathrm{T}} K_{\mathrm{e}}^{\mathrm{T}}\right) + K_{\mathrm{e}} C^{\mathrm{e}}_{\epsilon\epsilon} K_{\mathrm{e}}^{\mathrm{T}} \\
&= \left(C^{\mathrm{e}}_{\psi\psi}\right)^{\mathrm{f}} - K_{\mathrm{e}} M \left(C^{\mathrm{e}}_{\psi\psi}\right)^{\mathrm{f}} - \left(C^{\mathrm{e}}_{\psi\psi}\right)^{\mathrm{f}} M^{\mathrm{T}} K_{\mathrm{e}}^{\mathrm{T}} \\
&\quad + K_{\mathrm{e}}(M\left(C^{\mathrm{e}}_{\psi\psi}\right)^{\mathrm{f}} M^{\mathrm{T}} + C^{\mathrm{e}}_{\epsilon\epsilon}) K_{\mathrm{e}}^{\mathrm{T}} \\
&= (I - K_{\mathrm{e}} M)\left(C^{\mathrm{e}}_{\psi\psi}\right)^{\mathrm{f}}.
\end{aligned}$$

The last expression in this equation is the traditional result for the minimum error covariance found in the KF analysis scheme. This implies that the EnKF in the limit of an infinite ensemble size will give exactly the same result in the computation of the analysis as the KF and EKF. Note that this derivation clearly states that the observations $\boldsymbol{d}$ must be treated as random variables to get the measurement error covariance matrix $\boldsymbol{C}^{\mathrm{e}}_{\epsilon\epsilon}$ into the expression. It has been assumed that the distributions used to generate the model state ensemble and the observation ensemble are independent. In Chap. 13 we will see that it is also possible to derive deterministic analysis schemes where the perturbation of measurements is avoided. This reduces sampling errors but may introduce other problems.

Finally, it should be noted that the EnKF analysis scheme is approximate in the sense that it does not properly take into account non-Gaussian contributions in the prior for $\boldsymbol{\psi}$. In other words, it does not solve the Bayesian update equation for non-Gaussian pdfs. On the other hand, it is not a pure resampling of a Gaussian posterior distribution. Only the updates are linear and these are added to the prior non-Gaussian ensemble. Thus, the updated ensemble will inherit many of the non-Gaussian properties from the forecast ensemble. In summary, we have a very computational efficient analysis scheme where we avoid traditional resampling of the posterior, and the solution becomes something between a linear Gaussian update and a full Bayesian computation. This will be elaborated on in more detail in the following chapters.

4.3.4 Discussion

We now have a complete system of equations which constitutes the ensemble Kalman filter (EnKF), and the similarity with the standard Kalman filter is maintained both for the prediction of error covariances and in the analysis scheme. For linear dynamics the EnKF solution will converge exactly to the KF solution with increasing ensemble size.

We will now examine the forecast step a little further. In the EnKF each ensemble member evolves in time according to the stochastic model dynamics. The ensemble covariance matrix of the errors in the model equations, given by

$$\boldsymbol{C}^{\mathrm{e}}_{qq} = \overline{d\boldsymbol{q}_k d\boldsymbol{q}_k^{\mathrm{T}}}, \tag{4.42}$$

converges to $\boldsymbol{C}_{qq}$ in the limit of an infinite ensemble size.

The ensemble mean then evolves according to the equation

$$\begin{aligned}\overline{\boldsymbol{\psi}_{k+1}} &= \overline{\boldsymbol{G}(\boldsymbol{\psi}_k)} \\ &= \boldsymbol{G}(\overline{\boldsymbol{\psi}_k}) + \mathrm{n.l.},\end{aligned} \tag{4.43}$$

where n.l. represents the terms which may arise if $\boldsymbol{G}$ is nonlinear. Compare this equation with the approximate equation for the mean (4.27) used with the EKF, where only the first correction term is included. One of the advantages of the EnKF is that it models the exact equation for the mean and there is no closure assumption used since each ensemble member is integrated by the full nonlinear model. The only approximation is the limited size of the ensemble.

The error covariance of the ensemble evolves according to

$$\left(C^{\text{e}}_{\psi\psi}\right)^{k+1} = G'\left(C^{\text{e}}_{\psi\psi}\right)^{k} G'^{\text{T}} + C^{\text{e}}_{qq} + \text{n.l.}, \qquad (4.44)$$

where G' is the tangent linear operator evaluated at ψ in the current time step. This is again an equation of the same form as is used in the standard Kalman filter, except for the extra n.l.-terms that may appear if G is nonlinear as seen in (4.22). Implicitly, the EnKF retains all these terms also for the error covariance evolution and there is no closure approximation used.

For a linear dynamical model the sampled $C^{\text{e}}_{\psi\psi}$ converges to $C_{\psi\psi}$ for an infinite ensemble size, and independently of the model, $C^{\text{e}}_{\epsilon\epsilon}$ converges to $C_{\epsilon\epsilon}$ and C^{e}_{qq} converges to C_{qq}. Thus, in this limit the KF and the EnKF are equivalent.

For nonlinear dynamics the EnKF includes the full effect of these terms and there are no linearizations or closure assumptions used. In addition, there is no need for a tangent linear operator or its adjoint, and this makes the method very easy to implement for practical applications.

This leads to an interpretation of the EnKF as a purely statistical Monte Carlo method where the ensemble of model states evolves in state space with the mean as the best estimate and the spreading of the ensemble as the error variance. At measurement times each observation is represented by another ensemble, where the mean is the actual measurement and the variance of the ensemble represents the measurement errors. Thus, we combine a stochastic prediction step with a stochastic analysis step.

4.3.5 Example with a QG model

Evensen and van Leeuwen (1996) proved the EnKF's capabilities with nonlinear dynamics in an application where Geosat radar altimeter data were assimilated into a quasi geostrophic (QG) model to study the ring-shedding process in the Agulhas current flowing along the southeast coast of South Africa. This was the first real application of an advanced sequential assimilation method for estimating the ocean circulation. It proved that the EnKF with its fully nonlinear evolution of error statistics could be used with nonlinear and unstable dynamical models. In addition it showed that the low computational cost of the EnKF allows for reasonably sized model grids to be used.

A series of plots of the analyzed estimates for the upper layer stream function is given in Fig. 4.4 for different time steps. The results were in good agreement with the assimilated data and the assimilation run was well constrained by the data.

A conclusion from this work was that the assimilation of data helped compensate for neglected physics in the model. The QG model has a too slow final wave steepening and ring shedding, caused by the lack of ageostrophic effects in the model. This was accounted for by the assimilation of the data.

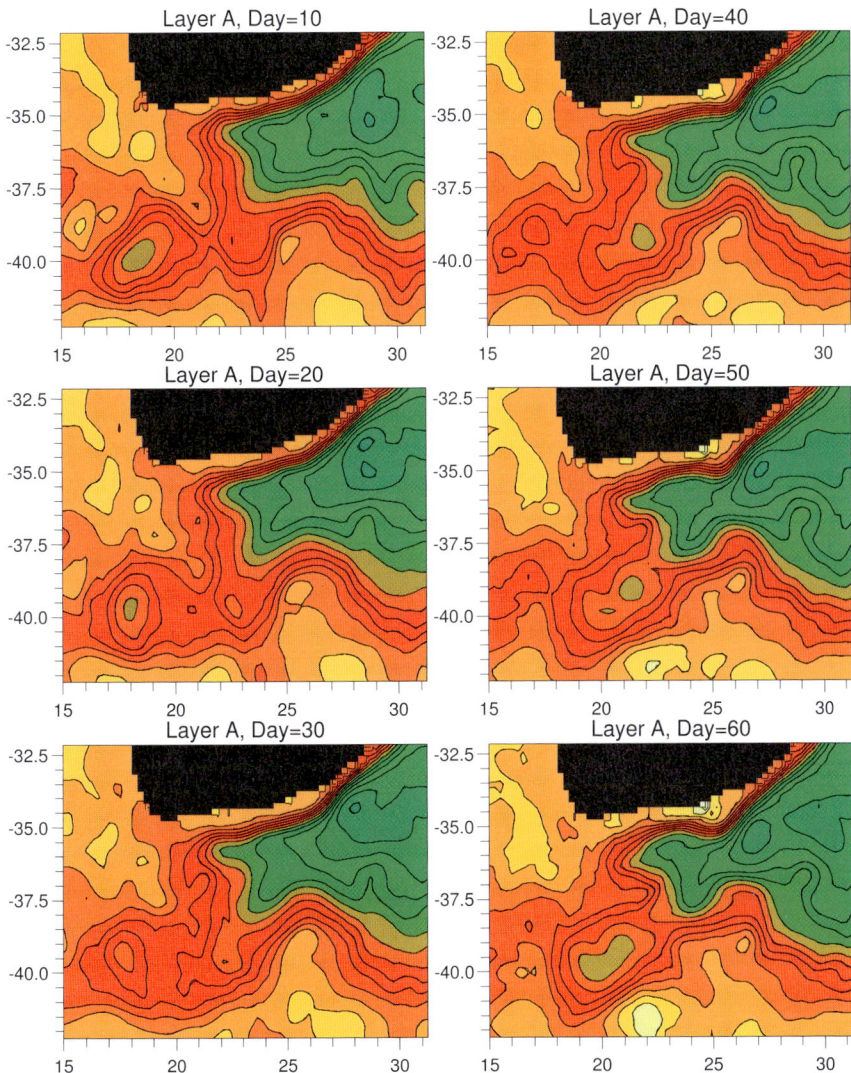

Fig. 4.4. Example of an EnKF experiment for the Agulhas current system from *Evensen and van Leeuwen* (1996)

In the experiment an ensemble size of 500 was used. The numerical grid consisted of two layers of 51×65 grid points, and the total number of unknowns was 6630, which is 13 times the number of ensemble members. The 500 ensemble members were sufficient to give a good representation of the gridded Geosat data and the space of possible model solutions.

5
Variational inverse problems

The purpose of this chapter is to introduce the basic formalism needed for properly formulating and solving linear variational inverse problems. Contrary to the sequential methods which update the model solution every time observations are available, variational methods seek an estimate in space and time where the estimate at a particular time is dependent on both past and future measurements.

We start by discussing a very simple example to illustrate the inverse problem and in particular the effect of including model errors. Thereafter a simple scalar model is used in a more typical illustration where the general formulation of the inverse problem is discussed and the Euler–Lagrange equations which determine the minimizing solution are derived.

Different methods are available for solving the Euler–Lagrange equations and we briefly discuss the popular representer method (see *Bennett*, 1992, 2002) which has proven extremely useful for solving linear and weakly nonlinear variational inverse problems.

5.1 Simple illustration

We will start with a very simple example to illustrate the mathematical properties of a variational problem and the difference between a weak and a strong constraint formulation. We define the simple model

$$\frac{d\psi}{dt} = 1, \qquad (5.1)$$
$$\psi(0) = 0, \qquad (5.2)$$
$$\psi(1) = 2, \qquad (5.3)$$

having one initial condition and one final condition. Clearly this is an overdetermined problem and it has no solution. However, if we relax the conditions by adding unknown errors to each of them the system becomes

5 Variational inverse problems

$$\frac{d\psi}{dt} = 1 + q, \tag{5.4}$$

$$\psi(0) = 0 + a, \tag{5.5}$$

$$\psi(1) = 2 + b. \tag{5.6}$$

The system is now under-determined since we can get whatever solution we want by choosing the different error terms. A statistical hypothesis $\mathcal{H}_0$, is now needed for the error terms,

$$\overline{q(t)} = 0, \quad \overline{q(t_1)q(t_2)} = C_0\delta(t_1 - t_2), \quad \overline{q(t)a} = 0,$$
$$\overline{a} = 0, \quad \overline{a^2} = C_0, \quad \overline{ab} = 0, \tag{5.7}$$
$$\overline{b} = 0, \quad \overline{b^2} = C_0, \quad \overline{q(t)b} = 0.$$

That is, we assume that we know the statistical behaviour of the error terms through their first and second order moments. In this example the variances are all set equal to C_0 for simplicity.

It is now possible to seek the solution, which is as close as possible to the initial and final conditions while at the same time it almost satisfies the model equations, by minimizing the error terms in the form of a weak constraint penalty function

$$\mathcal{J}[\psi] = W_0 \int_0^1 \left(\frac{d\psi}{dt} - 1\right)^2 dt + W_0\big(\psi(0) - 0\big)^2 + W_0\big(\psi(1) - 2\big)^2, \tag{5.8}$$

where W_0 is the inverse of the error variance C_0. Then ψ is an extremum of the penalty function if

$$\delta\mathcal{J}[\psi] = \mathcal{J}[\psi + \delta\psi] - \mathcal{J}[\psi] = \mathcal{O}(\delta\psi^2), \tag{5.9}$$

when $\delta\psi \to 0$. Now, using

$$\mathcal{J}[\psi + \delta\psi] = W_0 \int_0^1 \left(\frac{d\psi}{dt} - 1 + \frac{d\delta\psi}{dt}\right)^2 dt$$
$$+ W_0\big(\psi(0) - 0 + \delta\psi(0)\big)^2 + W_0\big(\psi(1) - 2 + \delta\psi(1)\big)^2 \tag{5.10}$$

in (5.9) and dropping the common nonzero factor $2W_0$, and all terms proportional to $\mathcal{O}(\delta\psi^2)$, we must have

$$\int_0^1 \frac{d\delta\psi}{dt}\left(\frac{d\psi}{dt} - 1\right) dt + \delta\psi(0)\big(\psi(0) - 0\big) + \delta\psi(1)\big(\psi(1) - 2\big) = 0, \tag{5.11}$$

or from integration by part,

$$\delta\psi\left(\frac{d\psi}{dt} - 1\right)\bigg|_0^1 - \int_0^1 \delta\psi \frac{d^2\psi}{dt^2} dt$$
$$+ \delta\psi(0)\big(\psi(0) - 0\big) + \delta\psi(1)\big(\psi(1) - 2\big) = 0. \tag{5.12}$$

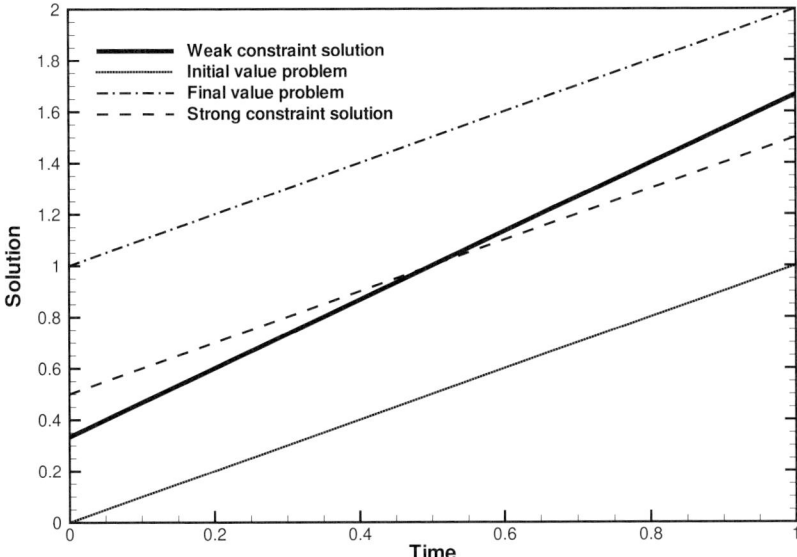

Fig. 5.1. Inverse solution from the simple example

This gives the following system of equations

$$\delta\psi(0)\left(-\frac{d\psi}{dt}+1+\psi\right)\bigg|_{t=0}=0, \qquad (5.13)$$

$$\delta\psi(1)\left(\frac{d\psi}{dt}-1+\psi-2\right)\bigg|_{t=1}=0, \qquad (5.14)$$

$$\delta\psi\left(\frac{d^2\psi}{dt^2}\right)=0, \qquad (5.15)$$

or since $\delta\psi$ is arbitrary

$$\frac{d\psi}{dt}-\psi=1 \quad \text{for } t=0, \qquad (5.16)$$

$$\frac{d\psi}{dt}+\psi=3 \quad \text{for } t=1, \qquad (5.17)$$

$$\frac{d^2\psi}{dt^2}=0. \qquad (5.18)$$

This is an elliptic boundary value problem in time with mixed Dirichlet and Neumann boundary conditions. The general solution is

$$\psi=c_1 t+c_2, \qquad (5.19)$$

and the constants in this case become $c_1=4/3$ and $c_2=1/3$.

In the case when we let the errors in the dynamical model go to zero, we approach the strong constraint limit where the dynamical model is assumed to be perfect. The strong constraint model solution is $\psi = t + c_2$ from (5.4), i.e. the slope is the one defined by the original model and no deviation of this is allowed. The free constant c_2 will take a value between 0 and 1, depending on the relative magnitude between the weights on the two conditions. In this case with equal weight we will have $c_2 = 0.5$.

By allowing for model errors to account for an imperfect model, we will through a weak constraint variational formulation also allow for a deviation from the exact model trajectory. This is important for the mathematical conditioning of the variational problem, and we will later see that the weak constraint problem can be solved as easily as the strong constraint problem. The results from this example are shown in Fig. 5.1. The upper and lower curves are the respective solutions of the final and initial value problems. The weak constraint inverse estimate is seen to have a steeper slope than the exact model would allow, in order to obtain an estimate in better agreement with the two conditions. The strong constraint estimate is shown for comparison.

Finally, it is interesting to examine what the KF solution becomes in this example. The KF starts by solving the initial value problem until $t = 1$, thus for $t \in [0, 1)$ the solution is just $\psi(t) = t$. The initial error variance is set to C_0 and the increase of error variance when integrating the model over one time unit is also C_0. Thus for the prediction at $t = 1$, the error variance equals $2C_0$. The update equation (3.14) then becomes

$$\begin{aligned} \psi^{\mathrm{a}} &= \psi^{\mathrm{f}} + \frac{C^{\mathrm{f}}_{\psi\psi}}{C_{\epsilon\epsilon} + C^{\mathrm{f}}_{\psi\psi}} (d - \psi^{\mathrm{f}}) \\ &= 1 + \frac{2C_0}{C_0 + 2C_0}(2 - 1) \\ &= 5/3. \end{aligned} \quad (5.20)$$

This is in fact identical to the weak constraint variational solution at $t = 1$. Thus, could there be some connection between the problem solved by a variational method and the KF? In fact it will be shown later that for linear inverse problems, the KF and the weak constraint variational method, when both formulated consistently and using the same prior error statistics, give identical solutions at the final time. Thus for forecasting purposes, it does not matter which method is used.

5.2 Linear inverse problem

In this section we will define the inverse problem for a simple linear model and derive the Euler–Lagrange equations for a weak constraint variational formulation.

5.2.1 Model and observations

Assume now that we have given a simple scalar model with an initial condition and a set of measurements, all subject to errors,

$$\frac{d\psi}{dt} = \psi + q, \qquad (5.21)$$

$$\psi(0) = \Psi_0 + a, \qquad (5.22)$$

$$\mathcal{M}[\psi] = \boldsymbol{d} + \boldsymbol{\epsilon}. \qquad (5.23)$$

The inverse problem can then be defined as finding an estimate which is close to the initial condition and the set of measurements, while at the same time it is almost satisfying the model equation.

5.2.2 Measurement functional

The linear measurement operator $\mathcal{M}[\psi]$, of dimension M equal to the number of measurements, relates the observations $\boldsymbol{d}$ to the model state variable $\psi(t)$.

As an example, a direct measurement of $\psi(t)$ will have a measurement functional of the form

$$\mathcal{M}_i[\psi(t)] = \int_0^T \psi(t)\delta(t - t_i)\, dt = \psi(t_i), \qquad (5.24)$$

where t_i is the measurement location in time and the subscript i denotes the component of the measurement functional.

Note for later use that the observation of the Dirac delta function becomes

$$\mathcal{M}_{i(2)}[\delta(t_1 - t_2)] = \int_0^T \delta(t_1 - t_2)\delta(t_2 - t_i)\, dt_2 = \delta(t_1 - t_i). \qquad (5.25)$$

The subscript (2) on $\mathcal{M}_i$ defines the variable that the functional is operating on. Multiplying this with $\delta\psi(t_1)$ and integrating with respect to t_1 gives

$$\int_0^T \delta\psi(t_1)\mathcal{M}_{i(2)}[\delta(t_1 - t_2)]\, dt_1 = \delta\psi(t_i) = \mathcal{M}_{i(1)}[\delta\psi(t_1)]. \qquad (5.26)$$

5.2.3 Comment on the measurement equation

In (3.2) and (3.23) we defined a measurement equation where we related the measurements to the true state, and ϵ became the real measurement errors. Let us now write

$$\boldsymbol{d} = \boldsymbol{d}^{\mathrm{t}} + \boldsymbol{\epsilon}_d, \qquad (5.27)$$

which defines $\boldsymbol{\epsilon}_d$ as the real measurement errors. In some cases we will also have that

$$\mathcal{M}[\psi^{\mathrm{t}}] = \boldsymbol{d}^{\mathrm{t}} + \boldsymbol{\epsilon}_{\mathcal{M}}, \qquad (5.28)$$

which states that there is an additional error associated with the measurement operator $\mathcal{M}$. An example of such an error could be related to the interpolation on a numerical grid when a measurement is located in the center of a grid cell. We can then write

$$\begin{aligned} d &= \mathcal{M}[\psi^t] + \epsilon_d - \epsilon_\mathcal{M} \\ &= \mathcal{M}[\psi^t] + \epsilon. \end{aligned} \quad (5.29)$$

Thus, we can say that the measurement is related to the true state through (5.29), where $\epsilon = \epsilon_d - \epsilon_\mathcal{M}$ accounts for both measurement errors and errors in the measurement operator.

In the measurement equation (5.23) there is no reference to the true value ψ^t. In fact (5.23) is an equation which relates an estimate ψ to the measurements d, allowing for a random error ϵ. Thus, we use this equation to impose an additional constraint to the model defined by (5.21) and (5.22). The random error ϵ that represents both the errors in the measurements and the measurement operator now defines the accuracy of the measurement equation (5.23), just as the random errors a and q define the accuracy of the model equation and the initial condition.

5.2.4 Statistical hypothesis

Again a statistical hypothesis $\mathcal{H}_0$ is needed for describing the unknown error terms and we adapt the following:

$$\begin{array}{lll} \overline{q} = 0, & \overline{q(t_1)q(t_2)} = C_{qq}(t_1, t_2), & \overline{q(t)a} = 0, \\ \overline{a} = 0, & \overline{a^2} = C_{aa}, & \overline{a\epsilon} = 0, \\ \overline{\epsilon} = 0, & \overline{\epsilon\epsilon^T} = C_{\epsilon\epsilon}, & \overline{q(t)\epsilon} = 0. \end{array} \quad (5.30)$$

In addition we will now define the functional inverse W_{qq} of the model error covariance C_{qq}, from the integral

$$\int_0^T C_{qq}(t_1, t_2) W_{qq}(t_2, t_3) \, dt_2 = \delta(t_1 - t_3), \quad (5.31)$$

and W_{aa} as the inverse of C_{aa}.

5.2.5 Weak constraint variational formulation

A weak constraint cost function can now be defined as

$$\begin{aligned} \mathcal{J}[\psi] = &\iint_0^T \left(\frac{d\psi(t_1)}{dt_1} - \psi(t_1)\right) W_{qq}(t_1, t_2) \left(\frac{d\psi(t_2)}{dt_2} - \psi(t_2)\right) dt_1 \, dt_2 \\ &+ W_{aa}\bigl(\psi(0) - \Psi_0\bigr)^2 \\ &+ \bigl(d - \mathcal{M}[\psi]\bigr)^T W_{\epsilon\epsilon} \bigl(d - \mathcal{M}[\psi]\bigr). \end{aligned} \quad (5.32)$$

Note that all first guesses, including the initial conditions, are penalized in (5.32). This is required in order to have a well-posed problem with a unique solution, as was shown by *Bennett and Miller* (1990).

The time-correlation in the model weight has a regularizing effect. Model errors are normally correlated in time, and the result of neglecting the time correlation is that the estimate will have discontinuous time derivatives at measurement locations.

5.2.6 Extremum of the penalty function

From standard variational calculus we know that ψ is an extremum if

$$\delta \mathcal{J} = \mathcal{J}[\psi + \delta\psi] - \mathcal{J}[\psi] = \mathcal{O}(\delta\psi^2), \tag{5.33}$$

when $\delta\psi \to 0$. Evaluating $\mathcal{J}[\psi + \delta\psi]$ we get

$$\begin{aligned}
\mathcal{J}[\psi + \delta\psi] = \iint_0^T &\left(\frac{d\psi}{dt} - \psi + \frac{d\delta\psi}{dt} - \delta\psi\right)_1 W_{qq}(t_1, t_2) \\
&\times \left(\frac{d\psi}{dt} - \psi + \frac{d\delta\psi}{dt} - \delta\psi\right)_2 dt_1\, dt_2 \\
+ W_{aa}&\bigl(\psi(0) + \delta\psi(0) - \Psi_0\bigr)^2 \\
+ \bigl(\boldsymbol{d} - &\boldsymbol{\mathcal{M}}[\psi] - \boldsymbol{\mathcal{M}}[\delta\psi]\bigr)^{\mathrm{T}} \boldsymbol{W}_{\epsilon\epsilon}\bigl(\boldsymbol{d} - \boldsymbol{\mathcal{M}}[\psi] - \boldsymbol{\mathcal{M}}[\delta\psi]\bigr),
\end{aligned} \tag{5.34}$$

where the subscripts 1 and 2 denote functions of t_1 and t_2. This can be rewritten as

$$\begin{aligned}
\mathcal{J}[\psi + \delta\psi] &= \mathcal{J}[\psi] \\
&+ 2\iint_0^T \left(\frac{d\delta\psi}{dt} - \delta\psi\right)_1 W_{qq}(t_1,t_2) \left(\frac{d\psi}{dt} - \psi\right)_2 dt_1\, dt_2 \\
&+ 2W_{aa}\delta\psi(0)\bigl(\psi(0) - \Psi_0\bigr) \\
&- 2\boldsymbol{\mathcal{M}}^{\mathrm{T}}[\delta\psi]\boldsymbol{W}_{\epsilon\epsilon}\bigl(\boldsymbol{d} - \boldsymbol{\mathcal{M}}[\psi]\bigr) + \mathcal{O}(\delta\psi^2).
\end{aligned} \tag{5.35}$$

Now, evaluating the variational derivative (5.33) and requiring that the remaining terms are proportional to $\delta\psi^2$, we must have

$$\begin{aligned}
\iint_0^T &\left(\frac{d\delta\psi}{dt} - \delta\psi\right)_1 W_{qq}(t_1, t_2) \left(\frac{d\psi}{dt} - \psi\right)_2 dt_1\, dt_2 \\
&+ W_{aa}\delta\psi(0)\bigl(\psi(0) - \Psi_0\bigr) \\
&- \boldsymbol{\mathcal{M}}^{\mathrm{T}}[\delta\psi]\boldsymbol{W}_{\epsilon\epsilon}\bigl(\boldsymbol{d} - \boldsymbol{\mathcal{M}}[\psi]\bigr) = 0.
\end{aligned} \tag{5.36}$$

This equation defines an extremum of the penalty function.

5.2.7 Euler–Lagrange equations

Start from (5.36) and define the "adjoint" variable λ as

$$\lambda(t_1) = \int_0^T W_{qq}(t_1,t_2)\left(\frac{d\psi}{dt} - \psi\right)_2 dt_2. \tag{5.37}$$

We now insert this in (5.36) and use integration by part to eliminate the derivative of the variation, i.e.

$$\int_0^T \frac{d\delta\psi}{dt}\lambda\, dt = \delta\psi\lambda\Big|_{t=0}^{t=T} - \int_0^T \delta\psi\frac{d\lambda}{dt} dt. \tag{5.38}$$

Then we use (5.26) to get the measurement term under the integral and proportional to $\delta\psi$.

Equation (5.36) then becomes

$$-\int_0^T \delta\psi\left(\frac{d\lambda}{dt} + \lambda + \mathcal{M}_{(2)}^{\text{T}}[\delta(t_1-t_2)]W_{\epsilon\epsilon}\left(d - \mathcal{M}[\psi]\right)\right)_1 dt_1$$
$$+ \delta\psi(0)\left(W_{aa}\left(\psi(0) - \Psi_0\right) - \lambda(0)\right) \tag{5.39}$$
$$+ \delta\psi(T)\lambda(T) = 0.$$

To obtain the final form of the Euler–Lagrange equations, we first multiply (5.37) with $W_{qq}(t,t_1)$ from the left, integrate in t_1 and use (5.31). This results in (5.40), given below. Further, assuming that the variation $\delta\psi$ in (5.39) is arbitrary, we get an equation for λ and conditions at time $t=0$ and $t=T$. Thus, we have the following Euler–Lagrange equations:

$$\frac{d\psi}{dt} - \psi = \int_0^T C_{qq}(t,t_1)\lambda(t_1)\, dt_1, \tag{5.40}$$

$$\psi(0) = \Psi_0 + C_{aa}\lambda(0), \tag{5.41}$$

$$\frac{d\lambda}{dt} + \lambda = -\mathcal{M}_{(2)}^{\text{T}}[\delta(t-t_2)]W_{\epsilon\epsilon}\left(d - \mathcal{M}[\psi]\right), \tag{5.42}$$

$$\lambda(T) = 0. \tag{5.43}$$

This system of Euler–Lagrange equations defines the extrema ψ of $\mathcal{J}$. The system consists of the original forward model forced by a term that is proportional to the adjoint variable λ in (5.40). The magnitude of this term is defined by the model error covariance C_{qq}, thus large model errors give a large contribution through the forcing term. The forward model is integrated from an initial condition which also contains a similar correction term proportional to the adjoint variable λ. The equation for λ can be integrated backward in time from a "final" condition, while forced by delta functions scaled by the residual between the measurement and forward model estimate ψ at each measurement location. Thus, the forward model requires knowledge of the adjoint variable to be integrated and the backward model uses the forward variable at measurement locations. We therefore have a coupled boundary value problem in time where the forward and backward models must be solved simultaneously. The system comprises a well-posed problem and as long as the model is linear, it will have one unique solution, ψ.

The simplest approach for solving the Euler–Lagrange equations, may be to define an iteration. An iteration for the system (5.40)–(5.43) can be defined by using the previous iterate of λ when integrating the forward model. However, this iteration will generally not converge as pointed out by *Bennett* (1992).

5.2.8 Strong constraint approximation

A much-used approach relies on the assumption that the model is perfect, i.e. $C_{qq} = 0$ in (5.40). This leads to the so-called adjoint method originally proposed by *Talagrand and Courtier* (1987), *Courtier and Talagrand* (1987) and later discussed in a number of publications, e.g. *Courtier et al.* (1994), *Courtier* (1997). This removes the coupling to λ in the forward model. However, the system is still coupled through the λ appearing in the initial condition. One is then seeking the initial condition resulting in the model trajectory which is closest to the measurements. The so-called adjoint method uses this approach and defines a solution method where the system may be iterated as follows:

$$\frac{d\psi^l}{dt} - \psi^l = 0, \tag{5.44}$$

$$\psi^l(0) = \psi^{l-1}(0) - \gamma\left(\psi^{l-1}(0) - \Psi_0 - C_{aa}\lambda^{l-1}(0)\right), \tag{5.45}$$

$$\frac{d\lambda^l}{dt} + \lambda^l = -\mathcal{M}_{(2)}^{\mathrm{T}}[\delta(t_1 - t_2)]\mathbf{W}_{\epsilon\epsilon}\left(\mathbf{d} - \mathcal{M}_{(4)}[\psi_4^l]\right), \tag{5.46}$$

$$\lambda^l(T) = 0. \tag{5.47}$$

The iteration defined for the initial condition uses that (5.41), or the expression in parantheses from (5.45), is the gradient of the penalty function with respect to the initial conditions. Thus, the iteration (5.45) is a standard gradient descent method where γ is the step length in the direction of the gradient. It should be noted that when $\psi = \psi(\mathbf{x})$, the dimension of the problem becomes infinite, and when $\psi(\mathbf{x})$ is discretized on a numerical grid, it becomes finite and equal to the number of grid nodes.

Note also that while the weak constraint formulation with proper knowledge of the error statistics defines a well-posed estimation problem where the estimate will be located within the statistical uncertainties of the first guesses, the strong constraint assumption violates this property of the inverse problem since one assumes that the model is better than it actually is.

5.2.9 Solution by representer expansions

For linear dynamics, it is possible to solve the Euler–Lagrange equations (5.40–5.43) exactly without using iterations. This can be done by assuming a solution of the form

$$\psi(t) = \psi_F(t) + \boldsymbol{b}^\mathrm{T} \boldsymbol{r}(t), \tag{5.48}$$
$$\lambda(t) = \lambda_F(t) + \boldsymbol{b}^\mathrm{T} \boldsymbol{s}(t), \tag{5.49}$$

as was previously also used for the time independent problem in (3.39). The dimensions of the vectors $\boldsymbol{b}$, $\boldsymbol{r}$ and $\boldsymbol{s}$ are all equal to the number of measurements, M. Assuming this form for the solution is equivalent to assuming that the minimizing solution is a first guess model solution ψ_F plus a linear combination of time dependent influence functions or representers $\boldsymbol{r}(t)$, one for each measurement. For a comprehensive discussion of this method see *Bennett* (1992, 2002). The practical implementation is discussed in great detail by *Chua and Bennett* (2001).

Inserting (5.48) and (5.49) into the Euler–Lagrange equations (5.40–5.43) and choosing first guesses ψ_F and λ_F that satisfy unforced exact equations

$$\frac{d\psi_F}{dt} - \psi_F = 0, \tag{5.50}$$
$$\psi_F(0) = \Psi_0, \tag{5.51}$$
$$\frac{d\lambda_F}{dt} + \lambda_F = 0, \tag{5.52}$$
$$\lambda_F(T) = 0, \tag{5.53}$$

gives us the following system of equations for the vector of representers $\boldsymbol{r}(t)$ and corresponding adjoints $\boldsymbol{s}(t)$:

$$\boldsymbol{b}^\mathrm{T}\left(\frac{d\boldsymbol{r}}{dt} - \boldsymbol{r} - C_{qq}\boldsymbol{s}\right) = 0, \tag{5.54}$$
$$\boldsymbol{b}^\mathrm{T}\left(\boldsymbol{r}(0) - C_{aa}\boldsymbol{s}\right) = 0, \tag{5.55}$$
$$\boldsymbol{b}^\mathrm{T}\left(\frac{d\boldsymbol{s}}{dt} + \boldsymbol{s}\right) = -\boldsymbol{\mathcal{M}}_{(2)}^\mathrm{T}[\delta(t-t_2)]\boldsymbol{W}_{\epsilon\epsilon}\left(\boldsymbol{d} - \boldsymbol{\mathcal{M}}[\psi_F + \boldsymbol{b}^\mathrm{T}\boldsymbol{r}]\right), \tag{5.56}$$
$$\boldsymbol{b}^\mathrm{T}\boldsymbol{s}(T) = 0. \tag{5.57}$$

If we define $\boldsymbol{b}$ as

$$\boldsymbol{b} = \boldsymbol{W}_{\epsilon\epsilon}\left(\boldsymbol{d} - \boldsymbol{\mathcal{M}}[\psi_F + \boldsymbol{b}^\mathrm{T}\boldsymbol{r}]\right), \tag{5.58}$$

then (5.56) becomes

$$\boldsymbol{b}^\mathrm{T}\left(\frac{d\boldsymbol{s}}{dt} + \boldsymbol{s} + \boldsymbol{\mathcal{M}}_{(2)}[\delta(t-t_2)]\right) = 0, \tag{5.59}$$

and the coupling to the solution on the right side of (5.56) is removed.

Equation (5.58) is exactly the same as (3.38) and the derivation in (3.42–3.45) leads to the same linear system for the coefficients $\boldsymbol{b}$,

$$\left(\mathcal{M}^\mathrm{T}[\boldsymbol{r}] + \boldsymbol{C}_{\epsilon\epsilon}\right)\boldsymbol{b} = \boldsymbol{d} - \mathcal{M}[\psi_F]. \tag{5.60}$$

Given that $\boldsymbol{b}$ in general is nonzero, we now have the following set of equations in addition to (5.50–5.53):

$$\frac{d\boldsymbol{r}}{dt} - \boldsymbol{r} = C_{qq}\boldsymbol{s}, \tag{5.61}$$

$$\boldsymbol{r}(0) = C_{aa}\boldsymbol{s}, \tag{5.62}$$

from (5.54) and (5.55) for the representers, and

$$\frac{d\boldsymbol{s}}{dt} + \boldsymbol{s} = -\mathcal{M}_{(2)}\bigl[\delta(t_1 - t_2)\bigr], \tag{5.63}$$

$$\boldsymbol{s}(T) = 0, \tag{5.64}$$

from (5.59) and (5.57) for the adjoints of the representers.

The equations for $\boldsymbol{s}$ can now be solved as a sequence of final value problems since they are decoupled from the forward equations for the representers. As soon as $\boldsymbol{s}$ is found the representers can be solved for. Together with the first guess solution ψ_F, found from solving (5.50)–(5.53), this provides the information needed for solving the system (5.60) for $\boldsymbol{b}$. The final estimate is then found by solving the Euler–Lagrange equation of the form

$$\frac{d\psi}{dt} - \psi = \int_0^T C_{qq}(t, t_1) \lambda(t_1) dt_1, \tag{5.65}$$

$$\psi(0) = \Psi_0 + C_{aa}\lambda(0), \tag{5.66}$$

$$\frac{d\lambda}{dt} + \lambda = -\mathcal{M}_{(1)}^\mathrm{T}\bigl[\delta(t - t_1)\bigr]\boldsymbol{b}, \tag{5.67}$$

$$\lambda(T) = 0. \tag{5.68}$$

The numerical load is then $2M+3$ model integrations, but note that only two model states need to be stored in space and time. If the solution is constructed directly from (5.48) all the representers need to be stored.

Thus, the representer expansion decouples the Euler–Lagrange equations for the weak constraint problem, which can now be solved exactly without any iterations. Further, the dimension of the problem is the number of measurements, which is normally much less than the number of unknowns in a discrete state vector.

5.3 Representer method with an Ekman model

In *Eknes and Evensen* (1997), the representer method was implemented with an Ekman flow model and used to solve an inverse problem with a long time series of real velocity measurements. In addition a parameter estimation problem was treated but this will be discussed later. The model used is very simple and allows for a simple interpretation and demonstration of the method.

5.3.1 Inverse problem

The Ekman model was written in a nondimensional form as

$$\frac{\partial \boldsymbol{u}}{\partial t} + \boldsymbol{k} \times \boldsymbol{u} = \frac{\partial}{\partial z}\left(A\frac{\partial \boldsymbol{u}}{\partial z}\right) + \boldsymbol{q}, \tag{5.69}$$

where $\boldsymbol{u}(z,t)$ is the horizontal velocity vector, $A = A(z)$ is the diffusion coefficient and $\boldsymbol{q}(z,t)$ is the stochastic model error. The initial conditions are given as

$$\boldsymbol{u}(z,0) = \boldsymbol{u}_0 + \boldsymbol{a}, \tag{5.70}$$

where $\boldsymbol{a}$ contains the stochastic errors in the first-guess initial condition $\boldsymbol{u}_0$. The boundary conditions for the model are

$$A\frac{\partial \boldsymbol{u}}{\partial z}\bigg|_{z=0} = \left(c_d\sqrt{u_a^2 + v_a^2}\right)\boldsymbol{u}_a + \boldsymbol{b}_0, \tag{5.71}$$

$$A\frac{\partial \boldsymbol{u}}{\partial z}\bigg|_{z=-H} = \boldsymbol{0} + \boldsymbol{b}_H, \tag{5.72}$$

where the position $z = 0$ is at the ocean surface and the lower boundary is at $z = -H$, c_d is the wind drag coefficient, $\boldsymbol{u}_a$ is the atmospheric wind speed, and $\boldsymbol{b}_0$ and $\boldsymbol{b}_H$ are the stochastic errors in the boundary conditions.

Now a set of measurements $\boldsymbol{d}$ of the true solution is assumed given and linearly related to the model variables by the measurement equation

$$\mathcal{M}[\boldsymbol{u}] = \boldsymbol{d} + \boldsymbol{\epsilon}. \tag{5.73}$$

5.3.2 Variational formulation

A convenient variational formulation is

$$\begin{aligned}
\mathcal{J}[\boldsymbol{u}] &= \int_0^T dt_1 \int_0^T dt_2 \int_{-H}^0 dz_1 \int_{-H}^0 dz_2 \, \boldsymbol{q}^{\mathrm{T}}(z_1,t_1) \boldsymbol{W}_{qq}(z_1,t_1,z_2,t_2) \boldsymbol{q}(z_2,t_2) \\
&+ \int_{-H}^0 dz_1 \int_{-H}^0 dz_2 \, \boldsymbol{a}^{\mathrm{T}}(z_1) \boldsymbol{W}_{aa}(z_1,z_2) \boldsymbol{a}(z_2) \\
&+ \int_0^T dt_1 \int_0^T dt_2 \, \boldsymbol{b}_0^{\mathrm{T}}(t_1) \boldsymbol{W}_{b_0 b_0}(t_1,t_2) \boldsymbol{b}_0(t_2) \\
&+ \int_0^T dt_1 \int_0^T dt_2 \, \boldsymbol{b}_H^{\mathrm{T}}(t_1) \boldsymbol{W}_{b_H b_H}(t_1,t_2) \boldsymbol{b}_H(t_2) \\
&+ \boldsymbol{\epsilon}^{\mathrm{T}} \boldsymbol{W}_{\epsilon\epsilon} \boldsymbol{\epsilon}.
\end{aligned} \tag{5.74}$$

A simpler way of writing this may be

$$\begin{aligned}\mathcal{J}[u] = {}& q^{\mathrm{T}} \bullet W_{qq} \bullet q \\ &+ a^{\mathrm{T}} \circ W_{aa} \circ a \\ &+ b_0^{\mathrm{T}} * W_{b_0 b_0} * b_0 \\ &+ b_H^{\mathrm{T}} * W_{b_H b_H} * b_H \\ &+ \epsilon^{\mathrm{T}} W_{\epsilon\epsilon} \epsilon,\end{aligned} \quad (5.75)$$

where the bullets mean integration both in space and time, the open circles mean integration in space, the asterisks mean integration in time. Here $W_{\epsilon\epsilon}$ is the inverse of the measurement error covariance matrix $C_{\epsilon\epsilon}$, while the other weights are functional inverses of the respective covariances. For the model weight, this can be expressed as $C_{qq} \bullet W_{qq} = \delta(z_1 - z_3)\delta(t_1 - t_3)I$, or written out,

$$\int_0^T dt_2 \int_{-H}^0 dz_2 \, C_{qq}(z_1, t_1, z_2, t_2) W_{qq}(z_2, t_2, z_3, t_3) \\ = \delta(z_1 - z_3)\delta(t_1 - t_3)I. \quad (5.76)$$

These weights determine the spatial and temporal scales for the physical problem and ensure smooth influences from the measurements.

5.3.3 Euler–Lagrange equations

Following the procedure outlined in the previous sections, we can derive the Euler–Lagrange equations. This leads to the forward model

$$\frac{\partial u}{\partial t} + k \times u = \frac{\partial}{\partial z}\left(A \frac{\partial u}{\partial z}\right) + C_{qq} \bullet \lambda, \quad (5.77)$$

with initial conditions

$$u|_{t=0} = u_0 + C_{aa} \circ \lambda, \quad (5.78)$$

and boundary conditions

$$A\frac{\partial u}{\partial z}\bigg|_{z=0} = c_d \sqrt{u_a^2 + v_a^2}\, u_a + C_{b_0 b_0} * \lambda, \quad (5.79)$$

$$A\frac{\partial u}{\partial z}\bigg|_{z=-H} = -C_{b_H b_H} * \lambda. \quad (5.80)$$

In addition we obtain the adjoint model

$$-\frac{\partial \lambda}{\partial t} - k \times \lambda = \frac{\partial}{\partial z}\left(A \frac{\partial \lambda}{\partial z}\right) + \mathcal{M}^{\mathrm{T}}\left[\delta(z-z_2)\delta(t-t_2)\right] W_{\epsilon\epsilon}\left(d - \mathcal{M}[u]\right), \quad (5.81)$$

subject to the "final" condition

5 Variational inverse problems

$$\lambda|_{t=T} = 0, \tag{5.82}$$

and the boundary conditions

$$\left.\frac{\partial \lambda}{\partial z}\right|_{z=0, z=-H} = 0. \tag{5.83}$$

The system (5.77) to (5.83) is the Euler–Lagrange equations which here comprise a two-point boundary value problem in space and time, and since they are coupled they must be solved simultaneously.

5.3.4 Representer solution

Assuming a solution in the standard form

$$\boldsymbol{u}(z,t) = \boldsymbol{u}_F(z,t) + \sum_{i=1}^{M} b_i \boldsymbol{r}_i(z,t), \tag{5.84}$$

$$\boldsymbol{\lambda}(z,t) = \boldsymbol{\lambda}_F(z,t) + \sum_{i=1}^{M} b_i \boldsymbol{s}_i(z,t), \tag{5.85}$$

we find the equations for the representers and their adjoints. The M representers are found by solving the initial value problems

$$\frac{\partial \boldsymbol{r}_i}{\partial t} + \boldsymbol{k} \times \boldsymbol{r}_i = \frac{\partial}{\partial z}\left(A \frac{\partial \boldsymbol{r}_i}{\partial z}\right) + \boldsymbol{C}_{qq} \bullet \boldsymbol{s}_i, \tag{5.86}$$

with initial condition

$$\boldsymbol{r}_i|_{t=0} = \boldsymbol{C}_{aa} \circ \boldsymbol{s}_i, \tag{5.87}$$

and boundary conditions

$$\left.A\frac{\partial \boldsymbol{r}_i}{\partial z}\right|_{z=0} = \boldsymbol{C}_{b_0 b_0} * \boldsymbol{s}_i, \tag{5.88}$$

$$\left.A\frac{\partial \boldsymbol{r}_i}{\partial z}\right|_{z=-H} = -\boldsymbol{C}_{b_H b_H} * \boldsymbol{s}_i. \tag{5.89}$$

These equations are coupled to the adjoints of the representers $\boldsymbol{s}_i$, which satisfy the "final" value problems

$$-\frac{\partial \boldsymbol{s}_i}{\partial t} - \boldsymbol{k} \times \boldsymbol{s}_i = \frac{\partial}{\partial z}\left(A\frac{\partial \boldsymbol{s}_i}{\partial z}\right) + \mathcal{M}_i[\delta(z-z_2)\delta(t-t_2)], \tag{5.90}$$

with "final" conditions

$$\boldsymbol{s}_i|_{t=T} = \boldsymbol{0}, \tag{5.91}$$

and boundary conditions

$$\left.\frac{\partial \boldsymbol{s}_i}{\partial z}\right|_{z=0,\ z=-H} = 0. \tag{5.92}$$

The coefficients $\boldsymbol{b}$ are again found by solving the system (5.60).

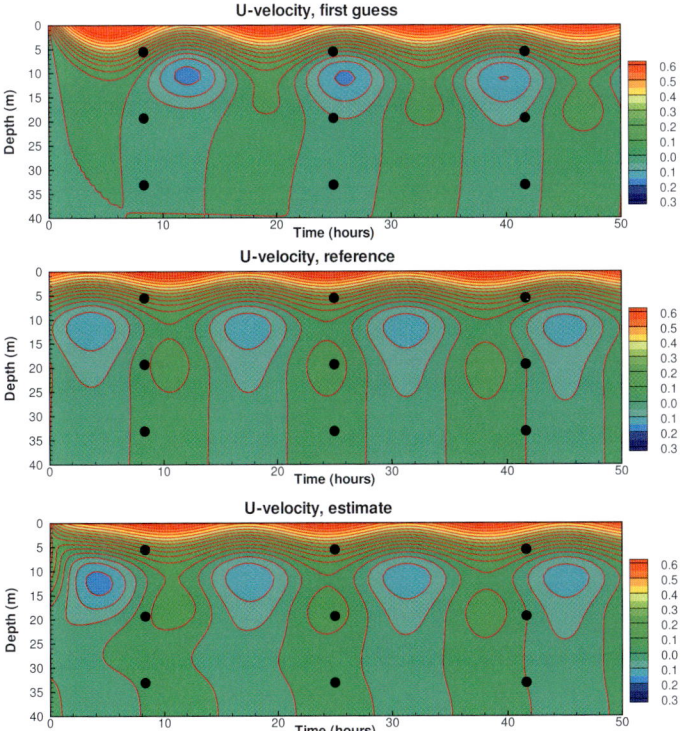

Fig. 5.2. The u components of *(from top to bottom)* the first-guess estimate $\boldsymbol{u}_F$, the reference case $\boldsymbol{u}$ and the inverse estimate of $\boldsymbol{u}$. The contour intervals are 0.05 m s^{-1} for all the velocity plots. The measurement locations are marked with a bullet. The v components are similar in structure and not shown. Reproduced from *Eknes and Evensen* (1997)

5.3.5 Example experiment

Here a simple example will be used to illustrate the method. A constant wind with $\boldsymbol{u}_a = (10 \text{ m s}^{-1}, 10 \text{ m s}^{-1})$ has been used to spin up the vertical velocity structure in the first-guess solution, starting with an initial condition $\boldsymbol{u}(z,0) = \boldsymbol{0}$ and then performing 50 hours of integration. The reference case, from which velocity data are extracted, is generated by continuing the integration for another 50 hours.

By measuring the reference case and adding Gaussian noise, nine simulated measurements of $\boldsymbol{u}$ were generated; that is, a total of 18 measurements of u and v components were used. The locations of the measurements are shown in Fig. 5.2.

All error terms are assumed to be unbiased and uncorrelated, and the error covariances were specified as follows:

62   5 Variational inverse problems

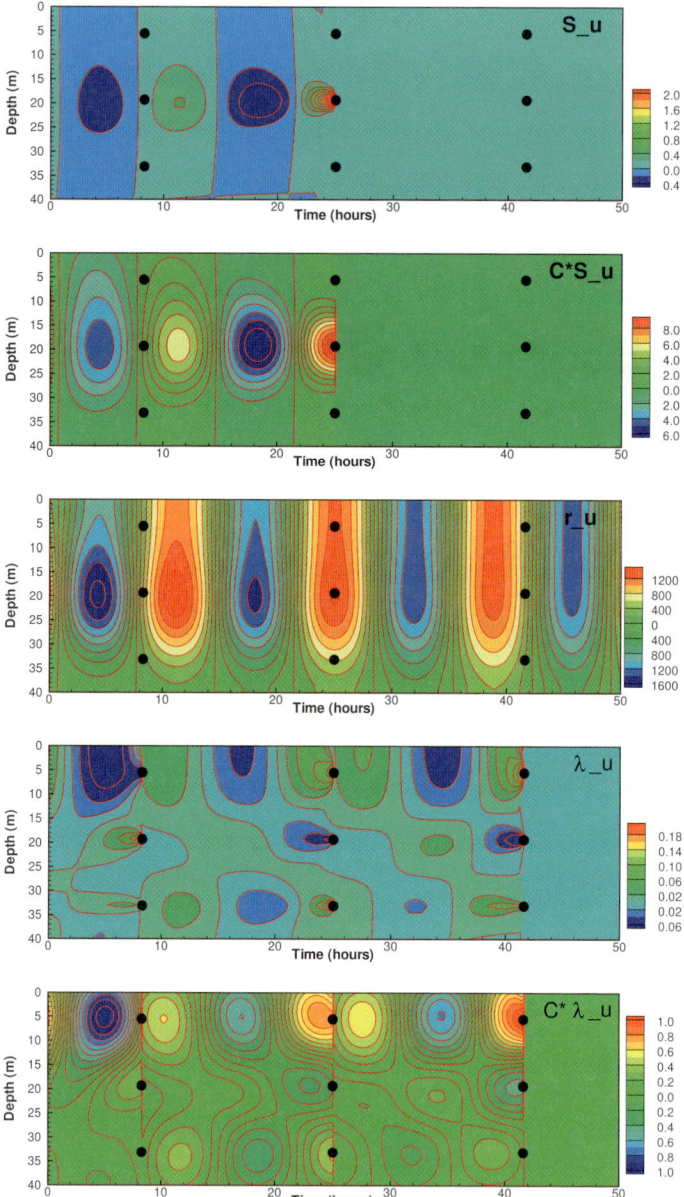

Fig. 5.3. The u component of *(top to bottom)* s_5, $C_{qq} \bullet s_5$, r_5, the adjoint $\boldsymbol{\lambda}$, and $C_{qq} \bullet \boldsymbol{\lambda}$. The measurement locations are marked with a bullet. The v components are similar in structure and not shown. Reproduced from *Eknes and Evensen* (1997)

$$\boldsymbol{C}_{aa}(z_1, z_2) = \sigma_a^2 \exp\left(-\left(\frac{z_1 - z_2}{l_a}\right)^2\right) \boldsymbol{I}, \tag{5.93}$$

$$\boldsymbol{C}_{b_0 b_0}(t_1, t_2) = \sigma_{b_0}^2 \delta(t_1 - t_2) \boldsymbol{I}, \tag{5.94}$$

$$\boldsymbol{C}_{b_H b_H}(t_1, t_2) = \sigma_{b_H}^2 \delta(t_1 - t_2) \boldsymbol{I}, \tag{5.95}$$

$$\boldsymbol{C}_{qq}(z_1, t_1, z_2, t_2) = \sigma_q^2 \exp\left(-\left(\frac{z_1 - z_2}{l_q}\right)^2\right) \delta(t_1 - t_2) \boldsymbol{I}, \tag{5.96}$$

$$\boldsymbol{C}_{\epsilon\epsilon} = \sigma_o^2 \boldsymbol{I}. \tag{5.97}$$

Here it has been assumed that the model and the boundary errors are uncorrelated in time. This is convenient for computational reasons, but for more realistic applications, such a correlation should probably be included. The error variances all correspond to a 5–10% standard deviation of the variables or terms they represent errors in. This means that all first guesses and the model dynamics are assumed to be reasonably accurate and they all have similar impact on the inverse solution. Small perturbations in the weights give only small perturbations in the inverse estimate. However, large perturbations may cause problems; for example, with zero weights on some of the first guesses, the inverse problem may become ill-posed. The de-correlation lengths are similar to the characteristic length scales of the dynamical system. This ensures that the representers also become smooth with similar length scales as the dynamical solution.

The first-guess, the reference solution, and the inverse estimate are given in Fig. 5.2. The reference solution is regenerated quite well, even though the first-guess solution is out of phase with the reference case and the measurements do not resolve the time period of the oscillation. In fact, a single measurement may suffice for reconstructing the correct phase since the corresponding representer will carry the information both forward and backward in time, although the errors will be larger with less measurements. Note that the quality of the inverse estimate is poorest near the initial time. This is probably caused by a poor choice of weights for the initial conditions relative to the initial condition that was actually used.

To illustrate the solution procedure using the representer method in more detail, the u-components of the variables $\boldsymbol{s}_5$, $\boldsymbol{r}_5$, $\boldsymbol{\lambda}$, and the right-hand sides $\boldsymbol{C}_{qq} \bullet \boldsymbol{s}_5$ and $\boldsymbol{C}_{qq} \bullet \boldsymbol{\lambda}$, are given in Fig. 5.3. These plots demonstrate how the information from the measurements is taken into account and influences the solution. Measurement number five corresponds to the u component at the location $(z, t) = (-20.0, 25.0)$.

The upper plot shows the u-component of $\boldsymbol{s}_5$ and it is clear from (5.90) that it is forced by the δ-function at the measurement location. This information is then propagated backward in time while the u and v components interact during the integration.

Thereafter, $\boldsymbol{s}_i$ is used on the right-hand side of the forward equation for the representer and is also used to generate the initial and boundary conditions.

The convolution $C_{qq} \bullet s_5$, is a smoothing of s_5 according to the covariance functions contained in C_{qq}, as can be observed from the second plot in Fig. 5.3.

The representer r_5 is smooth and is oscillating in time with a period reflecting the inertial oscillations described by the dynamical model. Note that the representers will have a discontinuous time derivative at the measurement location since the right-hand side $C_{qq} \bullet s_5$ is discontinuous there. However, if a correlation in time was allowed in C_{qq}, then $C_{qq} \bullet s_5$ would be continuous and the representer r_5 would be smooth.

After the representers have been calculated and measured to generate the representer matrix, the coefficient vector b is solved for and used in (5.81) to decouple the Euler–Lagrange equations. The u-component of λ (Fig. 5.3) illustrates how the various measurements have a different impact determined by values of the coefficients in b, which again are determined by the quality of the first-guess solution versus the quality of the measurements and the residual between the measurements and the first-guess solution. After λ is found, the right-hand side in the forward model equation can be constructed through the convolution $C_{qq} \bullet \lambda$, and this field is given at the bottom of Fig. 5.3. Clearly, the role of this term is to force the solution to smooth the measurements.

5.3.6 Assimilation of real measurements

The representer implementation will now be examined using the LOTUS–3 data set (*Bowers et al.*, 1986) in a similar setup to the one used by *Yu and O'Brien* (1991, 1992). The LOTUS–3 measurements were collected in the northwestern Sargasso Sea (34° N, 70° W) during the summer of 1982. Current meters were fixed at depths of 5, 10, 15, 20, 25, 35, 50, 65, 75 and 100 m and measured the in situ currents. A wind recorder mounted on top of the LOTUS–3 tower measured the wind speeds. The sampling interval was 15 min, and the data used by *Yu and O'Brien* (1991, 1992) were collected in the period from June 30 to July 9, 1982. Here, data from the same time period are used. However, while *Yu and O'Brien* (1991, 1992) used all data collected during the 10 days, we have used a sub-sampled data set consisting of measurements collected at a 5-hour time interval at the depths 5, 25, 35, 50 and 75 m. The reason for not using all the measurements is to reduce the size of the representer matrix $\mathcal{M}^{\mathrm{T}}[r]$, and thus the computational cost. The inertial period and the vertical length scale are still resolved, and it is expected that mainly small-scale noise is rejected by subsampling the measurements.

The model was initialized by the first measurements collected on June 30, 1982. The standard deviation of the small-scale variability of the velocity observations was estimated to be close to 0.025 m s^{-1}, and this value was used to determine the error variances for the observations and the initial conditions. A similar approach was also used for the surface boundary conditions by looking at small-scale variability of the wind data. The model error variance was specified after a few test runs to give a relatively smooth inverse estimate

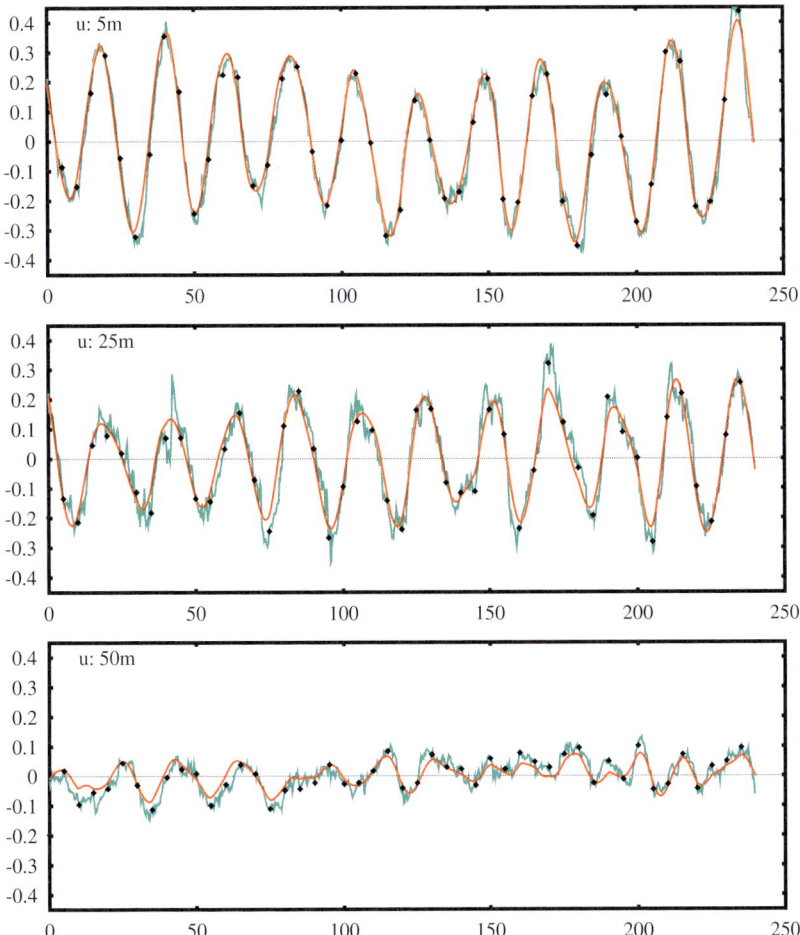

Fig. 5.4. Weak constraint results from the LOTUS–3 assimilation experiment from *Eknes and Evensen* (1997). Inverse estimate for the u component of velocity *(red lines)*, the time series of measurements *(blue lines)*, and the subsampled measurements *(bullets)*, at 5, 25 and 50 m

which seemed to be nearly consistent with the model dynamics and at the same time was close to the observations without over-fitting them.

The Ekman model describes wind driven currents and inertial oscillations only, while the measurements may also contain contributions from, e.g. pressure-driven currents. Therefore some drift in the measurements has been removed from the deepest moorings as was also done by *Yu and O'Brien* (1991, 1992).

The results from the inverse calculation are shown in Fig. 5.4 as time series of the u component of the velocity at various depths. The inverse esti-

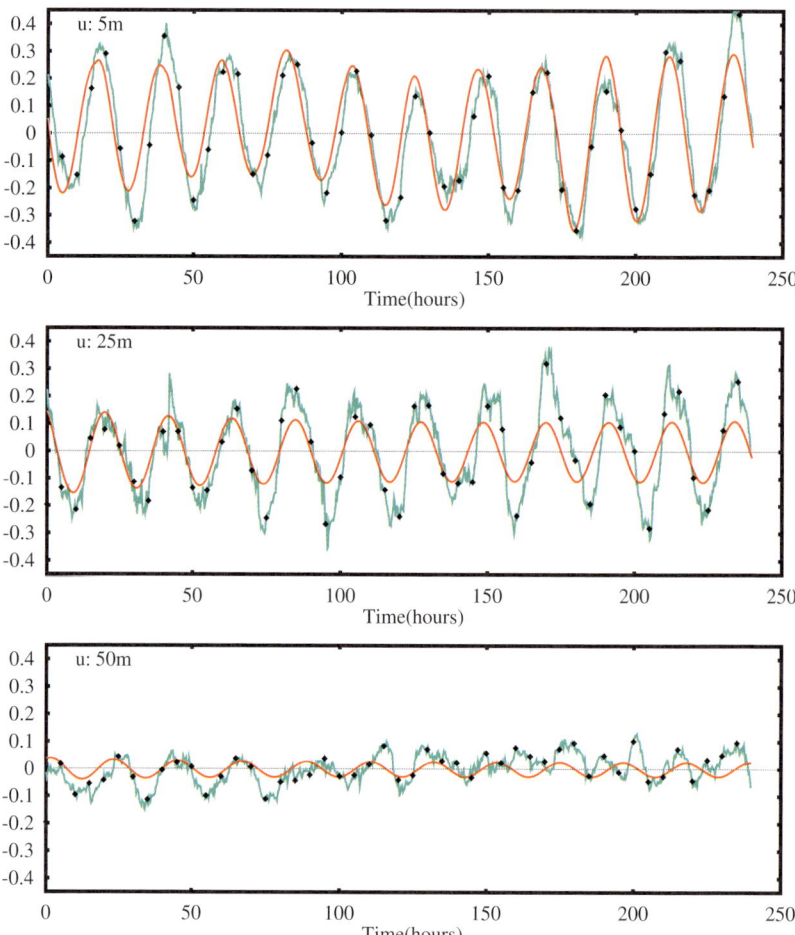

Fig. 5.5. Strong constraint results from the LOTUS–3 assimilation experiment from *Eknes and Evensen* (1997). Inverse estimate for the u component of velocity *(red lines)*, the time series of measurements *(blue lines)*, and the subsampled measurements *(bullets)*, at 5, 25 and 50 m

mate is plotted together with the full time series of the measurements. The measurements used in the inversion are shown as bullets.

It is first of all evident that both the amplitude and phase of the inverse estimate are in good agreement with the measurements at all times and depths. Note also that the inverse estimate is smooth and does not exactly interpolate the measurements. By a closer examination of the inverse estimate, it is possible to see that the time derivative of the inverse estimate is discontinuous at measurement locations. This is caused by neglecting the time correlation in the model error covariances.

For comparison, a strong constraint inversion was performed and the results are shown in Fig. 5.5. Note that the strong constraint inverse for a linear model is easily solved for without any iterations simply by calculating the representer solution with the model error covariance set to zero.

It is clear from comparisons that the strong constraint solution in the upper part of the ocean is in reasonable phase with the measurements, as determined by the initial conditions, while the amplitudes are not as good as in the weak constraint inverse. The only way the amplitudes can change when the model is assumed to be perfect is by vertical transfer of momentum from the surface. This is seen to work reasonably well near the surface, while in the deeper ocean, there is hardly any effect from the wind stress and the strong constraint inverse solution is also far from the measurements. The solution is actually rather close to a sine curve representing the pure inertial oscillations. The strong constraint results from *Yu and O'Brien* (1992) are similar to ours and also have the same problems with amplitude and phase. These results indicate that model deficiencies, such as neglected physics, should be accounted for through a weak constraint variational formulation to ensure an inverse solution in agreement with the measurements.

5.4 Comments on the representer method

Some important comments should be made regarding the representer method. For details we refer to the monographs by *Bennett* (1992, 2002).

1. As in (3.55) an inner product can be defined for the current time dependent problem, and a reproducing kernel for this inner product becomes the error covariance in time for the first guess state estimate. Thus, the same theory as was used in Chap. 3 can be used again to prove properties of the problem.

2. The representer solution provides the optimal minimizing solution of the linear inverse problem. It was shown by *Bennett* (1992) that by assuming a solution

$$\psi(t) = \psi_\mathrm{F}(t) + \boldsymbol{b}^\mathrm{T} \boldsymbol{r}(t) + g(t), \tag{5.98}$$

where $g(t)$ is an arbitrary function orthogonal to the space spanned by the representers, it can be shown that we must have $g(t) \equiv 0$, using a procedure similar to the one presented for the time independent problem in Sect. 3.2.6. This also shows that the solution is searched for in the M-dimensional space spanned by the representers. Thus, we have reduced the infinite dimensional problem defined by the penalty function to an M-dimensional problem.

3. The representer method can only be used to solve linear inverse problems. However, for nonlinear dynamical models, it can still be applied if one can define a convergent sequence of linear iterates of the nonlinear model,

where each linear iterate is solved for using the representer method. As an example consider the equation

$$\frac{\partial u}{\partial t} + u\frac{\partial u}{\partial x} = \cdots . \quad (5.99)$$

If the solution of this equation can be found from the iteration

$$\frac{\partial u^i}{\partial t} + u^{i-1}\frac{\partial u^i}{\partial x} = \cdots , \quad (5.100)$$

then one can also define a convergent sequence of linear inverse problems which can be solved exactly using representer expansions. This approach has been used with many realistic ocean and atmospheric circulation models by Bennett and coworkers and has proved to work well when the nonlinearities are not too strong. It was in fact used for an inversion of a global atmospheric primitive equation model by *Bennett et al.* (1996).

4. From the algorithm as described above, it may seem like one has to solve for a representer corresponding to each individual measurement, at the cost of two model integrations for each. However, it turns out that it is possible to solve the system (5.60) without first constructing the matrix $\mathcal{M}^\mathrm{T}[\boldsymbol{r}]$. This is possible since only the product of $\mathcal{M}^\mathrm{T}[\boldsymbol{r}]$ with an arbitrary vector $\boldsymbol{v}$ is required if an iterative solver such as the conjugate gradient method is used. This product can be evaluated by two model integrations by using a clever algorithm which is described by *Egbert et al.* (1994) and *Bennett* (2002). This is easily seen if we multiply the transposes of (5.61), (5.62), (5.63) and (5.64) with $\boldsymbol{v}$ to get

$$\frac{\partial \boldsymbol{r}^\mathrm{T}\boldsymbol{v}}{\partial t} + \boldsymbol{r}^\mathrm{T}\boldsymbol{v} = C_{qq} \bullet (\boldsymbol{s}^\mathrm{T}\boldsymbol{v}), \quad (5.101)$$

$$(\boldsymbol{r}^\mathrm{T}\boldsymbol{v})(0) = C_{aa}(\boldsymbol{s}^\mathrm{T}\boldsymbol{v})(0), \quad (5.102)$$

$$\frac{\partial (\boldsymbol{s}^\mathrm{T}\boldsymbol{v})}{\partial t} + \boldsymbol{s}^\mathrm{T}\boldsymbol{v} = -\mathcal{M}^\mathrm{T}[\delta]\boldsymbol{v}, \quad (5.103)$$

$$(\boldsymbol{s}^\mathrm{T}\boldsymbol{v})(t_k) = 0. \quad (5.104)$$

Here we note that $\boldsymbol{s}^\mathrm{T}\boldsymbol{v} = \boldsymbol{v}^\mathrm{T}\boldsymbol{s}$ is in this case a scalar function of time, just like the original model state. One backward integration of the final value problem defined by (5.103) and (5.104) results in the solution $(\boldsymbol{s}^\mathrm{T}\boldsymbol{v})$, which is then used to solve the initial value problem, (5.101) and (5.102), for the function $(\boldsymbol{r}^\mathrm{T}\boldsymbol{v})$. Since the measurement operator is linear, we then get

$$\mathcal{M}[(\boldsymbol{r}^\mathrm{T}\boldsymbol{v})] = \mathcal{M}^\mathrm{T}[\boldsymbol{r}]\boldsymbol{v}, \quad (5.105)$$

which is needed in the iterative solver.

Thus, for each linear iterate, the representer solution can be found by a number of model integrations equal to two times the number of conjugate gradient iterations to find $\boldsymbol{b}$, plus two integrations to find the final

5.4 Comments on the representer method

solution. The conjugate gradient iterations converge quickly if a good preconditioner is used and often a few selected representers are computed and measured first to construct the preconditioner (*Bennett*, 2002).

5. Finally, the convolutions appearing in the Euler–Lagrange equations can also be computed very efficiently if specific covariance functions are used. In particular it is explained in *Bennett* (2002) how one can compute the convolutions by solving simple differential equations using an approach developed by *Derber and Rosati* (1989) and *Egbert et al.* (1994).

6. Note that the equation for b, (5.60), is similar to the one solved in the analysis scheme in the standard Kalman filter. Furthermore, in the Kalman filter the representers or influence functions are defined as the measurements of the error covariance matrix at a particular time, while in the representer method the representers are functions of space and time. It can be shown that the representers correspond to the measurements of the space-time error covariance of the first guess solution. Thus, there are similarities between the analysis step in the Kalman filter and the representer method.

To summarize, the representer method is an extremely efficient methodology for solving linear inverse problems and it is also applicable to many nonlinear dynamical models. Note that the method requires knowledge of the dynamical equations and numerical code to derive the adjoint equations. Further, the actual derivation of the adjoint model and its implementation may be cumbersome for some models. This is contrary to the ensemble methods that will be discussed later. They only require the dynamical model as a black box for integrating model states forward in time.

6
Nonlinear variational inverse problems

This chapter considers highly nonlinear variational inverse problems and their properties. More general inverse formulations for nonlinear dynamical models will be treated extensively in the following chapters, but an introduction is in place here. The focus will be on some highly nonlinear problems which cannot easily be solved using the representer method. Examples are given were instead, so-called direct minimization methods are used.

6.1 Extension to nonlinear dynamics

It was pointed out in the previous chapter that, rather than solving one nonlinear inverse problem, one may define a convergent sequence of linear iterates for the nonlinear model equation, and then solve a linear inverse problem for each iterate using the representer method.

On the other hand, it is also possible to define a variational inverse problem for a nonlinear model. As an example, when starting from the system (5.21–5.23) but with the right-hand-side of (5.21) replaced by a nonlinear function, $G(\psi)$, we obtain Euler–Lagrange equations on the form

$$\frac{d\psi}{dt} - G(\psi) = \int_0^T C_{qq}(t, t_1)\lambda(t_1)\,dt_1, \tag{6.1}$$

$$\psi(0) = \Psi_0 + C_{aa}\lambda(0), \tag{6.2}$$

$$\frac{d\lambda}{dt} + G^*(\psi)\lambda = -\boldsymbol{\mathcal{M}}_{(2)}^{\mathrm{T}}\bigl[\delta(t-t_2)\bigr]\boldsymbol{W}_{\epsilon\epsilon}\Bigl(\boldsymbol{d} - \boldsymbol{\mathcal{M}}[\psi]\Bigr), \tag{6.3}$$

$$\lambda(T) = 0, \tag{6.4}$$

where $G^*(\psi)$ is the transpose of the tangent linear operator of $G(\psi)$ evaluated at ψ. Thus, like in the EKF we need to use linearized model operators, but this time for the backward or adjoint equation. We can expect that this may lead to similar problems as was found using the EKF.

Note that, for nonlinear dynamics the adjoint operator (or adjoint equation) does not exist, since the penalty function no longer defines an inner product for a Hilbert space. This is resolved by instead using the adjoint of the tangent linear operator.

In the following we will consider a variational inverse problem for the highly nonlinear and chaotic Lorenz equations and use this to illustrate typical problems that may show up when working with nonlinear dynamics.

6.1.1 Generalized inverse for the Lorenz equations

Several publications have examined assimilation methods with chaotic and unstable dynamics. In particular, the Lorenz model (*Lorenz*, 1963) has been examined with many different assimilation methods. Results have been used to suggest properties and possibilities of the methods for applications with oceanic and atmospheric models which may also be strongly nonlinear and chaotic.

The Lorenz model is a system of three first order coupled and nonlinear differential equations for the variables x, y and z,

$$\frac{dx}{dt} = \sigma(y - x) + q_x, \tag{6.5}$$

$$\frac{dy}{dt} = \rho x - y - xz + q_y, \tag{6.6}$$

$$\frac{dz}{dt} = xy - \beta z + q_z, \tag{6.7}$$

with initial conditions

$$x(0) = x_0 + a_x, \tag{6.8}$$
$$y(0) = y_0 + a_y, \tag{6.9}$$
$$z(0) = z_0 + a_z. \tag{6.10}$$

Here $x(t)$, $y(t)$ and $z(t)$ are the dependent variables and we have chosen the following commonly used values for the parameters in the equations: $\sigma = 10$, $\rho = 28$ and $\beta = 8/3$. We have also defined the error terms $\boldsymbol{q}(t)^{\mathrm{T}} = (q_x(t), q_y(t), q_z(t))$ and $\boldsymbol{a}^{\mathrm{T}} = (a_x, a_y, a_z)$ which have error statistics described by the 3×3 error covariance matrices $\boldsymbol{C}_{qq}(t_1, t_2)$ and $\boldsymbol{C}_{aa}$. The system leads to chaotic solutions where small perturbations of initial conditions lead to a completely different solution after a certain time integration.

Measurements of the solution are represented through the measurement equation

$$\mathcal{M}[\boldsymbol{x}] = \boldsymbol{d} + \boldsymbol{\epsilon}. \tag{6.11}$$

Further, by allowing the dynamical model equations (6.5–6.7) to contain errors, we obtain the standard weak constraint variational formulation,

$$\mathcal{J}[x,y,z] = \iint_0^T q(t_1)^{\mathrm{T}} W_{qq}(t_1,t_2) q(t_2) dt_1 dt_2 \qquad (6.12)$$
$$+ a^{\mathrm{T}} W_{aa} a + \epsilon^{\mathrm{T}} W_{\epsilon\epsilon} \epsilon.$$

The weight matrix, $W_{qq}(t_1,t_2) \in \Re^{3\times 3}$, is defined as the inverse of the model error covariance matrix, $C_{qq}(t_2,t_3) \in \Re^{3\times 3}$, from

$$\int_0^T W_{qq}(t_1,t_2) C_{qq}(t_2,t_3) dt_2 = \delta(t_1 - t_3) I, \qquad (6.13)$$

and we have the weight matrices, $W_{aa} = C_{aa}^{-1} \in \Re^{3\times 3}$ and $W_{\epsilon\epsilon} = C_{\epsilon\epsilon}^{-1} \in \Re^{M\times M}$.

6.1.2 Strong constraint assumption

The strong constraint assumption leads to the adjoint method which has proven to be efficient for linear dynamics, given that the strong constraint assumption is valid.

The strong constraint assumption, solved by the adjoint method, has been extensively used in the atmosphere and ocean communities. Particular effort has been invested in developing the adjoint method for use in weather forecasting systems, where it is named 4DVAR (4–dimensional variational method). 4DVAR implementations are today in operational or preoperational use at atmospheric weather forecasting centers, but common for these is that they still only works well for rather short assimilation time intervals of one day or less. The causes for this may be connected to the tangent linear approximation but also to the chaotic nature of the dynamical model.

The strong constraint inverse problem for the Lorenz equations is defined by assuming that the model is perfect, $q(t) \equiv 0$, and only the initial conditions contain errors. A number of papers have examined the adjoint method with the Lorenz model, see e.g. *Gauthier* (1992), *Stensrud and Bao* (1992), *Miller et al.* (1994), *Pires et al.* (1996). In these works it was found that there is a strong sensitivity of the penalty function with respect to the initial conditions. In particular there is a problem when the assimilation time interval exceeds a few times the predictability time of the model.

Miller et al. (1994) found that the penalty function changed from a nearly quadratic shape around the global minimum, for short assimilation time intervals, to a shape similar to a white noise process when the assimilation time interval was extended.

This is illustrated in Fig. 6.1 which plots values of the cost function with respect to variation in $x(0)$ while $y(0) = y_0$ and $z(0) = z_0$ are kept constant at their prior estimates. It is further assumed that all components of the solution $x(t)$ are observed at regular time intervals $t_j = j\Delta t_{\mathrm{obs}}$, for $j = 1,\ldots,m$, with $\Delta t_{\mathrm{obs}} = 1$. We can then define the measurement equation for each measurement time t_j, as

6 Nonlinear variational inverse problems

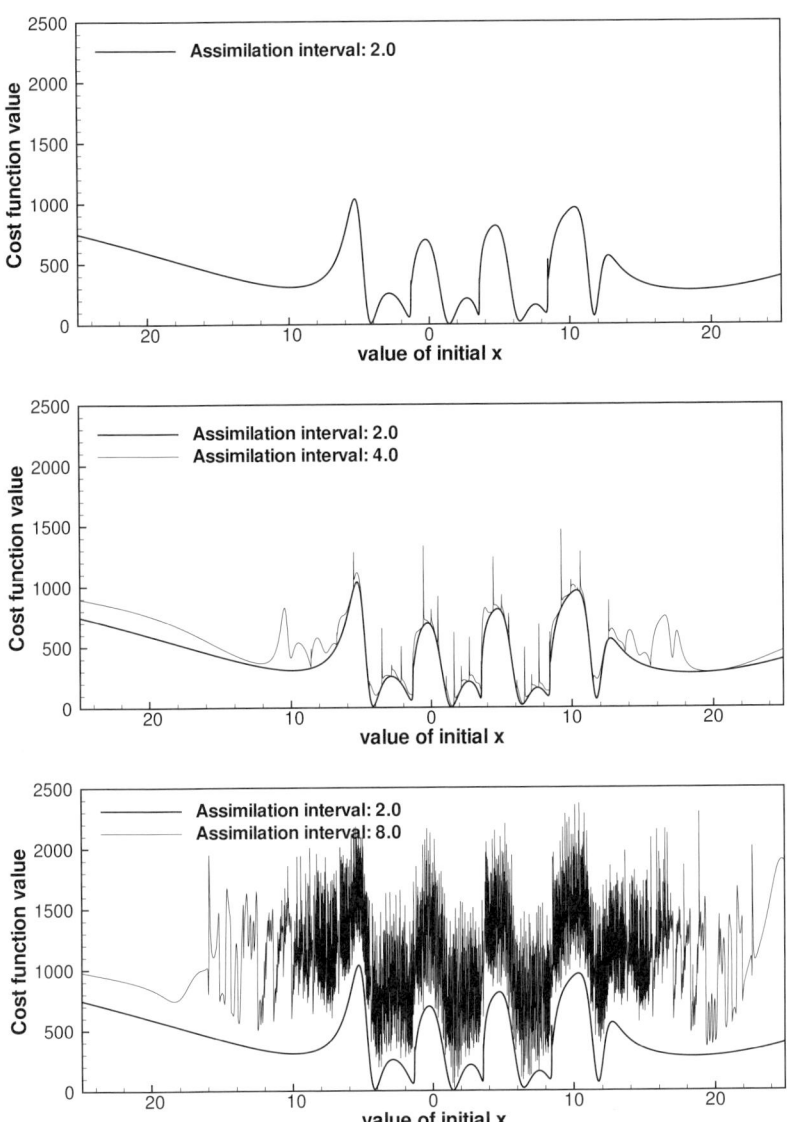

Fig. 6.1. Strong constraint penalty function for the Lorenz model as a function of the initial x-value, keeping y and z constant, when using data in the intervals $t \in [0, 2]$ *(upper plot)*, $t \in [0, 4]$ *(middle plot)*, and $t \in [0, 8]$ *(lower plot)*

6.1 Extension to nonlinear dynamics

$$\mathcal{M}_j[\boldsymbol{x}] = \boldsymbol{d}_j + \boldsymbol{\epsilon}_j, \tag{6.14}$$

where $\boldsymbol{\epsilon}_j$ represents the random errors in the measurements.

The value of the penalty function can be evaluated from

$$\mathcal{J}_J[\boldsymbol{x}(0)] = \big(\boldsymbol{x}(0) - \boldsymbol{x}_0\big)^{\mathrm{T}} \boldsymbol{W}_{aa}\big(\boldsymbol{x}(0) - \boldsymbol{x}_0\big) \\
+ \sum_{j=1}^{J}\big(\boldsymbol{d}_j - \mathcal{M}_j[\boldsymbol{x}]\big)^{\mathrm{T}} \boldsymbol{W}_{\epsilon\epsilon}(j)\big(\boldsymbol{d}_j - \mathcal{M}_j[\boldsymbol{x}]\big), \tag{6.15}$$

where the subscript J, defines the length of the assimilation time interval and indicates that measurements up to the J'th measurement time are included. The weights $\boldsymbol{W}_{aa}$ and $\boldsymbol{W}_{\epsilon\epsilon}(j)$ are three by three matrices and have the same interpretation as in the previous sections.

The upper plot of Fig. 6.1 is for a very short assimilation time interval of $t \in [0, 2]$, i.e. only twice the characteristic time scale of the model dynamics. It is clear that even for this short time-interval there are local minima in the cost function and a very good prior estimate of the initial state is needed for a gradient based method to converge to the global minimum near $x(0) = 1.5$. In the middle plot the assimilation interval is extended to $t \in [0, 4]$ and we see that even though the basic shape is the same there now appear some additional spikes and local minima in the cost function. When the assimilation time interval is extended to $t \in [0, 8]$ in the lower plot, the shape of the cost function appears nearly as a white noise process. It is obvious that these cost functions cannot be minimized using traditional gradient based methods, and obviously, the strong constraint problem for the Lorenz equations becomes practically impossible to solve for long assimilation time intervals, independent of the method used.

It should at this time be noted that this is mainly a result of the formulation of the problem, i.e. the assumption that the model is an exact representation of unstable and chaotic dynamics. It is not unlikely that similar problems can occur in models of the ocean and atmosphere which resolves the chaotic mesoscale circulation, and this may be one of the reasons why 4DVAR appears to be limited to short assimilation time intervals in these applications.

The approach for resolving this problem in the atmospheric community has been to solve a sequence of strong constraint inverse problems, of the form (6.15), defined for separate subintervals in time. To illustrate this, assume that we have divided the assimilation time interval into one-day sub-intervals, and we define a strong constraint inverse problem for each one-day time interval on the form (6.15). Thus:

1. We start by solving the first sub-problem for day one which results in an estimate for the initial conditions at day one.
2. Integration of the model from this initial condition provides the strong constraint inverse solution for day one.
3. We then use the inverse solution from the end of day one to specify the prior estimate of the initial conditions for day two.

4. The problem now is that, for day two, one cannot easily compute an estimate of a new updated prior error statistics $\boldsymbol{W}_{aa}$, for the initial conditions, that accounts for the new information introduced in the previous inverse calculation. Thus, the original prior $\boldsymbol{W}_{aa}$ is used repeatedly for each sub-interval.

Using this procedure, there is no proper time evolution of the error covariances, thus a different problem than the originally posed strong constraint problem is solved. Estimation of the proper error covariance matrix would require the computation of the inverse of the Hessian of the penalty function, which equals the error covariance matrix for the estimated initial conditions, followed by the evolution of this error covariance matrix through the assimilation interval using an approximate error covariance equation like in the EKF.

6.1.3 Solution of the weak constraint problem

We already saw that if the dynamical model is not too nonlinear, a convergent sequence of linear iterates may be defined, and each iterate can be optimally solved using the representer method. For dynamical models with stronger nonlinearities the sequence of linear iterates may not converge and alternative methods need to be used.

Another class of methods for minimizing (6.12) is named substitution methods. These are methods that guess candidates for the minimizing solution and then evaluate the value of the penalty function. Dependent of the algorithm used the new candidate may be accepted with a specified probability if it results in a lower value for the penalty function.

A discrete version of the penalty function is now needed and we represent the model variables $x(t)$, $y(t)$, and $z(t)$ on a numerical grid in time. The variables are stored in the state vectors $\boldsymbol{x}$, $\boldsymbol{y}$, and $\boldsymbol{z}$, all belonging to $\Re^n$, i.e. we have the vector $\boldsymbol{x}^{\mathrm{T}} = (x_1, x_2, \ldots, x_n)$, and similarly for $\boldsymbol{y}$ and $\boldsymbol{z}$, where n is the number of grid points in time. The discrete analog to (6.12) then becomes

$$\mathcal{J}[\boldsymbol{x}, \boldsymbol{y}, \boldsymbol{z}] = \Delta t^2 \sum_{i=1}^{n} \sum_{j=1}^{n} \boldsymbol{q}(i)^{\mathrm{T}} \boldsymbol{W}_{qq}(i,j) \boldsymbol{q}(j) + \boldsymbol{a}^{\mathrm{T}} \boldsymbol{W}_{aa} \boldsymbol{a} + \boldsymbol{\epsilon}^{\mathrm{T}} \boldsymbol{W}_{\epsilon\epsilon} \boldsymbol{\epsilon}, \quad (6.16)$$

where $\boldsymbol{q}(i)^{\mathrm{T}} = (q_x(t_i), q_y(t_i), q_z(t_i))$. Furthermore, there will be no integration of the model equations required using the substitution methods and simple numerical discretizations based on second order centered differences for the time derivatives can be used, i.e.

$$\begin{aligned}
\frac{x_{i+1} - x_{i-1}}{2\Delta t} &= \sigma(y_i - x_i) + q_x(t_i), \\
\frac{y_{i+1} - y_{i-1}}{2\Delta t} &= \rho x_i - y_i - x_i z_i + q_y(t_i), \\
\frac{z_{i+1} - z_{i-1}}{2\Delta t} &= x_i y_i - \beta z_i + q_z(t_i),
\end{aligned} \quad (6.17)$$

where $i = 2, \ldots, n-1$ is the time-step index, with n the total number of time steps.

Note that the evaluation of the double sum in (6.16) is costly. Here, an alternative method like the one used for the convolutions in the representer method could be used.

An even more efficient approach was used by *Evensen and Fario* (1997). It is assumed that the model weight can be written as

$$\boldsymbol{W}_{qq}(t_1, t_2) = \boldsymbol{W}_{qq}\delta(t_1 - t_2), \tag{6.18}$$

where $\boldsymbol{W}_{qq}$ is a constant 3×3 matrix. This eliminates one of the summations in the model term in (6.16) and allows for more efficient computational algorithms. However, the correlation in time of the model errors has a time regularizing effect on the inverse estimate which has now been lost.

To ensure a smooth solution in time the regularization is instead accounted for by a smoothing term

$$\mathcal{J}_S[\boldsymbol{x}, \boldsymbol{y}, \boldsymbol{z}] = \Delta t \sum_{i=1}^{n} \boldsymbol{\eta}_i^{\mathrm{T}} \boldsymbol{W}_{\eta\eta} \boldsymbol{\eta}_i, \tag{6.19}$$

where $\boldsymbol{\eta}_i^{\mathrm{T}} = \big(\eta_x(t_i), \eta_y(t_i), \eta_z(t_i)\big)$, with

$$\eta_x(t_i) = \frac{x_{i+1} - 2x_i + x_{i-1}}{\Delta t^2}, \tag{6.20}$$

and $\boldsymbol{W}_{\eta\eta}$ is a weight matrix determining the relative impact of the smoothing term.

It would have been more consistent to actually smooth the model errors instead of the inverse estimate, since it can be shown that such a smoothing constraint, used together with the penalty term for the model errors, would define a norm. Moreover, there is a unique correspondence between such a smoothing norm and a covariance matrix, as shown by *McIntosh* (1990). On the other hand, the smoothing term as included here, will improve the conditioning of the method since only smooth functions are searched for.

The penalty function now becomes

$$\begin{aligned}\mathcal{J}[\boldsymbol{x},\boldsymbol{y},\boldsymbol{z}] = {}& \Delta t \sum_{i=1}^{n} \boldsymbol{q}_i^{\mathrm{T}} \boldsymbol{W}_{qq} \boldsymbol{q}_i + \boldsymbol{a}^{\mathrm{T}} \boldsymbol{W}_{aa} \boldsymbol{a} + \boldsymbol{\epsilon}^{\mathrm{T}} \boldsymbol{w} \boldsymbol{\epsilon} \\ & + \Delta t \sum_{i=1}^{n} \boldsymbol{\eta}_i^{\mathrm{T}} \boldsymbol{W}_{\eta\eta} \boldsymbol{\eta}_i.\end{aligned} \tag{6.21}$$

For $\boldsymbol{q}_1$, $\boldsymbol{q}_n$, $\boldsymbol{\eta}_1$ and $\boldsymbol{\eta}_n$ we use second order one-sided difference formulas.

6.1.4 Minimization by the gradient descent method

A very simple approach for minimizing the penalty function (6.21) is to use a gradient descent algorithm as was done by *Evensen* (1997), *Evensen and Fario*

(1997). The gradient $\nabla_{(\boldsymbol{x},\boldsymbol{y},\boldsymbol{z})}\mathcal{J}[\boldsymbol{x},\boldsymbol{y},\boldsymbol{z}]$, with respect to the full state vector in time $(\boldsymbol{x},\boldsymbol{y},\boldsymbol{z})$, is easily derived. When the gradient is known it can be used in a descent algorithm to search for the minimizing solution. Thus, for the Lorenz model we solve the iteration

$$\begin{pmatrix}\boldsymbol{x}\\\boldsymbol{y}\\\boldsymbol{z}\end{pmatrix}^{i+1} = \begin{pmatrix}\boldsymbol{x}\\\boldsymbol{y}\\\boldsymbol{z}\end{pmatrix}^{i} - \gamma \begin{pmatrix}\nabla_{\boldsymbol{x}}\mathcal{J}[\boldsymbol{x},\boldsymbol{y},\boldsymbol{z}]\\\nabla_{\boldsymbol{y}}\mathcal{J}[\boldsymbol{x},\boldsymbol{y},\boldsymbol{z}]\\\nabla_{\boldsymbol{z}}\mathcal{J}[\boldsymbol{x},\boldsymbol{y},\boldsymbol{z}]\end{pmatrix}^{i}. \tag{6.22}$$

with γ being a step length. Given a first guess estimate, the gradient of the cost function is evaluated and a new state estimate can be searched for in the direction of the gradient.

The required storage for the gradient descent method is of order the size of the state vector in space and time, which is the same as for the adjoint and representer methods.

Note that, using a gradient descent method there is no need for any model integrations. This is contrary to the representer and adjoint methods which integrate both the forward model and the adjoint model, and to the Kalman filter where the forward model is needed.

As long as the penalty function does not contain any local minima, the gradient method will eventually converge to the minimizing solution. However, the obvious drawback is that the dimension of the problem becomes huge for high dimensional problems, i.e. the number of dependent variables times the grid points in time and space. For the Lorenz model this becomes $3n$. This is normally much larger than the number of measurements which defines the dimension of the problem as solved by the representer method. Thus, a proper conditioning may be needed to ensure that high dimensional problems converge in an acceptable number of iterations.

6.1.5 Minimization by genetic algorithms

With nonlinear dynamics the penalty function is clearly not convex in general due to the first term in (6.21) containing the model residuals. However, both the measurement penalty term and the smoothing norm will give a quadratic contribution to the penalty function and if the weights, $\boldsymbol{W}_{\epsilon\epsilon}$ and $\boldsymbol{W}_{\eta\eta}$, are large enough compared to the dynamical weight $\boldsymbol{W}_{qq}$, one can expect a nearly quadratic penalty function. On the contrary, if the model residuals are the dominating terms in the penalty function, clearly a pure descent algorithm may get trapped in local minima and the solution found may depend on the first guess in the iteration.

A special class of substitution methods contains the so-called genetic algorithms. These are typically statistical methods which guess new candidates for the minimizing solution at random or using some wise candidate selection algorithm. Then an acceptance algorithm is used to decide whether the new candidate is accepted or not. The acceptance algorithm is dependent on the value of the penalty function but also has a random component which allows it to escape local minima.

Statistical versions of the genetic methods exploit the fact that the minimizing solution can be interpreted as the maximum likelihood estimate of a probability density function,

$$f(\boldsymbol{x}, \boldsymbol{y}, \boldsymbol{z}) \propto \exp\Big(-\mathcal{J}[\boldsymbol{x}, \boldsymbol{y}, \boldsymbol{z}]\Big). \tag{6.23}$$

Moments of $f(\boldsymbol{x}, \boldsymbol{y}, \boldsymbol{z})$ could be estimated using standard numerical integration based on Monte Carlo methods using points selected at random from some distribution. However, this would be extremely inefficient due to the huge state space associated with many high dimensional models, such as models of the ocean and atmosphere.

Metropolis algorithm

Instead a method by *Metropolis et al.* (1953) is useful, and we now illustrate it for the variable $\boldsymbol{\psi}^\mathrm{T} = (\boldsymbol{x}, \boldsymbol{y}, \boldsymbol{z})$. The algorithm samples a pdf by performing a random walk through the space of interest. At each sample position $\boldsymbol{\psi}$, a perturbation is added to generate a new candidate $\boldsymbol{\psi}_1$, and this candidate is accepted according to a probability

$$p = \min\left(1, \frac{f(\boldsymbol{\psi}_1)}{f(\boldsymbol{\psi})}\right). \tag{6.24}$$

The mechanism for accepting the candidate with probability p, is implemented by drawing a random number ξ, from the uniform distribution on the interval $[0, 1]$ and then accepting $\boldsymbol{\psi}_1$ if $\xi \leq p$. The conditional uphill climb, based on the value of p and ξ, is due to *Metropolis et al.* (1953) and is named the Metropolis algorithm. They also gave a proof that the method was ergodic, i.e. any state can be reached from any other, and that the trials would sample the probability distribution $f(\boldsymbol{\psi})$. Clearly, in a high dimensional space with strongly nonlinear dynamics, the random trials may be too random and most of the time lead to candidates $\boldsymbol{\psi}_1$, with very low probabilities, which are only occasionally accepted. Thus, the algorithm becomes very inefficient.

Hybrid Monte Carlo algorithm

In *Bennett and Chua* (1994) an alternative to a random walk, which provided a significantly faster convergence, was used when solving for the inverse of a nonlinear open ocean shallow water model. The algorithm which is due to *Duane et al.* (1987) ensures that candidates with acceptable probabilities are constructed. It is based on constructing the Hamiltonian

$$\mathcal{H}[\boldsymbol{\psi}, \boldsymbol{\pi}] = \mathcal{J}[\boldsymbol{\psi}] + \frac{1}{2}\boldsymbol{\pi}^\mathrm{T}\boldsymbol{\pi}, \tag{6.25}$$

and then deriving the canonical equations of motion in $(\boldsymbol{\psi}, \boldsymbol{\pi})$ phase space, with respect to a pseudo time variable τ,

$$\frac{\partial \psi_i}{\partial \tau} = \frac{\partial \mathcal{H}}{\partial \pi_i} = \pi_i, \qquad (6.26)$$

$$\frac{\partial \pi_i}{\partial \tau} = -\frac{\partial \mathcal{H}}{\partial \psi_i} = -\frac{\partial \mathcal{J}}{\partial \psi_i}. \qquad (6.27)$$

This system is integrated for a pseudo time interval, $\tau \in [0, \tau_1]$, using the previously accepted value of $\boldsymbol{\psi}$ and a random guess for $\boldsymbol{\pi}(0)$ as initial conditions. The Metropolis algorithm can then be used for the new guess $\boldsymbol{\psi}(\tau_1)$. In *Duane et al.* (1987), it was proved that this algorithm also preserved detailed balance, i.e.

$$f(\boldsymbol{\psi}_1, \boldsymbol{\psi}_2) = f(\boldsymbol{\psi}_2|\boldsymbol{\psi}_1)f(\boldsymbol{\psi}_1) = f(\boldsymbol{\psi}_1|\boldsymbol{\psi}_2)f(\boldsymbol{\psi}_2), \qquad (6.28)$$

which is needed for showing that a long sequence of random trials will converge towards the distribution (6.23).

The interpretation of the method is clear. In the Hamiltonian (6.25), the penalty function defines a potential energy while a kinetic energy is represented by the last term. The canonical equations describe motion along lines of constant total energy. Thus, with a finite and random initial momentum, the integration of the canonical equation over a pseudo time interval will result in a new candidate with a different distribution of potential and kinetic energy. Unless the initial momentum is very large this will always result in a candidate which has a reasonable probability. If the initial momentum is zero, it will result in a candidate with less potential energy and higher probability. If the initial candidate is a local minimum, the random initial momentum may provide enough energy to escape the local minimum.

Note that, after a minimum of the variational problem has been found, the posterior error statistics can be estimated by collecting samples of nearby states. Thus, by using the hybrid Monte Carlo method to generate a Markov chain that samples the probability function, a statistical variance estimate can be generated. This method may be used to generate error estimates independently of the minimization technique used to solve the weak constraint problem. Hence, it could also be used in combination with the representer method which does not easily provide error estimates.

Simulated annealing

When working with a penalty function which has many local minima, the so-called simulated annealing technique may be used to improve the convergence to the stationary distribution, based on the method's capability of escaping local minima.

6.1 Extension to nonlinear dynamics

The simulated annealing method (see *Kirkpatrick et al., 1983, Azencott, 1992*) is extremely simple in its basic formulation and can be illustrated using an example where a penalty function $\mathcal{J}[\boldsymbol{\psi}]$, which may be nonlinear and discontinuous, is to be minimized with respect to the variable $\boldsymbol{\psi}$:

$\boldsymbol{\psi}$ first guess
for $i = 1 : \ldots$
　$\boldsymbol{\psi}_1 = \boldsymbol{\psi} + \Delta\boldsymbol{\psi}$
　if $(\mathcal{J}[\boldsymbol{\psi}_1] < \mathcal{J}[\boldsymbol{\psi}])$ **then**
　　$\boldsymbol{\psi} = \boldsymbol{\psi}_1$
　else
　　$\xi \in [0,1]$ random number
　　$p = \exp\bigl((\mathcal{J}[\boldsymbol{\psi}] - \mathcal{J}[\boldsymbol{\psi}_1])/\theta\bigr) \in [0,1]$
　　if $p > \xi$ **then** $\boldsymbol{\psi} = \boldsymbol{\psi}_1$
　end
　$\theta = f(\theta, i, \mathcal{J}_{\min})$
end

Here $\Delta\boldsymbol{\psi}$ might be a normal distributed random vector, but it is more efficient to simulate it using the hybrid Monte Carlo technique just described.

The temperature scheme $\theta = \theta(\theta, i, \mathcal{J}_{\min})$ is used to cool or relax the system and is normally a decreasing function of iteration counter i.

The trials will then converge towards a distribution

$$f(\boldsymbol{\psi}) \propto \exp\bigl(-\mathcal{J}[\boldsymbol{\psi}]/\theta\bigr), \qquad (6.29)$$

By slowly decreasing the value of θ the distribution will approach the delta function at the minimizing value of $\boldsymbol{\psi}$. The clue is then to choose a temperature scheme where one avoids getting trapped in local minima for too many iterations, or where too many uphill climbs are accepted. In *Bohachevsky et al.* (1986), it was suggested that the temperature should be chosen so that $p \in [0.5, 0.9]$. Here also a generalized algorithm was proposed where p was calculated according to $p = \exp\bigl(\beta(\mathcal{J}[\boldsymbol{\psi}] - \mathcal{J}[\boldsymbol{\psi}_1])/(\mathcal{J}[\boldsymbol{\psi}] - \mathcal{J}_{\min})\bigr)$, where β is approximately 3.5 and $\mathcal{J}_{\min}$ is an estimate of the normally unknown minimum value of the penalty function. Then the probability of accepting a detrimental step tend to zero as the random walk approaches the global minimum. If a value of the cost function is found which is less than $\mathcal{J}_{\min}$ this value will replace $\mathcal{J}_{\min}$.

Simulated annealing was previously used by *Barth and Wunsch* (1990) to optimize an oceanographic data collection scheme. The use of the hybrid Monte Carlo method in combination with simulated annealing has been extensively discussed by *Neal* (1992, 1993) in the context of Bayesian training of back-propagation networks. The method was also used to invert an inverse for a primitive equation model on a domain with ill-posed open boundaries by *Bennett and Chua* (1994). An application with the Lorenz equations was discussed by *Evensen and Fario* (1997) and will be illustrated below.

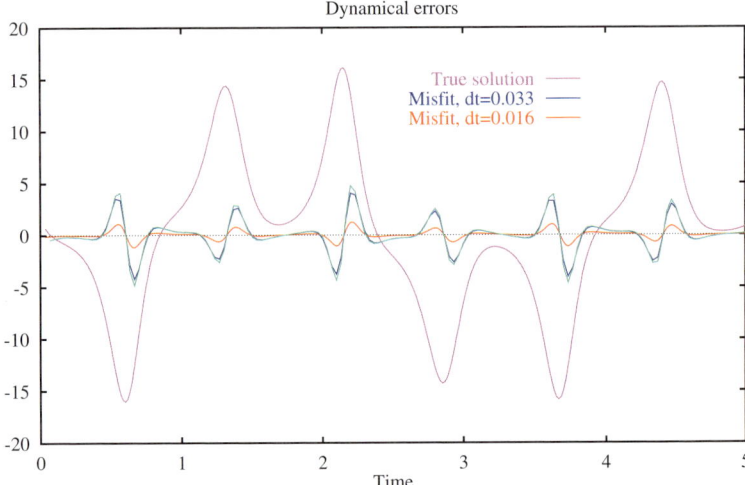

Fig. 6.2. Errors in the difference approximation used for the time derivative, plotted together with the reference solution used in the calculation of the errors. The two similar curves for $\Delta t = 0.033$ are comparing the actual calculated misfits and the lowest-order error term in the discrete time derivative. Reproduced from *Evensen and Fario* (1997)

6.2 Example with the Lorenz equations

We will now present an example where the gradient descent and the simulated annealing algorithm are used with the Lorenz equations. This example is similar to the one discussed by *Evensen and Fario* (1997).

6.2.1 Estimating the model error covariance

In an identical twin experiment it is possible to generate accurate estimates of the model error covariance. First the reference or true solution is computed using a highly accurate ordinary differential equation solver. Then the only significant contribution to the dynamical error term $\boldsymbol{q}_n$, is the error introduced in the approximate time discretization (6.17). These misfits can be evaluated and used to determine the weight matrices $\boldsymbol{W}_{qq}$ and $\boldsymbol{W}_{\eta\eta}$, which are needed in the inverse calculation.

An alternative is to evaluate the first order error term in the centered first derivative approximation used in the discrete model equations (6.17), i.e. we write for the time derivative of $x(t)$,

$$\frac{\partial x}{\partial t} = \frac{x(t+\Delta t) - x(t-\Delta t)}{2\Delta t} + \frac{1}{6}\frac{\partial^3 x}{\partial t^3}\Delta t^2 + \ldots, \qquad (6.30)$$

and evaluate the error term given the true solution.

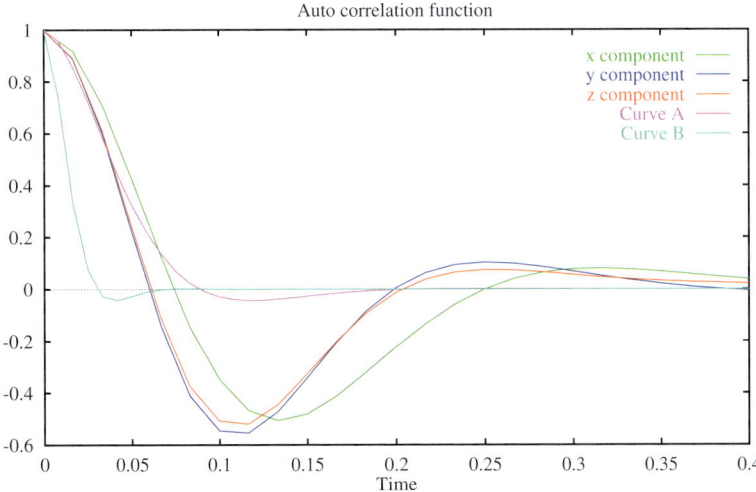

Fig. 6.3. Auto-correlation functions calculated for the computed dynamical misfits for the x, y, and z component of the solution, and two auto-correlation functions corresponding to the smoothing norm with $\gamma = 0.0008$ *(curve A)* and $\gamma = 0.00001$ *(curve B)*. Reproduced from *Evensen and Fario* (1997)

In Fig. 6.2 the dynamical misfits are plotted using two different time steps. Clearly, the errors increase with the length of the time step and the maximum errors are located at the peaks of the reference solution. The two almost identical curves for $\Delta t = 0.033$ are generated using the two different approaches just described.

The error covariance matrix $\boldsymbol{C}_{qq}$ can be estimated from a long time series of these errors, and is of course dependent on the time step used. In the experiments presented here we use a time step of $\Delta t = 0.01667$ and the corresponding error covariance matrix then becomes

$$\boldsymbol{C}_{qq} = \begin{bmatrix} 0.1491 & 0.1505 & 0.0007 \\ 0.1505 & 0.9048 & 0.0014 \\ 0.0007 & 0.0014 & 0.9180 \end{bmatrix}, \qquad (6.31)$$

where the integration has been performed for a long time interval $t \in [0, 1667]$, i.e. 100 000 time steps. The inverse of this matrix is used for $\boldsymbol{W}_{qq}$ in the penalty function (6.21).

6.2.2 Time correlation of the model error covariance

The errors are also clearly correlated in time. In Fig. 6.3 the auto-correlation functions for the x, y, and z components of the dynamical errors are plotted. Since it is inconvenient to use a full space and time covariance matrix, we introduce the smoothing term (6.19), which act as a regularization term on the minimizing solution.

It can be shown that a smoothing norm of the type

$$||\psi|| = \int_0^T \psi^2 + \gamma \psi_{tt}^2 \, dt \tag{6.32}$$

has a Fourier transform equal to

$$\hat{\psi} = \left(1 + \gamma \omega^4\right)^{-1}. \tag{6.33}$$

The limiting behaviour for increasing frequency ω is then proportional to $(\gamma \omega^4)^{-1}$; thus high frequencies are penalized most strongly in the smoothing norm. The ψ^2 term is added here, as a first guess penalty, for illustrational purposes. Without this term, the limiting behaviour for $\omega \to 0$ would be singular and the corresponding auto-correlation function would become very flat. In the actual inverse formulation, the dynamical and initial residual will provide the first guess penalty, ensuring a well-behaved limiting behaviour when $f \to 0$.

An inverse Fourier transform of the spectrum (6.33) gives an auto-correlation function which is shown in Fig. 6.3 for two values of γ, i.e. $\gamma = 0.0008$ for *curve A* and $\gamma = 0.00001$ for *curve B*. For $\gamma = 0.0008$ the auto-correlation function has a similar half width to the auto-correlation functions of the dynamical errors. However, it turned out that for this value of γ the inverse estimate became too smooth, i.e. the peaks in the solutions were to low compared to the reference solution. We decided to use $\gamma = 0.00001$ which gave an inverse estimate more in agreement with the reference solution. Based on the time series of dynamical misfits in Fig. 6.2, it is also clear that the errors are rather smooth for most of the time while they have sudden changes close to the peaks of the reference solution. The computed auto-correlation function will describe an "average" smoothness of the dynamical misfits which is too smooth near the peaks in the reference solution. This can then justify the use of the smaller smoothing weight $\gamma = 0.00001$.

The error covariance matrix $\boldsymbol{C}_{aa}$ for the errors in the initial conditions, and the measurement error covariance matrix $\boldsymbol{C}_{\epsilon\epsilon}$, are both assumed to be diagonal and with the same error variance equal to 0.5. The model error covariance matrix is given by (6.31) and the smoothing weight matrix is chosen to be diagonal and given by $\boldsymbol{W}_{\eta\eta} = \gamma \boldsymbol{I}$ with $\gamma = 0.00001$.

6.2.3 Inversion experiments

For all the cases to be discussed the initial condition for the reference case is given by $(x_0, y_0, z_0) = (1.508870, -1.531271, 25.46091)$ and the observations and first guess initial conditions are simulated by adding normal distributed noise, with zero mean and variance equal to 0.5, to the reference solution.

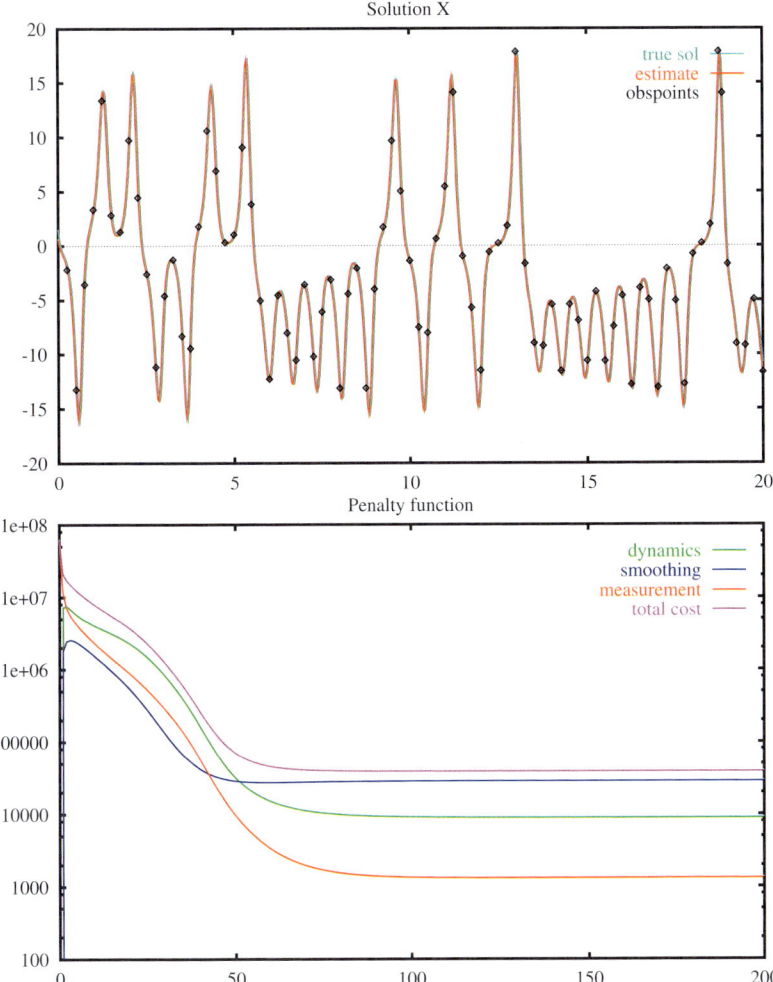

Fig. 6.4. Case A: The inverse estimate for x *(top)* and the terms in the penalty function *(bottom)*. The estimated solution is given by the solid line. The dashed line is the true reference solution, and the diamonds show the simulated observations. The same line types will be used also in the following figures. Reproduced from *Evensen and Fario* (1997)

These are lower values than the variances equal to 2.0, used in *Miller et al.* (1994) and *Evensen and Fario* (1997).

The first guess used in the gradient descent method was initially chosen as the mean of the reference solution, i.e. about $(0, 0, 23)$. However, there seems to be a possibility for a local minima close to the zero solution where both the dynamical penalty term and the smoothing penalty vanish. It is therefore

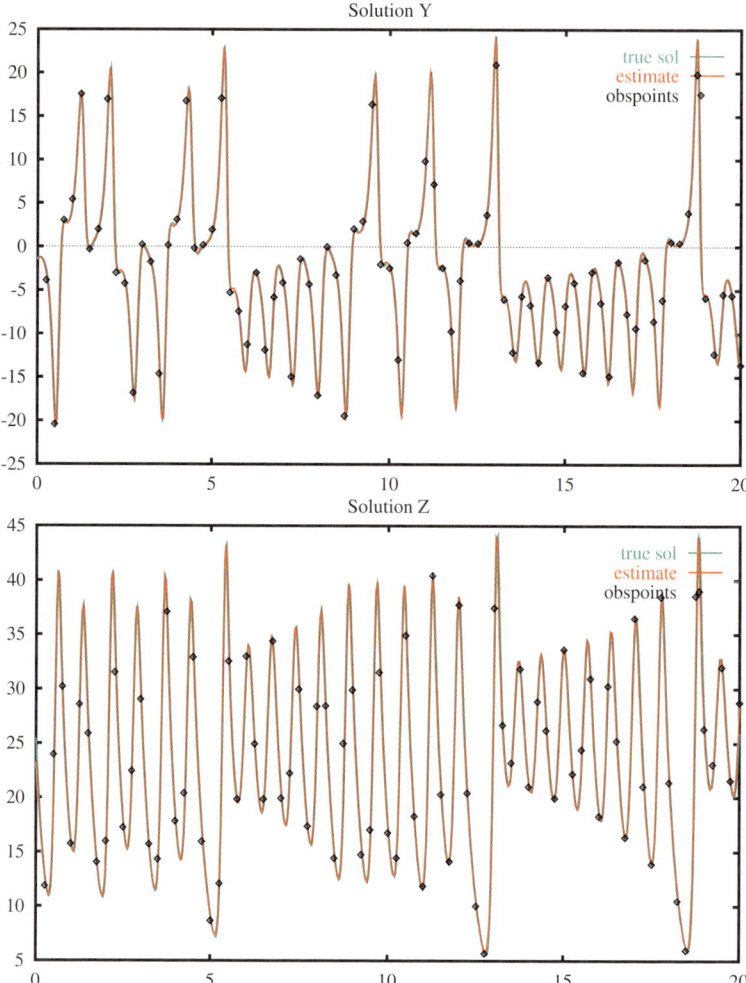

Fig. 6.5. Case A: The inverse estimate for y (top) and z (bottom). Reproduced from *Evensen and Fario* (1997)

not wise to use an estimate close to the zero solution as the first guess in the descent algorithm. To reduce the probability of getting trapped in eventual local minima, an objective analysis estimate, consistent with the measurements, was used as a first guess in the descent algorithm. It was calculated using a smoothing spline minimization algorithm which is equivalent to objective analysis (*McIntosh*, 1990). This could easily be done by replacing the dynamical misfit term with a penalty of a first-guess estimate in the inverse formulation (6.21). Some examples will now be discussed.

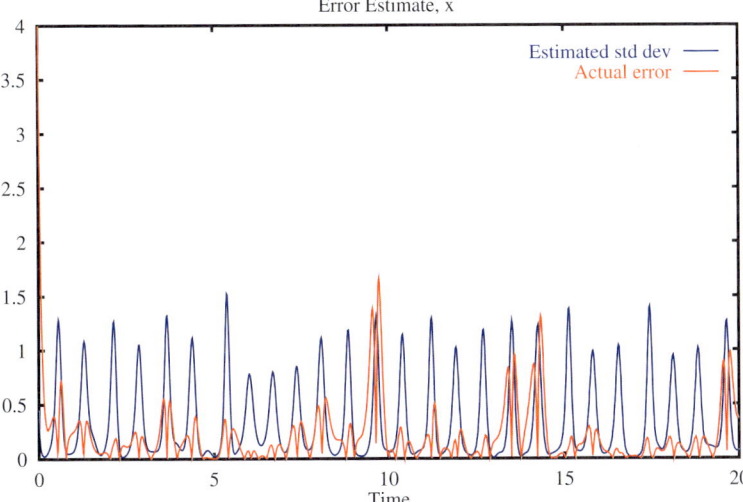

Fig. 6.6. Case A: Statistical error estimates (standard deviations) for x together with the absolute value of the actual errors. Reproduced from *Evensen and Fario* (1997)

Case A

This case can be considered as a base case and is, except for the lower measurement errors, similar to the case discussed by *Miller et al.* (1994); i.e. the time interval is $t \in [0, 20]$ and the distance between the measurements is $\Delta t_{\text{obs}} = 0.25$. The gradient descent method was in this case capable of finding the global minimum when starting from the objective analysis estimate. The minimizing solution for the three variables is given in Figs. 6.4 and 6.5 together with the terms in the penalty function as a function of iteration. We find it amazing how close the inverse estimate is to the reference solution. The quality of this inverse estimate is clearly superior to previous inverse calculations using the extended Kalman filter or a strong constraint formulation.

From the terms in the penalty function given in Fig. 6.4, it is seen that the first guess is close to the measurements and rather smooth, while the dynamical residuals are large and contribute with more than 99 % of the total value of the cost function. During the iterations, the dynamical misfit is reduced while there is an initial increase in the smoothing and measurement terms, which indicates that the final inverse solution is further from the measurements and less smooth than the first guess.

The hybrid Monte Carlo method was used to estimate the standard deviations of the errors in the minimizing solution. These are plotted together with the true differences between the estimate and the reference solution in Fig. 6.6 for the x-component. The largest errors appear around the peaks of the solution and the statistical and true errors are similar.

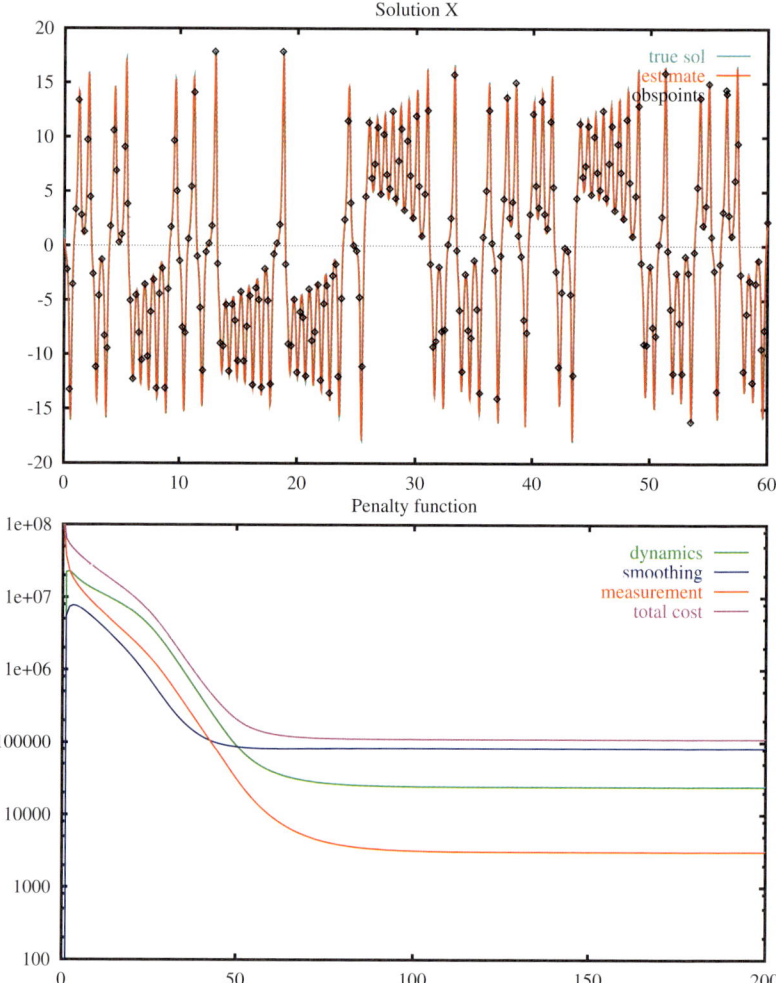

Fig. 6.7. Case B: The inverse estimate for x *(top)* and the penalty function *(bottom)*. Reproduced from *Evensen and Fario* (1997)

Case B

Here, we extended the time interval to $T = 60$, to test the sensitivity of the inverse estimate with respect to a longer time interval. The number of measurements is increased by a factor of 3 to give the same data density as in Case A. Note that the value of the cost function is also increased by about a factor of 3. This case behaves similarly to Case A, with convergence to the global minimum at a similar rate as in Case A. In Fig. 6.7 the x-component of the solution is given together with the terms in the penalty function.

6.2 Example with the Lorenz equations

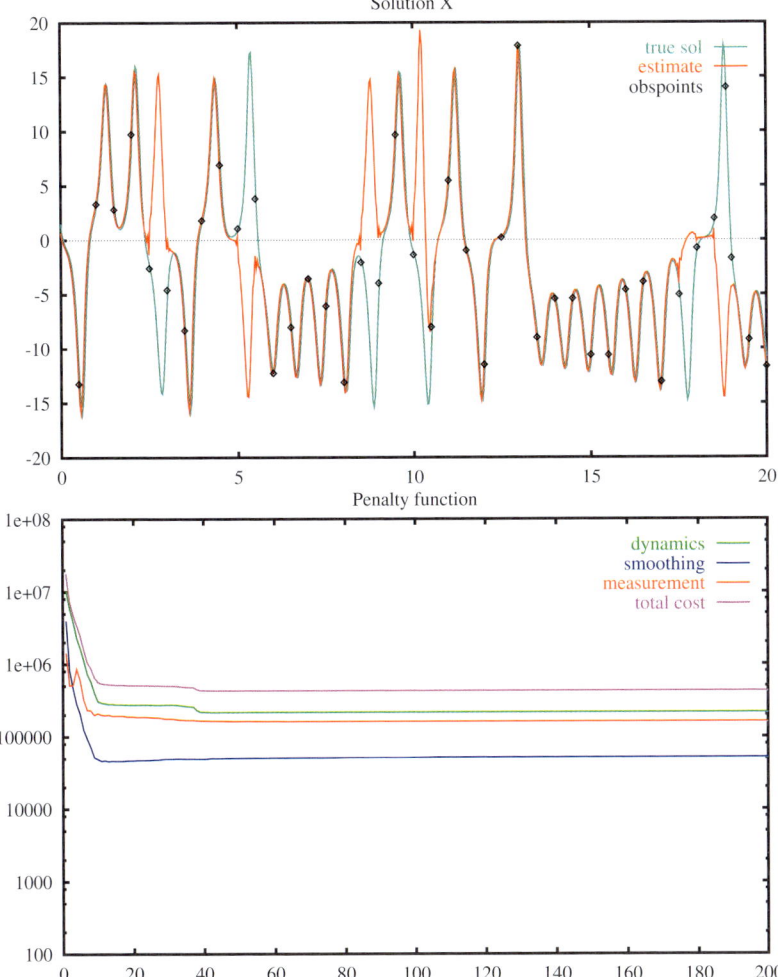

Fig. 6.8. Case C: The inverse estimate for x *(top)* and the penalty function *(bottom)*. Reproduced from *Evensen and Fario* (1997)

An important conclusion from this example is that by using a weak constraint variational formulation for the inverse, the strong sensitivity with respect to perturbations in initial conditions which is observed for strong constraint variational formulations, is completely removed. The weak constraint formulation allows the dynamical model to "forget" very past and future information. The convergence of the inverse calculation therefore has a "local" behaviour where the current estimate at two distant locations have vanishing influence on each other.

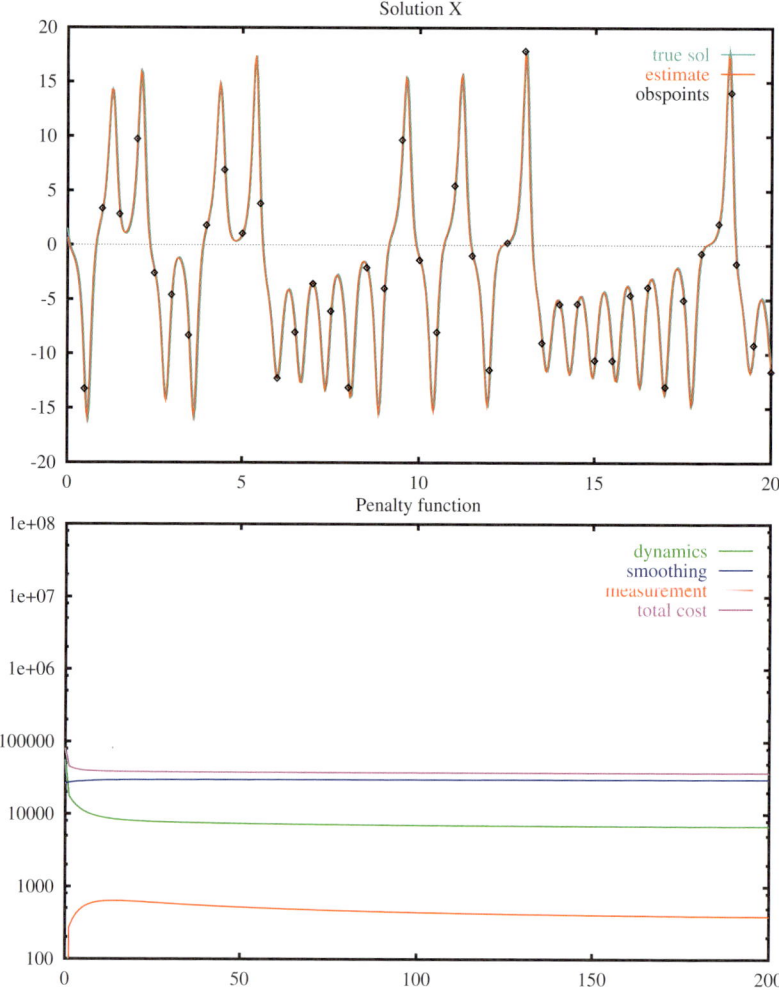

Fig. 6.9. Case C: The inverse estimate for x *(top)* and the penalty function *(bottom)* when the reference solution is used as the first guess in the gradient descent algorithm. Reproduced from *Evensen and Fario (1997)*

Case C

When the distance between the measurements is increased to $\Delta t_{\text{obs}} = 0.50$, a solution is found which misses several of the transitions, as seen in the solution for the x-component given in Fig. 6.8 together with the terms in the penalty function. This is an indication that the gradient algorithm converged to a local minimum. We can verify that this is in fact the case by running another minimization where the true reference solution is used as the first guess for the gradient method. The result is given in Fig. 6.9 where, after

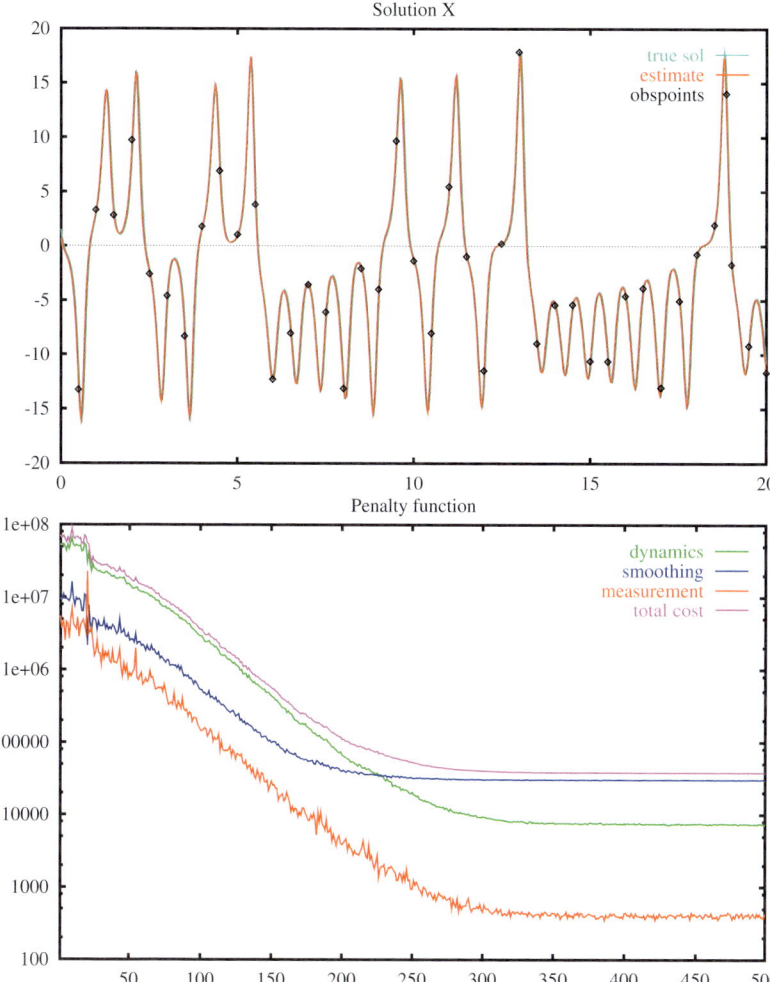

Fig. 6.10. Case C1: The inverse estimate for x *(top)* and the penalty function *(bottom)* where a genetic algorithm based on simulated annealing is used. Reproduced from *Evensen and Fario* (1997)

a minor initial adjustment, the algorithm converges to the global minimum which has a significantly lower value of the cost function and which captures all the transitions. Thus, we can conclude that when the measurement density is lowered the measurement term will give a smaller quadratic contribution to the cost function and at some stage local minima start to appear.

Case C1

This case is similar to Case C, but now the hybrid Monte Carlo method is used in combination with simulated annealing for minimizing the penalty function. The minimizing solution is in this case given in Fig. 6.10. Note that the number of iterations required for convergence is higher in this case than in the previous ones. This is due to perturbations caused by the annealing process that allows uphill moves to migrate out of local minima. The method used here is actually not proper annealing but should be denoted quenching, since the system is cooled too fast to guarantee that the global minimum will be found. In fact, in a similar case in *Evensen and Fario* (1997) a local minimum was found.

6.2.4 Discussion

A weak constraint variational formulation for the Lorenz model has been minimized using a gradient descent method.

It has been illustrated that by imposing the dynamical model as a weak constraint, by allowing the dynamics to contain errors, this leads to a better posed problem than the strong constraint formulation. The weak constraint formulation eliminates the sensitivity with respect to the initial conditions since, by allowing for model errors, the estimate can deviate from an exact model trajectory and thereby forget very past and future information. Further, there are no limitations on the length of the assimilation interval.

The inverse was calculated using the full state in "space" and time as control variables. The huge state space associated with such a formulation is the main objection against using a gradient descent method for a weak constraint inverse calculation. It could be compared to the mathematically very appealing representer method (Bennett, 1992), where the solution is searched for in a space with dimension equal to the number of measurements. On the other hand, with a gradient descent approach there is no need to integrate any dynamical equations, since a new candidate for the solution in space and time is substituted in every iteration. This gives rise to the notation substitution methods, where the important issue is the method used for proposing the solution candidates.

A gradient descent method will always provide a solution. However, it may be a local minimum if the penalty function is not convex. Statistical methods based on simulated annealing in combination with a hybrid Monte Carlo method for generating the candidates are much more expensive than a gradient descent approach but has a higher probability of finding the global minimum. The genetic methods will, for practical problems, only lead to a marginal improvement since they can only solve a slightly more difficult problem to a much larger cost. Thus, one should rather try to define a better posed problem, e.g. by introducing additional measurements.

It should be noted that with reasonable good measurement coverage the penalty function is essentially convex, but when either the number of measurements is decreased or with poorer quality of the measurements, the quadratic contribution to the penalty function from the measurement term has less influence and nonlinearities in the dynamics may give raise to local minima. Thus, the success of the substitution methods is strongly dependent on the measurement density. With sufficient number of measurements the algorithms converged to the global minimum of the weak constraint problem. When the number of measurements decreased, this resulted in a penalty function with multiple local minima and the gradient descent method was unable to converge to the global minimum.

It should also be pointed out that the gradient descent method does not directly provide error estimates for the minimizing solution. However, if the gradient descent method is first used to find the solution then the hybrid Monte Carlo method can be used to sample from the posterior distribution function and error variance estimates can be calculated.

An example of this method was used by *Natvik et al.* (2001) with a simple but nonlinear three component marine ecosystem model. In this case the dimension of the problem was equal to three variables times the number of grid nodes in time. Results similar to those found by *Evensen* (1997) were obtained, and the global minimum was found in the cases with sufficient measurement density. With a small number of measurements the gradient method converged to a local minimum.

The substitution methods solve for a state vector which consists of the model state vector in space and time. Clearly, this can be very large for realistic models and it does not appear to be a smart approach since we noted that the real dimension of the linear inverse problem equals the number of measurements. If the number of grid nodes is large, slow convergence is expected, and this was indeed a result from these studies.

In a final case, similar to case A, but only using measurements of the x-component of the solution, the global minimum was still found using the gradient descent method. In this case the estimates for y and z were entirely determined by the choice of model error covariance matrix and interactions through the dynamical equations. However, this case converged significantly slower. This is a result of poor conditioning and can be expected since the quadratic contribution from the measurement term is lower when only the x-component of the solution is measured. It also indicates that if the method is used with high dimensional problems, or with to sparse measurements, convergence problems may become crucial.

7
Probabilistic formulation

In the previous chapters we have discussed some traditional data assimilation methods and illustrated these with some simple examples. We will now present a mathematically and statistically consistent formulation of the combined parameter and state estimation problem. The starting point is Bayes' theorem which defines the posterior probability density function of the poorly known parameters and the model solution conditioned on a set of observations.

In the following chapters it will be seen that both the generalized inverse formulation and the EnKF as well as ensemble smoothers can be derived from Bayes' theorem. In addition it will be possible to properly interpret different assimilation methods and understand the assumption and approximations they relay on, and what they solve for.

The introduction of poorly known parameters does not complicate the discussion much. It is done since the parameter estimation problem is closely related to the state estimation problem, and it should in fact be treated as a combined parameter and state estimation problem. This is a fact that many works on parameter estimation have ignored, probably because the theoretical foundation for these problems and their solution methods have not previously been worked out.

7.1 Joint parameter and state estimation

The parameter estimation problem for a dynamical model can in a general form be formulated as *"how to find the joint pdf of the parameters and model state, given a set of measurements and a dynamical model with known uncertainties."*

This is vastly different from the traditional approach which is normally formulated as either *"how to find an estimate of the parameters which is as close as possible to the first guess values of the parameters and which results in a model solution which is as close as possible to a set of measurements"* or even simpler *"how to find the parameters resulting in a model solution* which is as close as possible to a set of measurements". Using these definitions, the dynamical model is considered to be perfect except for the errors in the poorly known parameters. A cost function, which measures the distance between the model solution and the measurements plus the deviation between the estimated parameter and its prior with some relative weight, is normally minimized with respect to the parameters.

Alternatively, a pure state estimation problem as was considered in the previous chapters can be defined. One is then searching for *the pdf of the model solution given a number of measurements related to the model solution*.

7.2 Model equations and measurements

We define a model with associated initial and boundary conditions on the spatial domain $\mathcal{D}$ with boundary $\partial \mathcal{D}$, and a set of observations,

$$\frac{\partial \boldsymbol{\psi}(\boldsymbol{x}, t)}{\partial t} = \boldsymbol{G}\big(\boldsymbol{\psi}(\boldsymbol{x}, t), \boldsymbol{\alpha}(\boldsymbol{x})\big) + \boldsymbol{q}(\boldsymbol{x}, t), \tag{7.1}$$

$$\boldsymbol{\psi}(\boldsymbol{x}, t_0) = \boldsymbol{\Psi}_0(\boldsymbol{x}) + \boldsymbol{a}(\boldsymbol{x}), \tag{7.2}$$

$$\boldsymbol{\psi}(\boldsymbol{x}, t)|_{\partial \mathcal{D}} = \boldsymbol{\Psi}_b(\boldsymbol{\xi}, t) + \boldsymbol{b}(\boldsymbol{\xi}, t), \tag{7.3}$$

$$\boldsymbol{\alpha}(\boldsymbol{x}) = \boldsymbol{\alpha}_0(\boldsymbol{x}) + \boldsymbol{\alpha}'(\boldsymbol{x}), \tag{7.4}$$

$$\mathcal{M}[\boldsymbol{\psi}, \boldsymbol{\alpha}] = \boldsymbol{d} + \boldsymbol{\epsilon}. \tag{7.5}$$

The model state $\boldsymbol{\psi}(\boldsymbol{x}, t) \in \Re^{n_\psi}$ is a vector consisting of the n_ψ model variables where each variable is a function of space and time. The nonlinear model is defined by (7.1) where $\boldsymbol{G}(\boldsymbol{\psi}, \boldsymbol{\alpha}) \in \Re^{n_\psi}$ is the nonlinear model operator. More general forms can be used for the nonlinear model operator, but the present one will suffice to demonstrate the methodologies considered here.

The model state is assumed to evolve in time from the initial state $\boldsymbol{\Psi}_0(\boldsymbol{x}) \in \Re^{n_\psi}$ defined in (7.2), under the constraints of the boundary conditions $\boldsymbol{\Psi}_b(\boldsymbol{\xi}, t) \in \Re^{n_\psi}$ defined in (7.3). The coordinate $\boldsymbol{\xi}$ is running over the surface $\partial \mathcal{D}$ where the boundary condition is defined.

We have defined $\boldsymbol{\alpha}(\boldsymbol{x}) \in \Re^{n_\alpha}$ as a set of n_α poorly known parameters of the model. These can be both a vector of spatial fields, in the form they are written here, or a vector of scalars, and they are assumed to be constant in time. A first guess value $\boldsymbol{\alpha}_0(\boldsymbol{x}) \in \Re^{n_\alpha}$, of the vector of parameters $\boldsymbol{\alpha}(\boldsymbol{x}) \in \Re^{n_\alpha}$, is introduced through (7.4).

Additional conditions are present in the form of the measurements $\boldsymbol{d} \in \Re^M$. These can be direct point measurements of the model solution or more complex parameters which are nonlinearly related to the model state. For the time being we will restrict ourselves to the case with linear measurements. An example of a direct measurement functional is then

$$\mathcal{M}_i[\boldsymbol{\psi}] = \int \int \boldsymbol{\psi}^{\mathrm{T}}(\boldsymbol{x}, t) \boldsymbol{\delta}_{\psi_i} \delta(t - t_i) \delta(\boldsymbol{x} - \boldsymbol{x}_i) dt\, d x, \tag{7.6}$$

where the integration is over the space and time domain of the model. The measurement d_i, is related to the model state variable as selected by the vector $\boldsymbol{\delta}_{\psi_i}$, and evaluated at the space and time location $(\boldsymbol{x}_i, t_i)$. If a three-variable model is used and the second variable is measured, then $\boldsymbol{\delta}_{\psi_i}$ becomes the vector $(0, 1, 0)^{\mathrm{T}}$ while $\delta(t - t_i)$ and $\delta(\boldsymbol{x} - \boldsymbol{x}_i)$ are Dirac delta functions.

In (7.1–7.5) we have also included unknown error terms which are representing the errors in the model equations, the initial and boundary conditions, the first guess for the model parameters and the measurements. Without these error terms the system as given above is over-determined and has no solution. On the other hand, when we introduce these error terms without additional conditions there are infinitively many solutions of the system. The way to proceed is to introduce a statistical hypothesis about the errors, e.g. assuming that they are normally distributed with means equal to zero and known error covariances.

7.3 Bayesian formulation

We now consider the model variables, the poorly known parameters, the initial and boundary conditions and the measurements as random variables which can be described by pdfs.

The joint pdf for the model state as a function of space and time and the parameters is $f(\boldsymbol{\psi}, \boldsymbol{\alpha})$. Further, for the measurements we can define the likelihood function $f(\boldsymbol{d}|\boldsymbol{\psi}, \boldsymbol{\alpha})$, thus we can measure both the model state and the parameters. Using Bayes' theorem the parameter estimation problem can be written as

$$f(\boldsymbol{\psi}, \boldsymbol{\alpha}|\boldsymbol{d}) \propto f(\boldsymbol{\psi}, \boldsymbol{\alpha}) f(\boldsymbol{d}|\boldsymbol{\psi}, \boldsymbol{\alpha}). \tag{7.7}$$

We have not included a denominator which normalizes the right-hand-side, thereby writing proportional to, $\propto$, rather than equal to.

Parameter estimation problems normally do not include the model state as a variable to be estimated. It is more common to first solve for the poorly known parameters alone, and then rerun the model to find the model solution. This implicitly assumes that the model, with the new estimates of the parameters, does not contain any errors. Generally, this is not a valid assumption.

In the dynamical model, we have specified initial and boundary conditions as random variables and we have included prior information about the parameters. Thus, we define the pdfs $f(\boldsymbol{\psi}_0)$, $f(\boldsymbol{\psi}_b)$ and $f(\boldsymbol{\alpha})$, for the estimates $\boldsymbol{\psi}_0$, $\boldsymbol{\psi}_b$ and $\boldsymbol{\alpha}$, of the initial and boundary conditions, and the parameters. We then write instead of $f(\boldsymbol{\psi}, \boldsymbol{\alpha})$,

$$\begin{aligned} f(\boldsymbol{\psi}, \boldsymbol{\alpha}, \boldsymbol{\psi}_0, \boldsymbol{\psi}_b) &= f(\boldsymbol{\psi}, \boldsymbol{\alpha}|\boldsymbol{\psi}_0, \boldsymbol{\psi}_b) f(\boldsymbol{\psi}_0) f(\boldsymbol{\psi}_b) \\ &= f(\boldsymbol{\psi}|\boldsymbol{\alpha}, \boldsymbol{\psi}_0, \boldsymbol{\psi}_b) f(\boldsymbol{\psi}_0) f(\boldsymbol{\psi}_b) f(\boldsymbol{\alpha}). \end{aligned} \tag{7.8}$$

Equation (7.7) should accordingly be written as

7 Probabilistic formulation

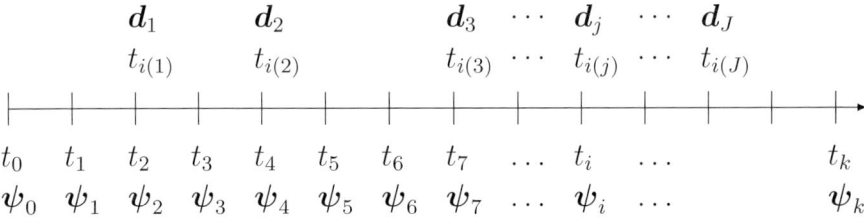

Fig. 7.1. Discretization in time. The time interval is discretized into $k+1$ nodes, at the times t_0 to t_k, where the model state vector $\boldsymbol{\psi}_i = \boldsymbol{\psi}(t_i)$ is defined. The measurement vectors $\boldsymbol{d}_j$ are available at the discrete subset of times $t_{i(j)}$, where $j = 1, \ldots, J$

$$f(\boldsymbol{\psi}, \boldsymbol{\alpha}, \boldsymbol{\psi}_0, \boldsymbol{\psi}_b | \boldsymbol{d}) \propto f(\boldsymbol{\psi}|\boldsymbol{\alpha}, \boldsymbol{\psi}_0, \boldsymbol{\psi}_b) f(\boldsymbol{\psi}_0) f(\boldsymbol{\psi}_b) f(\boldsymbol{\alpha}) f(\boldsymbol{d}|\boldsymbol{\psi}, \boldsymbol{\alpha}), \quad (7.9)$$

where it is also assumed that the boundary conditions and initial conditions are independent, although, this may not be true for their intersection at t_0. Here the pdf $f(\boldsymbol{\psi}|\boldsymbol{\alpha}, \boldsymbol{\psi}_0, \boldsymbol{\psi}_b)$ is the prior density for the model solution given the parameters and initial and boundary conditions.

7.3.1 Discrete formulation

In the following discussion it is convenient to work with a model state which is discretized in time, i.e. $\boldsymbol{\psi}(\boldsymbol{x}, t)$ is represented at fixed time intervals as $\boldsymbol{\psi}_i(\boldsymbol{x}) = \boldsymbol{\psi}(\boldsymbol{x}, t_i)$ with $i = 0, 1, \ldots, k$. Please refer to Fig. 7.1 for further illustration.

Furthermore, we define the pdf for the model integration from time t_{i-1} to t_i as $f(\boldsymbol{\psi}_i|\boldsymbol{\psi}_{i-1}, \boldsymbol{\alpha}, \boldsymbol{\psi}_b(t_i))$, which assumes that the model is a first order Markov process. In the general case when model errors are time correlated this could be written as $f(\boldsymbol{\psi}_i|\boldsymbol{\psi}_k, \ldots, \boldsymbol{\psi}_{i+1}, \boldsymbol{\psi}_{i-1}, \ldots, \boldsymbol{\psi}_0, \boldsymbol{\alpha}, \boldsymbol{\psi}_b(t_i))$ which for simplicity is written as $f(\boldsymbol{\psi}_i| \{\boldsymbol{\psi}_{l \neq i}\}, \boldsymbol{\alpha}, \boldsymbol{\psi}_b(t_i))$.

The joint pdf for the model solution and the parameters in (7.8) can now be written

$$f(\boldsymbol{\psi}_1, \ldots, \boldsymbol{\psi}_k, \boldsymbol{\alpha}, \boldsymbol{\psi}_0, \boldsymbol{\psi}_b) \propto f(\boldsymbol{\alpha}) f(\boldsymbol{\psi}_b) f(\boldsymbol{\psi}_0) \prod_{i=1}^{k} f(\boldsymbol{\psi}_i|\boldsymbol{\psi}_{i-1}, \boldsymbol{\alpha}, \boldsymbol{\psi}_b). \quad (7.10)$$

We now assume that the measurements $\boldsymbol{d} \in \Re^M$ can be divided into subsets of measurement vectors $\boldsymbol{d}_j \in \Re^{m_j}$, collected at times $t_{i(j)}$, with $j = 1, \ldots, J$ and $0 < i(1) < i(2) < \ldots < i(J) < k$. The subset $\boldsymbol{d}_j$ will only depend on $\boldsymbol{\psi}(t_{i(j)}) = \boldsymbol{\psi}_{i(j)}$ or $\boldsymbol{\alpha}$. Further, it is assumed that the measurement errors are uncorrelated in time. We can then write

$$f(\boldsymbol{d}|\boldsymbol{\psi}, \boldsymbol{\alpha}) = \prod_{j=1}^{J} f(\boldsymbol{d}_j|\boldsymbol{\psi}_{i(j)}, \boldsymbol{\alpha}). \quad (7.11)$$

7.3 Bayesian formulation

From Bayes' theorem, we now get

$$f(\boldsymbol{\psi}_1,\ldots,\boldsymbol{\psi}_k,\boldsymbol{\alpha},\boldsymbol{\psi}_0,\boldsymbol{\psi}_b|\boldsymbol{d}) \propto \\ f(\boldsymbol{\alpha})f(\boldsymbol{\psi}_0)f(\boldsymbol{\psi}_b)\prod_{i=1}^{k}f(\boldsymbol{\psi}_i|\boldsymbol{\psi}_{i-1},\boldsymbol{\alpha})\prod_{j=1}^{J}f(\boldsymbol{d}_j|\boldsymbol{\psi}_{i(j)},\boldsymbol{\alpha}). \tag{7.12}$$

The general case, when the model is not a first order Markov process, becomes

$$f(\boldsymbol{\psi}_1,\ldots,\boldsymbol{\psi}_k,\boldsymbol{\alpha},\boldsymbol{\psi}_0,\boldsymbol{\psi}_b|\boldsymbol{d}) \propto \\ f(\boldsymbol{\alpha})f(\boldsymbol{\psi}_0)f(\boldsymbol{\psi}_b)\prod_{i=1}^{k}f\bigl(\boldsymbol{\psi}_i\big|\{\boldsymbol{\psi}_{l\neq i}\},\boldsymbol{\alpha}\bigr)\prod_{j=1}^{J}f(\boldsymbol{d}_j|\boldsymbol{\psi}_{i(j)},\boldsymbol{\alpha}), \tag{7.13}$$

i.e. the model state at time t_i is dependent on the model state at all other times. This is the case when time correlated model errors are used. The previous equations constitute the most general formulation of the state and parameter estimation problem.

7.3.2 Sequential processing of measurements

We will now assume that the model can be written as a first order Markov process. This is not a strong assumption or simplification. It was shown by *Reichle et al.* (2002) and *Evensen* (2003) that in the case of time correlated model errors, it is still possible to reformulate the problem as a first order Markov process by augmenting the model errors to the model state vector. A simple equation forced by white noise can be used to simulate the time evolution of the model errors.

Evensen and van Leeuwen (2000) showed that a general smoother and filter could be derived from the Bayesian formulation given in (7.12). We now rewrite (7.12) as follows:

$$f(\boldsymbol{\psi}_1,\ldots,\boldsymbol{\psi}_k,\boldsymbol{\alpha},\boldsymbol{\psi}_0,\boldsymbol{\psi}_b|\boldsymbol{d}) \propto f(\boldsymbol{\alpha})f(\boldsymbol{\psi}_0)f(\boldsymbol{\psi}_b) \\ \prod_{i=1}^{i(1)} f(\boldsymbol{\psi}_i|\boldsymbol{\psi}_{i-1},\boldsymbol{\alpha})f(\boldsymbol{d}_1|\boldsymbol{\psi}_{i(1)},\boldsymbol{\alpha}) \\ \vdots \\ \prod_{i=i(J-1)+1}^{i(J)} f(\boldsymbol{\psi}_i|\boldsymbol{\psi}_{i-1},\boldsymbol{\alpha})f(\boldsymbol{d}_J|\boldsymbol{\psi}_{i(J)},\boldsymbol{\alpha}) \\ \prod_{i=i(J)+1}^{k} f(\boldsymbol{\psi}_i|\boldsymbol{\psi}_{i-1},\boldsymbol{\alpha}). \tag{7.14}$$

This expression can be evaluated sequentially in time as shown below, and the result will be identical to the one obtained by direct evaluation of (7.12),

$$f(\pmb{\psi}_1,\ldots,\pmb{\psi}_{i(1)},\pmb{\alpha},\pmb{\psi}_0,\pmb{\psi}_b|\pmb{d}_1) \propto$$
$$f(\pmb{\alpha})f(\pmb{\psi}_0)f(\pmb{\psi}_b)$$
$$\prod_{i=1}^{i(1)} f(\pmb{\psi}_i|\pmb{\psi}_{i-1},\pmb{\alpha})f(\pmb{d}_1|\pmb{\psi}_{i(1)},\pmb{\alpha}), \qquad (7.15)$$

$$f(\pmb{\psi}_1,\ldots,\pmb{\psi}_{i(2)},\pmb{\alpha},\pmb{\psi}_0,\pmb{\psi}_b|\pmb{d}_1,\pmb{d}_2) \propto$$
$$f(\pmb{\psi}_1,\ldots,\pmb{\psi}_{i(1)},\pmb{\alpha},\pmb{\psi}_0,\pmb{\psi}_b|\pmb{d}_1)$$
$$\prod_{i=i(1)+1}^{i(2)} f(\pmb{\psi}_i|\pmb{\psi}_{i-1},\pmb{\alpha})f(\pmb{d}_2|\pmb{\psi}_{i(2)},\pmb{\alpha}), \qquad (7.16)$$

$$\vdots$$

$$f(\pmb{\psi}_1,\ldots,\pmb{\psi}_{i(J)},\pmb{\alpha},\pmb{\psi}_0,\pmb{\psi}_b|\pmb{d}_1,\ldots,\pmb{d}_J) \propto$$
$$f(\pmb{\psi}_1,\ldots,\pmb{\psi}_{i(J-1)},\pmb{\alpha},\pmb{\psi}_0,\pmb{\psi}_b|\pmb{d}_1,\ldots,\pmb{d}_{J-1})$$
$$\prod_{i=i(J-1)+1}^{i(J)} f(\pmb{\psi}_i|\pmb{\psi}_{i-1},\pmb{\alpha})f(\pmb{d}_J|\pmb{\psi}_{i(J)},\pmb{\alpha}), \qquad (7.17)$$

$$f(\pmb{\psi}_1,\ldots,\pmb{\psi}_k,\pmb{\alpha},\pmb{\psi}_0,\pmb{\psi}_b|\pmb{d}_1,\ldots,\pmb{d}_J) \propto$$
$$f(\pmb{\psi}_1,\ldots,\pmb{\psi}_{i(J)},\pmb{\alpha},\pmb{\psi}_0,\pmb{\psi}_b|\pmb{d}_1,\ldots,\pmb{d}_J)$$
$$\prod_{i=i(J)+1}^{k} f(\pmb{\psi}_i|\pmb{\psi}_{i-1},\pmb{\alpha}). \qquad (7.18)$$

From these equations it is clear that, as long as the model is a first order Markov process and the measurements are available at discrete times with errors uncorrelated in time, we can process the measurements sequentially in time.

In (7.15) we compute the joint conditional pdf for the solution in the interval $[t_1, t_{i(1)}]$, the parameter $\pmb{\alpha}$ and the initial and boundary condition, given the measurements $\pmb{d}_1$.

This joint conditional pdf becomes the prior in (7.16) where the information from the measurements $\pmb{d}_2$ are introduced and the time interval is extended to $[t_1, t_{i(2)}]$. Thus, we compute the joint conditional pdf for the solution in the interval $[t_1, t_{i(2)}]$, the parameter $\pmb{\alpha}$ and the initial and boundary condition, given the measurements $\pmb{d}_1$ and $\pmb{d}_2$.

We can continue this sequential updating until all measurements have been processed and we get the pdf in (7.17). Thereafter, (7.18) is the prediction of $\pmb{\psi}_{i(m)+1},\ldots,\pmb{\psi}_k$, starting from the joint conditional pdf from (7.17).

We note again that these equations do not introduce any important approximations and thus describe the full inverse problem. Further, we claim that for many problems this sequential procedure provides a better posed approach for solving the inverse problem than trying to process all the measurements simultaneously as is normally done in variational formulations. The sequential processing is also very convenient for typical forecasting problems where new measurements can be processed when they arrive without recomputing the full inversion.

7.4 Summary

We have formulated the combined parameter and state estimation problem using Bayesian statistics and have seen that, under a condition of measurement errors being independent in time and the dynamical model being a Markov processes, a recursive formulation can be used for Bayes' theorem where measurements are processed sequentially in time.

The assumption of the model being a Markov process can be relaxed by defining a first order auto-regressive formula for the model errors and augmenting the model errors to the model state. In this case the Bayesian formulation also solves for the model errors.

It is seen that by augmenting the poorly known parameters to the model state we obtain a formulation where the model state and the parameters are solved for simultaneously. Hence, we have a combined parameter and state estimation problem.

In the next chapter we will use the standard Bayesian formulation as given by either (7.12) or (7.13) to derive the generalized variational inverse formulation for the combined parameter and state estimation problem.

Then in Chap. 9 the Ensemble Smoother (ES) is derived from the standard Bayesian formulation while the recursive form of Bayes' theorem, given by (7.15–7.18), is used to derive the Ensemble Kalman Smoother (EnKS) and the Ensemble Kalman Filter (EnKF).

8
Generalized Inverse

The variational inverse problems discussed in Chap. 5 can be derived from the Bayesian formulation presented in the previous Chapter by assuming Gaussian statistics for the priors. This was previously demonstrated by *van Leeuwen and Evensen* (1996) using the results from *Jazwinski* (1970). We will now derive the generalized inverse formulation for the combined parameter and state estimation problem starting from Bayes' theorem. Further, the resulting Euler–Lagrange equations are derived and we discuss some solution methods which also allow for the estimation of poorly known model parameters.

8.1 Generalized inverse formulation

We start from (7.13) and define Gaussian statistics for all the priors, transition densities and likelihoods which occur on the right-hand-side.

8.1.1 Prior density for the poorly known parameters

Assume that we have available a prior estimate $\boldsymbol{\alpha}_0(\boldsymbol{x}) \in \Re^{n_\alpha}$, of $\boldsymbol{\alpha}(\boldsymbol{x}) \in \Re^{n_\alpha}$, as defined in (7.4). Furthermore, the poorly known parameters $\boldsymbol{\alpha}(\boldsymbol{x})$ are assumed to be smooth functions of the spatial coordinates with Gaussian distributed errors. These conditions have the impact of a regularization of the inverse problem since they effectively reduce the degrees of freedom of the problem.

The smoothness of the estimated parameters is controlled by the definition of an error covariance $\boldsymbol{C}_{\alpha\alpha}(\boldsymbol{x}_1, \boldsymbol{x}_2) \in \Re^{n_\alpha \times n_\alpha}$. Here indices on $\boldsymbol{x}$, i.e. $\boldsymbol{x}_1, \boldsymbol{x}_2, \ldots$, denote dummy variables in $\mathcal{D}$. We can then define the inverse of $\boldsymbol{C}_{\alpha\alpha}(\boldsymbol{x}_1, \boldsymbol{x}_2)$, as $\boldsymbol{W}_{\alpha\alpha}(\boldsymbol{x}_1, \boldsymbol{x}_2)$, from

$$\int_{\mathcal{D}} \boldsymbol{C}_{\alpha\alpha}(\boldsymbol{x}_1, \boldsymbol{x}_3) \boldsymbol{W}_{\alpha\alpha}(\boldsymbol{x}_3, \boldsymbol{x}_2) d\boldsymbol{x}_3 = \delta(\boldsymbol{x}_1 - \boldsymbol{x}_2)\boldsymbol{I}, \qquad (8.1)$$

where $I \in \Re^{n_\alpha \times n_\alpha}$ is the diagonal identity matrix.

Note that a discretization of the parameter on a spatial grid leads to the use of matrices $C_{\alpha\alpha}$ and $W_{\alpha\alpha}$. Equation (8.1) is then replaced by a matrix-matrix multiplication, defining $C_{\alpha\alpha}$ as the matrix inverse of $W_{\alpha\alpha}$.

The prior pdf for α then becomes

$$f(\alpha) \propto \exp\left(-\frac{1}{2} \iint_\mathcal{D} \big(\alpha(x_1) - \alpha_0(x_1)\big)^\mathrm{T} \right.$$
$$\left. W_{\alpha\alpha}(x_1,x_2)\big(\alpha(x_2) - \alpha_0(x_2)\big) dx_1\, dx_2 \right). \tag{8.2}$$

8.1.2 Prior density for the initial conditions

The errors in the initial conditions are also assumed to have a Gaussian distribution, where $\Psi_0(x) \in \Re^{n_\psi}$ is the prior for the initial state, and $C_{aa}(x_1,x_2) \in \Re^{n_\psi \times n_\psi}$ defines the error covariance of the initial condition. As above we define the inverse of the error covariance $W_{aa}(x_1,x_2)$, from

$$\int_\mathcal{D} C_{aa}(x_1,x_3) W_{aa}(x_3,x_2)\, dx_3 = \delta(x_1 - x_2) I, \tag{8.3}$$

with $I \in \Re^{n_\psi \times n_\psi}$.

The prior pdf for the initial state then becomes

$$f(\psi_0) \propto \exp\left(-\frac{1}{2} \iint_\mathcal{D} \big(\psi_0(x_1) - \Psi_0(x_1)\big)^\mathrm{T} \right.$$
$$\left. W_{aa}(x_1,x_2)\big(\psi_0(x_2) - \Psi_0(x_2)\big) dx_1\, dx_2 \right). \tag{8.4}$$

8.1.3 Prior density for the boundary conditions

For the boundary condition which is defined on $\partial \mathcal{D}$ for all times $t \in [t_0, t_k]$, we define the covariance $C_{bb}(\xi_1,t_1,\xi_2,t_2) \in \Re^{n_\psi \times n_\psi}$ which has the inverse $W_{bb}(\xi_1,t_1,\xi_2,t_2)$ defined as

$$\int_{t_0}^{t_k} \int_{\partial\mathcal{D}} C_{bb}(\xi_1,t_1,\xi_3,t_3) W_{bb}(\xi_3,t_3,\xi_2,t_2)\, d\xi_3\, dt_3$$
$$= \delta(\xi_1 - \xi_2)\delta(t_1 - t_2) I, \tag{8.5}$$

where x_b is a coordinate over the surface $\partial \mathcal{D}$ and $I \in \Re^{n_\psi \times n_\psi}$. The prior pdf for the boundary conditions then becomes

$$f(\psi_b) \propto \exp\left(-\frac{1}{2} \iint_{\partial\mathcal{D}} \iint_{t_0}^{t_k} \big(\psi(\xi_1,t_1) - \psi_b(\xi_1,t_1)\big)^\mathrm{T} \right.$$
$$\left. W_{bb}(\xi_1,t_1,\xi_2,t_2)\big(\psi(\xi_2,t_2) - \psi_b(\xi_2,t_2)\big) dt_1\, dt_2\, d\xi_1\, d\xi_2 \right). \tag{8.6}$$

8.1.4 Prior density for the measurements

We will continue using the assumption that measurement errors are uncorrelated in time, although at least for the variational formulation this assumption is not required. With $\boldsymbol{C}_{\epsilon\epsilon}(t_{i(j)}) = \boldsymbol{W}_{\epsilon\epsilon}^{-1}(t_{i(j)}) \in \Re^{m_j \times m_j}$, with m_j being the number of measurements at time $t_{i(j)}$, we can write

$$f(\boldsymbol{d}_j|\boldsymbol{\psi}_{i(j)}, \boldsymbol{\alpha}) \propto \exp\left(-\frac{1}{2}\Big(\boldsymbol{d}_j - \mathcal{M}_j[\boldsymbol{\psi}_{i(j)}, \boldsymbol{\alpha}]\Big)^\mathrm{T} \boldsymbol{W}_{\epsilon\epsilon}(t_{i(j)}) \Big(\boldsymbol{d}_j - \mathcal{M}_j[\boldsymbol{\psi}_{i(j)}, \boldsymbol{\alpha}]\Big)\right), \quad (8.7)$$

for the prior information on the measurements. Here, we have used the vector of measurement functionals $\mathcal{M}_j \in \Re^{m_j}$, which corresponds to the vector of measurements $\boldsymbol{d}_j \in \Re^{m_j}$, and which takes the model state vector at the time $t_{i(j)}$, and possibly the parameter $\boldsymbol{\alpha}$, as arguments.

For the further discussion we write

$$\begin{aligned}
f(\boldsymbol{d}|\boldsymbol{\psi}, \boldsymbol{\alpha}) &\propto \prod_{j=1}^{m} f(\boldsymbol{d}_j|\boldsymbol{\psi}_{i(j)}, \boldsymbol{\alpha}) \\
&= \exp\left(-\frac{1}{2}\sum_{j=1}^{m}\Big(\boldsymbol{d}_j - \mathcal{M}_j[\boldsymbol{\psi}_{i(j)}, \boldsymbol{\alpha}]\Big)^\mathrm{T} \boldsymbol{W}_{\epsilon\epsilon}(t_{i(j)}) \Big(\boldsymbol{d}_j - \mathcal{M}_j[\boldsymbol{\psi}_{i(j)}, \boldsymbol{\alpha}]\Big)\right) \\
&= \exp\left(-\frac{1}{2}\Big(\boldsymbol{d} - \mathcal{M}[\boldsymbol{\psi}, \boldsymbol{\alpha}]\Big)^\mathrm{T} \boldsymbol{W}_{\epsilon\epsilon} \Big(\boldsymbol{d} - \mathcal{M}[\boldsymbol{\psi}, \boldsymbol{\alpha}]\Big)\right),
\end{aligned} \quad (8.8)$$

where $\boldsymbol{W}_{\epsilon\epsilon}$ is a matrix with the J sub-matrices, $\boldsymbol{W}_{\epsilon\epsilon}(t_{i(j)})$, on the diagonal.

8.1.5 Prior density for the model errors

Given a dynamical model we define the probability density functions for the model error using an assumption of Gaussian statistics. The model residual term is obtained from a short derivation and for simplicity we use a scalar model. The extension to a more general model like (7.1) is straight-forward.

We start by defining the discrete dynamical scalar model as

$$\psi_{i+1} = \psi_i + G(\psi_i, \alpha)\Delta t + q_i. \quad (8.9)$$

Here the function $G(\psi_i, \alpha)$ is a nonlinear model operator and q_i is an additive stochastic noise process. More general noise processes such as $G(\psi_i, q_i)$ can be treated as additive if we augment q_i to the state vector and define an additional equation which models q_i as an additive noise process.

It is useful to represent the noise as

$$q_i = \sigma\sqrt{\Delta t}\,\omega_i, \quad (8.10)$$

where $\overline{\omega_i \omega_j} = \Omega_{i,j}$ has unit variance and further defines correlations in time. Then σ is the standard deviation of the stochastic noise and the factor $\sqrt{\Delta t}$

ensures that the increase of variance with time will be independent of the time step used.

We can define the error covariance of the model noise as

$$C_{qq}(i,j) = \overline{q_i q_j} = \sigma^2 \Delta t \overline{\omega_i \omega_j}, \qquad (8.11)$$

or

$$\boldsymbol{C}_{qq} = \sigma^2 \Delta t \boldsymbol{\Omega}, \qquad (8.12)$$

i.e. for white model noise the increase in variance over a time unit is σ^2. The case with coloured noise is further treated in Chap. 12.

Now define the inverse $\boldsymbol{W}_{qq}$ of $\boldsymbol{C}_{qq}$ such that $\boldsymbol{W}_{qq}\boldsymbol{C}_{qq} = \boldsymbol{I}$, thus

$$\boldsymbol{W}_{qq} = \sigma^{-2}\Delta t^{-1} \boldsymbol{\Omega}^{-1}. \qquad (8.13)$$

We can now define the squared and weighted model residual terms, $q_i W_{qq}(i,j) q_j$, and the sum over i and j defines the measure of the total model misfit. In the limit when $\Delta t \to 0$ we can write

$$\sum_{ij} \frac{q_i}{\Delta t} \Delta t W_{qq}(i,j) \Delta t \frac{q_j}{\Delta t}$$
$$= \sum_{ij} \left(\frac{\psi_{i+1}-\psi_i}{\Delta t} - G_i\right) \Delta t W_{qq}(i,j) \Delta t \left(\frac{\psi_{i+1}-\psi_i}{\Delta t} - G_i\right) \qquad (8.14)$$
$$\to \int\!\!\int_{t_0}^{t_k} \left(\frac{\partial \psi}{\partial t} - G(\psi,\alpha)\right)_{t_1} W_{qq}(t_1,t_2) \left(\frac{\partial \psi}{\partial t} - G(\psi,\alpha)\right)_{t_2} dt_1 dt_2,$$

where $G_i = G(\psi_i, \alpha)$. If model errors are uncorrelated in time then $W_{qq}(i,j) = \sigma^{-2}\Delta t^{-1}\delta(i-j)$, and the sum will be over $q_i W_{qq}(i) q_i$, thus we get,

$$\sum_i q_i W_{qq}(i) q_i \to \int_{t_0}^{t_k} \left(\frac{\partial \psi}{\partial t} - G(\psi)\right) W_{qq}(t) \left(\frac{\partial \psi}{\partial t} - G(\psi)\right) dt. \qquad (8.15)$$

The relation to the transition densities in (7.12) and (7.13), when we assume Gaussian statistics, is

$$f(\boldsymbol{\psi}_i | \{\boldsymbol{\psi}_{l \neq i}\}, \boldsymbol{\alpha}) \propto \exp\left(-\frac{1}{2} \sum_j q_i W_{ij} q_j\right), \qquad (8.16)$$

and

$$\prod_{i=1}^{k} f(\boldsymbol{\psi}_i | \{\boldsymbol{\psi}_{l \neq i}\}, \boldsymbol{\alpha}) \propto \exp\left(-\frac{1}{2} \sum_{ij} q_i W_{ij} q_j\right), \qquad (8.17)$$

where we can replace the summations with the integrals from (8.14) and (8.15) in the limit when $\Delta t \to 0$.

8.1.6 Conditional joint density

Now, introducing the scalar products

$$\bullet \equiv \int_{t_0}^{t_k} \int_{\mathcal{D}} d\boldsymbol{x}\, dt, \quad \circ \equiv \int_{\mathcal{D}} d\boldsymbol{x}, \quad \star \equiv \int_{t_0}^{t_k} \int_{\partial \mathcal{D}} d\boldsymbol{\xi}\, dt, \qquad (8.18)$$

we can write the conditional pdf (7.13) as

$$f(\boldsymbol{\psi}_1, \ldots, \boldsymbol{\psi}_k, \boldsymbol{\alpha}, \boldsymbol{\psi}_0, \boldsymbol{\psi}_b | \boldsymbol{d}) \propto \exp\left(-\frac{1}{2}\mathcal{J}[\boldsymbol{\psi}, \boldsymbol{\alpha}]\right), \qquad (8.19)$$

where we have defined the function

$$\begin{aligned}
\mathcal{J}[\boldsymbol{\psi}, \boldsymbol{\alpha}] =\ & \left(\frac{\partial \boldsymbol{\psi}}{\partial t} - \boldsymbol{G}(\boldsymbol{\psi}, \boldsymbol{\alpha})\right)^{\mathrm{T}} \bullet \boldsymbol{W}_{qq} \bullet \left(\frac{\partial \boldsymbol{\psi}}{\partial t} - \boldsymbol{G}(\boldsymbol{\psi}, \boldsymbol{\alpha})\right) \\
& + (\boldsymbol{\psi}_0 - \boldsymbol{\Psi}_0)^{\mathrm{T}} \circ \boldsymbol{W}_{aa} \circ (\boldsymbol{\psi}_0 - \boldsymbol{\Psi}_0) \\
& + (\boldsymbol{\psi} - \boldsymbol{\psi}_b)^{\mathrm{T}} \star \boldsymbol{W}_{bb} \star (\boldsymbol{\psi} - \boldsymbol{\psi}_b) \\
& + (\boldsymbol{\alpha} - \boldsymbol{\alpha}_0)^{\mathrm{T}} \circ \boldsymbol{W}_{\alpha\alpha} \circ (\boldsymbol{\alpha} - \boldsymbol{\alpha}_0) \\
& + \Big(\boldsymbol{d} - \mathcal{M}[\boldsymbol{\psi}]\Big)^{\mathrm{T}} \boldsymbol{W}_{\epsilon\epsilon} \Big(\boldsymbol{d} - \mathcal{M}[\boldsymbol{\psi}]\Big).
\end{aligned} \qquad (8.20)$$

Thus, for Gaussian priors, maximization of the conditional joint density in (7.13) is equivalent to minimization of $\mathcal{J}$ as defined in (8.20). The minimum of $\mathcal{J}$ is also the maximum likelihood solution for $\boldsymbol{\psi}$ and $\boldsymbol{\alpha}$ as defined by the conditional joint pdf in (8.19).

The penalty function as defined by $\mathcal{J}$ will have a global minimum, but it may not be unique if the model is nonlinear. It can also possess several local minima and there is a risk of converging to one of these. It is also clear that in the case with no measurements there is a unique solution. This is the prior model solution, or central forecast, from (7.1–7.4) with all error terms set to zero, which then gives a value of $\mathcal{J} \equiv 0$. It corresponds to the maximum likelihood solution of the prior joint pdf and is therefore also named the *modal trajectory* (see Jazwinski, 1970).

The generalized inverse problem as defined by (8.20) may appear very complex at first. The introduction of parameters to be estimated, in addition to the state variables, leads to a strongly nonlinear problem even if the dynamical model is linear. However, iterative schemes have been used for the parameters in connection with the representer method by *Eknes and Evensen* (1997) and more recently by *Muccino and Bennett* (2001). This methodology will be further discussed and illustrated with an example from *Eknes and Evensen* (1997) in the following sections. The formulation of the combined parameter and state estimation problem was also discussed by *Evensen et al.* (1998).

From these studies, it became clear that the parameter estimation problem is difficult to solve using standard minimization algorithms due to the inherent nonlinearities. Other approaches for minimizing the penalty function (8.20) may use the direct iterative methods from Chap. 6 where candidates for a solution are generated, e.g. using the gradient of $\mathcal{J}$ with respect to the parameters and state variables, or even using genetic algorithms. Common for the direct methods is that they are extremely time consuming. The gradient methods may get trapped in local minima. The genetic algorithms should converge to a global minimum but are orders of magnitude more costly than the gradient methods. Because of this, other approaches have introduced assumptions of, e.g. zero model errors and sometimes also zero errors in the initial and/or the boundary conditions. It is clear that one then solves a different problem than the one originally posed and one will not find the correct solution unless these approximations are valid. In fact, one can find unphysical values of parameters which compensate for neglected errors in the model or conditions.

The state space associated with the variables $\boldsymbol{\psi}(\boldsymbol{x})$ and $\boldsymbol{\alpha}(\boldsymbol{x})$ can be huge. This has motivated some approaches for parameter estimation where $\boldsymbol{\alpha}(\boldsymbol{x})$ is approximated by a set of parameters with a smaller effective dimension. It should be noted that the use of a prior like (8.2) correctly reduces the effective dimension of $\boldsymbol{\alpha}(\boldsymbol{x})$ in a statistically consistent manner, and the problem with large state spaces is significantly reduced.

8.2 Solution methods for the generalized inverse problem

We will now use a simple scalar model formulation to illustrate some of the methods that may be used for minimizing (8.20). The use of a scalar model simplifies the notation and we avoid the specification of boundary conditions.

8.2.1 Generalized inverse for a scalar model

With $\psi(t)$ being a scalar model state, the system of equations now becomes

$$\frac{\partial \psi}{\partial t} = G(\psi, \alpha) + q, \tag{8.21}$$

$$\psi(t_0) = \Psi_0 + a, \tag{8.22}$$

$$\alpha = \alpha_0 + \alpha, \tag{8.23}$$

$$\mathcal{M}[\psi] = \boldsymbol{d} + \boldsymbol{\epsilon}. \tag{8.24}$$

The penalty function then simplifies to

$$\begin{aligned}\mathcal{J}[\psi, \alpha] = & \left(\frac{\partial \psi}{\partial t} - G(\psi, \alpha)\right) \bullet W_{qq} \bullet \left(\frac{\partial \psi}{\partial t} - G(\psi, \alpha)\right) \\ & + \left(\psi(t_0) - \Psi_0\right) W_{aa} \left(\psi(t_0) - \Psi_0\right) \\ & + (\alpha - \alpha_0) W_{\alpha\alpha} (\alpha - \alpha_0) \\ & + \left(\boldsymbol{d} - \mathcal{M}[\psi]\right)^{\mathrm{T}} \boldsymbol{W}_{\epsilon\epsilon} \left(\boldsymbol{d} - \mathcal{M}[\psi]\right).\end{aligned} \tag{8.25}$$

8.2 Solution methods for the generalized inverse problem

Note that, since there is no spatial dimension, we now have

$$\bullet \equiv \int_{t_0}^{t_k} dt, \tag{8.26}$$

and the product ∘, is replaced by scalar multiplication.

8.2.2 Euler–Lagrange equations

Note first that ψ is a function of α, since changing α will result in a different ψ. From standard variational calculus we know that $(\psi(\alpha), \alpha)$ defines an extremum if

$$\delta\mathcal{J} = \mathcal{J}\big[\psi(\alpha + \delta\alpha) + \delta\psi', \alpha + \delta\alpha\big] - \mathcal{J}\big[\psi(\alpha), \alpha\big] = \mathcal{O}(\delta\alpha^2, \delta\psi'^2), \tag{8.27}$$

when $\delta\alpha \to 0$ and $\delta\psi' \to 0$. Here, $\delta\alpha$ is a perturbation of the parameters, which also results in a perturbation ψ which becomes $\psi(\alpha + \delta\alpha) - \psi(\alpha)$. The perturbation $\delta\psi'$ is a perturbation of ψ which is independent of any perturbation of α.

Note that

$$\begin{aligned}\psi(\alpha + \delta\alpha) + \delta\psi' &= \psi(\alpha) + \psi_\alpha \delta\alpha + \delta\psi' + \mathcal{O}(\delta\alpha^2, \delta\psi'^2) \\ &= \psi(\alpha) + \delta\psi + \mathcal{O}(\delta\alpha^2, \delta\psi'^2),\end{aligned} \tag{8.28}$$

where we have defined

$$\psi_\alpha = \frac{\partial\psi}{\partial\alpha}, \tag{8.29}$$

and the total perturbation of ψ,

$$\delta\psi = \psi_\alpha \delta\alpha + \delta\psi'. \tag{8.30}$$

The nonlinear model operator can be expanded as

$$\begin{aligned}G\big(\psi(\alpha + \delta\alpha) &+ \delta\psi', \alpha + \delta\alpha\big) \\ &= G\big(\psi(\alpha), \alpha\big) + \frac{\partial G}{\partial \psi}(\psi_\alpha \delta\alpha + \delta\psi') + \frac{\partial G}{\partial \alpha}\delta\alpha + \mathcal{O}(\delta\alpha^2, \delta\psi'^2) \\ &= G\big(\psi(\alpha), \alpha\big) + \frac{\partial G}{\partial \psi}\delta\psi + \frac{\partial G}{\partial \alpha}\delta\alpha + \mathcal{O}(\delta\alpha^2, \delta\psi'^2).\end{aligned} \tag{8.31}$$

Evaluating $\delta\mathcal{J}$ from (8.27) we get

$$\begin{aligned}\frac{\delta\mathcal{J}}{2} &= \delta\alpha W_{\alpha\alpha}(\alpha - \alpha_0) \\ &+ \delta\psi(t_0) W_{aa}\big(\psi(t_0) - \Psi_0\big) \\ &+ M^T[\delta\psi] W_{\epsilon\epsilon}\big(d - M[\psi]\big) \\ &+ \int_{t_0}^{t_k}\left(\frac{\partial\delta\psi}{\partial t} - \frac{\partial G}{\partial \psi}\delta\psi - \delta\alpha\frac{\partial G}{\partial \alpha}\right)\lambda(t)\,dt \\ &+ \mathcal{O}(\delta\alpha^2, \delta\psi'^2),\end{aligned} \tag{8.32}$$

where we have defined the "adjoint" variable

$$\lambda(t_1) = \int_{t_0}^{t_k} W_{qq}(t_1, t_2) \left(\frac{\partial \psi}{\partial t} - G(\psi, \alpha) \right)_2 dt_2, \tag{8.33}$$

where the subscript 2 denotes function of t_2. Multiplying this equation with $\int_{t_0}^{t_k} dt_1 C_{qq}(t, t_1)$ from the left gives the equation

$$\frac{\partial \psi}{\partial t} - G(\psi, \alpha) = C_{qq} \bullet \lambda, \tag{8.34}$$

which is the original model with a representation of the model error involving a product between the model error covariance and the adjoint variable on the right hand side.

We now have from integration by part

$$\int_{t_0}^{t_k} \frac{\partial \delta \psi}{\partial t} \lambda \, dt = \delta \psi \lambda \Big|_{t_0}^{t_k} - \int_{t_0}^{t_k} \delta \psi \frac{\partial \lambda}{\partial t} dt. \tag{8.35}$$

Furthermore,

$$\mathcal{M}^{\mathrm{T}}[\delta \psi] = \int_{t_0}^{t_k} \delta \psi \mathcal{M}^{\mathrm{T}}\left[\delta(t - t_1)\right] dt_1, \tag{8.36}$$

which is easy to demonstrate, e.g. using a direct measurement functional,

$$\mathcal{M}_i[\delta(t - t_1)] = \int_{t_0}^{t_k} \delta(t - t_1)\delta(t_1 - t_i) dt_1 = \delta(t - t_i). \tag{8.37}$$

We can then write the variation (8.32) as

$$\begin{aligned}\frac{\delta \mathcal{J}}{2} &= \delta \alpha W_{\alpha\alpha}(\alpha - \alpha_0) \\ &\quad + \delta \psi(t_0) W_{aa}\big(\psi(t_0) - \Psi_0\big) \\ &\quad + \delta \psi(t_k)\lambda(t_k) - \delta \psi(t_0)\lambda(t_0) \\ &\quad - \int_{t_0}^{t_k} \delta \psi \frac{\partial \lambda}{\partial t} + \delta \psi \frac{\partial G}{\partial \psi}\lambda + \delta \alpha \frac{\partial G}{\partial \alpha}\lambda + \delta \psi \mathcal{M}^{\mathrm{T}}[\delta] \boldsymbol{W}_{\epsilon\epsilon}\big(\boldsymbol{d} - \mathcal{M}[\psi]\big) \, dt \\ &\quad + \mathcal{O}(\delta \alpha^2, \delta \psi'^2).\end{aligned} \tag{8.38}$$

We then reorder the terms to be proportional to either one of the variations $\delta \alpha$, $\delta \psi$, $\delta \psi(t_0)$ and $\delta \psi(t_k)$, to get

$$\begin{aligned}\frac{\delta \mathcal{J}}{2} &= \delta \alpha \left(W_{\alpha\alpha}(\alpha - \alpha_0) - \int_{t_0}^{t_k} \frac{\partial G}{\partial \alpha}\lambda \, dt \right) \\ &\quad + \delta \psi(t_0)\Big(W_{aa}\big(\psi(t_0) - \Psi_0\big) - \lambda(t_0)\Big) \\ &\quad + \delta \psi(t_k)\lambda(t_k) \\ &\quad - \int_{t_0}^{t_k} \delta \psi \left(\frac{\partial \lambda}{\partial t} + \frac{\partial G}{\partial \psi}\lambda + \mathcal{M}^{\mathrm{T}}[\delta]\boldsymbol{W}_{\epsilon\epsilon}\big(\boldsymbol{d} - \mathcal{M}[\psi]\big) \right) dt \\ &\quad + \mathcal{O}(\delta \alpha^2, \delta \psi'^2).\end{aligned} \tag{8.39}$$

8.2 Solution methods for the generalized inverse problem

If we require that $\delta\mathcal{J} = \mathcal{O}(\delta\alpha^2, \delta\psi'^2)$ we must have

$$\frac{\partial\psi}{\partial t} = G(\psi,\alpha) + C_{qq} \bullet \lambda, \tag{8.40}$$

$$\psi(t_0) = \Psi_0 + C_{aa}\lambda(t_0), \tag{8.41}$$

$$\frac{\partial\lambda}{\partial t} = -\frac{\partial G}{\partial\psi}\lambda - \mathcal{M}^{\mathrm{T}}[\delta]\boldsymbol{W}_{\epsilon\epsilon}\Big(\boldsymbol{d} - \mathcal{M}[\psi]\Big), \tag{8.42}$$

$$\lambda(t_k) = 0, \tag{8.43}$$

$$\alpha = \alpha_0 + C_{\alpha\alpha} \int_{t_0}^{t_k} \frac{\partial G}{\partial\alpha}\lambda\, dt. \tag{8.44}$$

These equations define the Euler–Lagrange equations for the weak constraint problem. They constitute a *coupled two point boundary value problem in time* for ψ and λ. The forward model is forced by a term representing model errors while the backward model is forced by *impulses* at measurement locations. The model operator of the backward model is the adjoint of the tangent linear forward model.

8.2.3 Iteration in α

It is common to define an iteration in α as follows

$$\alpha_{l+1} = \alpha_l - \gamma\left(\alpha_l - \alpha_0 - C_{\alpha\alpha}\int_{t_0}^{t_k}\frac{\partial G}{\partial\alpha}\bigg|_{\alpha_l}^{\psi_l}\lambda_l\, dt\right). \tag{8.45}$$

Here, the expression in the parentheses is just the gradient of the penalty function with respect to α, and γ is a step length. Thus, the iteration (8.45) is just the gradient descent method.

8.2.4 Strong constraint problem

A majority of previous works on parameter estimation solve a simpler version of the variational problem defined by (8.20) or (8.25). The parameter is still iterated as in (8.45), but an additional common simplification is to assume that the dynamical model has zero model errors, i.e. the prior for the model error covariance C_{qq} is set to zero. This corresponds to an infinite weight on the dynamical model which then must be satisfied exactly. From the Euler–Lagrange equations (8.40–8.43), it is seen that this eliminates the coupling of the dynamical model to the adjoint variable λ, although the initial condition still depends on λ. The so-called adjoint method solves this strong constraint problem by iteration of the initial conditions, using an equation similar to

$$\psi_{l+1}(t_0) = \psi_l(t_0) - \gamma\Big(\psi_l(t_0) - \Psi_0 + C_{aa}\lambda_l(t_0)\Big), \tag{8.46}$$

where the step length γ, may differ from the one used in (8.45). One can also choose to iterate both (8.45) and (8.46) simultaneously, or use an outer iteration of (8.45) and inner iteration for (8.46).

A further simplification is to assume that the initial conditions also are perfect, i.e. $C_{aa} \equiv 0$. This is equivalent to introducing an infinite weight on the term for the initial conditions in (8.20) and it will be exactly satisfied. This additional simplification completely decouples the dynamical model from the adjoint variable. The solution is then an exact model trajectory given the estimated parameter α. This is a commonly used form for the parameter estimation problem and it corresponds to minimizing a cost function containing the data misfit term and the prior term for the parameters. It is efficiently solved using the adjoint method and iterating the parameter, i.e. solve (8.40–8.44) with C_{qq} and C_{aa} set to zero.

The Euler–Lagrange equations for the strong constraint problem is most commonly derived from a Lagrangian function where the model and initial conditions are included using Lagrangian multipliers, i.e.

$$\mathcal{L}[\alpha, \lambda, \mu] = (\alpha - \alpha_0)W_{\alpha\alpha}(\alpha - \alpha_0)$$
$$+ \big(\psi(t_0) - \Psi_0\big)\mu$$
$$+ \big(d - \mathcal{M}[\psi]\big)^{\mathrm{T}} W_{\epsilon\epsilon}\big(d - \mathcal{M}[\psi]\big) \qquad (8.47)$$
$$+ \int_{t_0}^{t_k} \left(\frac{\partial \psi}{\partial t} - G(\psi, \alpha)\right) \lambda \, dt.$$

Variation with respect to μ returns the initial condition while variation with respect to λ returns the model. The variation with respect to α returns the Euler–Lagrange equations for the strong constraint problem as found above, i.e. (8.40–8.44) with C_{qq} and C_{aa} equal to zero. Thus, the Euler–Lagrange equations are decoupled and a solution can be found for α if the iteration (8.45) converges. This approach is normally named the adjoint method or 4DVAR method for parameter estimation.

An alternative approach for solving the strong constraint problem can be derived as follows. Evaluating the variation of (8.47) with respect to α when realizing that ψ is a function of α gives

$$\frac{\delta \mathcal{L}}{2} = \delta\alpha W_{\alpha\alpha}(\alpha - \alpha_0)$$
$$+ \delta\alpha \psi_\alpha(t_0)\mu$$
$$+ \delta\alpha \mathcal{M}^{\mathrm{T}}[\psi_\alpha] W_{\epsilon\epsilon}\big(d - \mathcal{M}[\psi]\big) \qquad (8.48)$$
$$+ \delta\alpha \int_{t_0}^{t_k} \left(\frac{\partial \psi_\alpha}{\partial t} - \frac{\partial G}{\partial \psi}\psi_\alpha - \frac{\partial G}{\partial \alpha}\right)\lambda \, dt$$
$$+ \mathcal{O}(\delta\alpha^2),$$

where we have used that $\delta\alpha$ is independent of time and that the measurement operator is linear. Since in addition λ and μ are arbitrary multipliers, we must have

$$\frac{\partial \psi}{\partial t} = G(\psi, \alpha), \tag{8.49}$$

$$\psi(t_0) = \Psi_0, \tag{8.50}$$

$$\frac{\partial \psi_\alpha}{\partial t} = \frac{\partial G}{\partial \psi}\psi_\alpha - \frac{\partial G}{\partial \alpha}, \tag{8.51}$$

$$\psi_\alpha(t_0) = 0, \tag{8.52}$$

$$\alpha = \alpha_0 + C_{\alpha\alpha}\mathcal{M}^{\mathrm{T}}[\psi_\alpha]\,\boldsymbol{W}_{\epsilon\epsilon}\Big(\boldsymbol{d} - \mathcal{M}[\psi]\Big). \tag{8.53}$$

Thus, we have derived a system of equations which consists of the original dynamical model with initial condition and an equation and initial condition for the sensitivity of ψ with respect to α, i.e. ψ_α. An equation for α includes the first guess value and an update term which includes the impact of measurements. It may be convenient to define an iteration in α as

$$\alpha_{l+1} = \alpha_l - \gamma\bigg(\alpha_l - \alpha_0 - C_{\alpha\alpha}\mathcal{M}^{\mathrm{T}}[\psi_{\alpha l}]\,\boldsymbol{W}_{\epsilon\epsilon}\Big(\boldsymbol{d} - \mathcal{M}[\psi_l]\Big)\bigg). \tag{8.54}$$

For each iteration in α we can solve the system (8.49–8.52) by forward integrations. There is no adjoint equation or backward integration involved. The forward models (8.49) and (8.51) should be integrated in parallel since the tangent linear operator in (8.51) is evaluated at the current estimate of the solution ψ. Note that the size of ψ_α, and the cost of solving (8.51), is proportional to the number of parameters included. In this example, we only have a single parameter and ψ_α becomes a scalar. Thus, with a low number of parameters this may be a more efficient approach than the adjoint method for solving the strong constraint parameter estimation problem. On the other hand, the adjoint method finds the gradient from one forward and one backward integration, independent of the number of parameters involved, but requires the model solution as a function of space and time to be stored and used for evaluation of the adjoint model operator.

8.3 Parameter estimation in the Ekman flow model

In Sect. 5.3 the representer method was used to solve the generalized inverse problem for an Ekman flow model. The discussion was taken from *Eknes and Evensen* (1997) which also considered the estimation of poorly known parameters in the model. In particular the first guesses of the wind drag and the vertical diffusion coefficient, c_{d_0} and $A_0(z)$, were allowed to contain errors, i.e.

$$c_d = c_{d_0} + p_{c_d}, \tag{8.55}$$

$$A(z) = A_0(z) + p_A(z), \tag{8.56}$$

where p_{c_d} and $p_A(z)$ are the unknown error terms. Thus a combined state estimation and parameter estimation problem was formulated and the penalty function for the state estimation problem given in (5.75) was extended to include two terms which penalize the deviation of estimated parameters from the first guess. Using the notation from Sect. 5.3, the generalized inverse for the combined parameter and state estimation problem was formulated as

$$\begin{aligned}\mathcal{J}[\boldsymbol{u}, c_d, A] = {} & \boldsymbol{q}^T \bullet \boldsymbol{W}_{qq} \bullet \boldsymbol{q} \\ & + \boldsymbol{a}^T \circ \boldsymbol{W}_{aa} \circ \boldsymbol{a} \\ & + \boldsymbol{b}_0^T * \boldsymbol{W}_{b_0 b_0} * \boldsymbol{b}_0 \\ & + \boldsymbol{b}_H^T * \boldsymbol{W}_{b_H b_H} * \boldsymbol{b}_H \\ & + p_A \circ W_{AA} \circ p_A \\ & + p_{c_d} W_{c_d c_d} p_{c_d} \\ & + \boldsymbol{\epsilon}^T \boldsymbol{W}_{\epsilon\epsilon} \boldsymbol{\epsilon},\end{aligned} \quad (8.57)$$

where the weight $W_{c_d c_d}$ is the inverse of the error variance $C_{c_d c_d}$ of p_{c_d}, and W_{AA} is the inverse of the error covariance C_{AA} of p_A. Since the wind drag coefficient and the vertical diffusion are allowed to contain errors, the variation of the penalty function with respect to these parameters must also be taken. This results in the additional equations

$$c_d = c_{d_0} + C_{c_d c_d} \int_0^T \boldsymbol{\lambda}^T(0, t)\, \boldsymbol{u}_a dt, \quad (8.58)$$

$$A = A_0 - C_{AA} \bullet \frac{\partial \boldsymbol{\lambda}^T}{\partial z} \frac{\partial \boldsymbol{u}}{\partial z}, \quad (8.59)$$

for the wind drag coefficient and the diffusion parameter. The addition of the two equations (8.58) and (8.59) to the system of Euler–Lagrange equations (5.77) to (5.83) makes the overall inverse problem nonlinear.

In Sect. 5.3 it was illustrated how the representer method could be used to solve exactly the Euler–Lagrange equations for the weak constraint inverse problem when $A(z)$ and c_d are known. When the parameters are allowed to contain errors, the inverse problem becomes nonlinear and therefore an iteration was used for $A(z)$ and c_d in (8.58) and (8.59). In each iteration, the representer technique was used to solve for the corresponding inverse estimate.

The equations (8.58) and (8.59) were iterated using a gradient descent method, i.e.

$$c_d^{l+1} = c_d^l - \gamma \left(c_d^l - c_{d_0} - C_{c_d c_d} \int_{t_0}^{t_k} (\boldsymbol{\lambda}^l)^{\mathrm{T}} \sqrt{u_a^2 + v_a^2}\, \boldsymbol{u}_a\, dt \right), \quad (8.60)$$

$$A^{l+1}(z) = A^l(z) - \gamma \left(A^l(z) - A_0(z) + C_{AA} \bullet \left(\frac{\partial \boldsymbol{u}^l}{\partial z} \right)^{\mathrm{T}} \frac{\partial \boldsymbol{\lambda}^l}{\partial z} \right). \quad (8.61)$$

8.3 Parameter estimation in the Ekman flow model

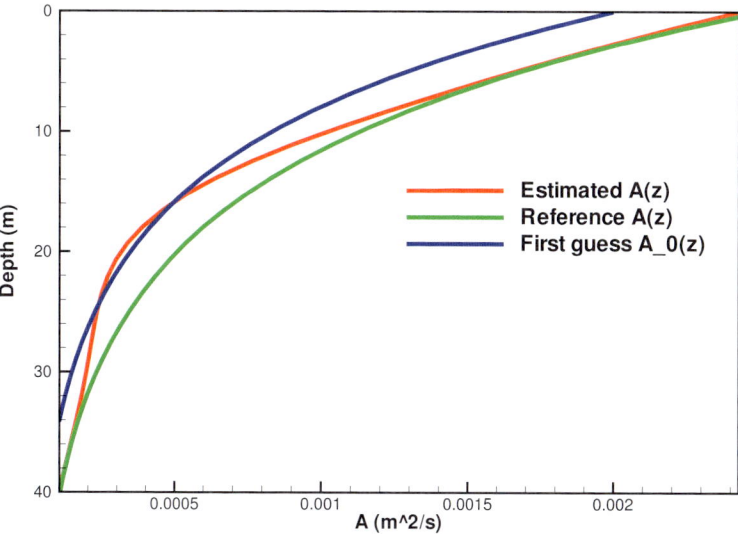

Fig. 8.1. The estimation of the eddy viscosity profile $A(z)$, from the identical twin experiment. Reproduced from *Eknes and Evensen* (1997)

Note that the expressions inside the parantheses are the actual gradients used in the gradient descent algorithm. The constant γ determines the length of the steps in the direction of the gradient in the parameter space and has an important impact on the convergence. The equations (8.60) and (8.61) are now iterated to generate new guesses c_d^{l+1} and A^{l+1}, which are used to solve for $\boldsymbol{u}^{l+1}$ and $\boldsymbol{\lambda}^{l+1}$ using the representer method.

The identical twin experiment from *Eknes and Evensen* (1997) resulted in estimates of the parameters as shown in Figs. 8.1 and 8.2. For the statistical priors used in this experiment we refer to *Eknes and Evensen* (1997). The estimation of the diffusion parameter $A(z)$ is illustrated in Fig. 8.1 where the first-guess $A_0(z)$ and the reference $A(z)$ are shown together with the estimate of $A(z)$. The weak signal below the Ekman layer makes it difficult to correct an erroneous first-guess of the diffusion parameter in the deep ocean. Note also that the estimate of $A(z)$ does not coincide with the reference diffusion parameter but is located somewhere in between the first-guess $A_0(z)$ and the exact $A(z)$ at most of the depths. At some depths the estimate is located to the left of both the first guess and the reference diffusion. This is not unexpected for this nonlinear problem where the minimum of the penalty function determines both the inverse solution and estimated parameters simultaneously, and these are mutually dependent. The estimation of the wind drag coefficient C_d is shown in Fig. 8.2. It converges to a value somewhere in between the first-guess and the reference value.

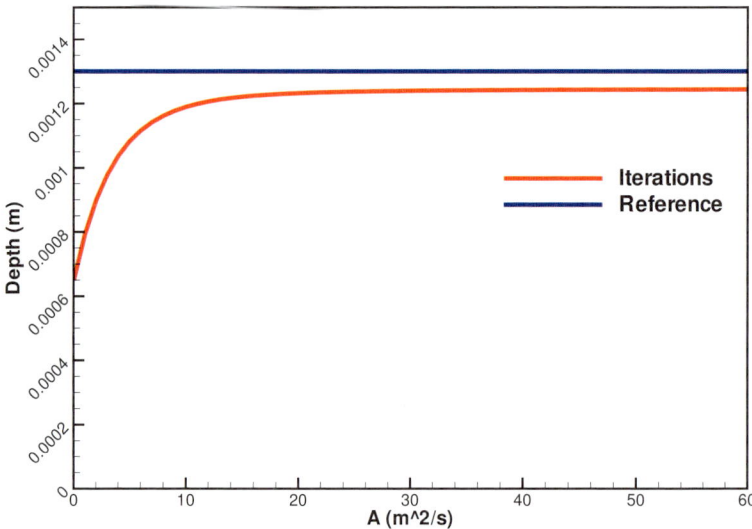

Fig. 8.2. The estimation of the wind-drag coefficient c_d, from the identical twin experiment. The number of iterations is given along the x axis. Reproduced from *Eknes and Evensen* (1997)

It was pointed out by *Bennett* (1992) and *Yu and O'Brien* (1991) that without a smoothing regularization on the diffusion coefficient $A(z)$, it is not clear if there is any difference in varying $A(0)$ or c_d in the surface condition (5.79), since $A(z)$ may then become discontinuous. However here, the non-diagonal weight will ensure a smooth $A(z)$. It is therefore expected that a vertical profile of the solution for $\boldsymbol{u}$, which is consistent with the measurements, will determine the profile for $A(z)$, while c_d will adjust to provide the correct surface forcing.

This illustration of a methodology for solving the combined state and parameter estimation problem considered a fairly simple dynamical model and it was shown that a better solution could be obtained both for the state and the parameters. The same methodology has later been examined by *Muccino and Bennett* (2001) with a nonlinear dynamical model (Korteweg-de Vries equation) containing several parameters.

They also defined an outer iteration of the parameter. Since the model dynamics is nonlinear, a sequence of linear inverse problems is next defined for each iterate of the parameter and each of these is solved using the representer method. It was found that the parameter estimation skill was limited due to the nonlinear and dispersive properties of the dynamical system. Further, they observed problems with convergence of the parameters, in particular when several parameters were estimated simultaneously. They had a fairly negative conclusion and suggested that one should rather admit errors in the dynamical equations than fiddle with the empirical formulas in the dynamical equations.

8.4 Summary

In this chapter we have derived the generalized inverse formulation for the combined state and parameter estimation problem. The starting point was the Bayes' theorem on the form (7.13) where all the data are introduced simultaneously together with an assumption of Gaussian priors. This led to the generalized inverse formulation in the form of a penalty function which is quadratic in the errors. From the generalized inverse, we derived the Euler–Lagrange equations which, in the parameter estimation case, pose a nonlinear problem even if the dynamical model is linear. We showed how we could resolve this nonlinearity by defining an iteration for the parameters to be estimated and then use the representer method to solve for the state for each iterate of the parameters.

Note that it is also possible to define a sequence of variational problems for each of (7.15–7.18) and the solution of one variational problem would then become the prior for the next. This could be a sensible approach except that the variational methods, such as the representer and adjoint methods, do not easily provide statistical information about the errors of the estimate, which is needed when the estimate is used as a prior for the next inversion. On the other hand, the genetic algorithms result in a sample of the posterior distribution, which might be used as the prior for the next inversion.

9
Ensemble methods

The focus in this Chapter will be on three methods, the Ensemble Smoother (ES), the Ensemble Kalman Smoother (EnKS) and the Ensemble Kalman Filter (EnKF). They belong to a general class of so-called particle methods which use a Monte Carlo or ensemble representation for the pdfs, an ensemble integration using stochastic models to model the time evolution of the pdfs, and different schemes for conditioning the predicted pdf given the observations.

Specific for the ES, EnKS and EnKF is the introduction of an assumption of a Gaussian pdf for the model prediction. This makes it possible to represent the pdf for the model prediction using only the mean and covariance of the pdf and a linear update equation can then be used. The discussion below will also allow for the estimation of poorly known model parameters.

9.1 Introductory remarks

Going back to the original Bayes' problem formulated as (7.12) or (7.13), we now assume that all the prior densities are known. The joint pdf for the model prediction until t_k is given by (7.10).

In Sect. 4.3 we derived the EnKF on the assumption that errors statistics could be described by error covariances represented by an ensemble of model states. The same approach can also be used when working with general pdfs.

Given a large sample of realizations for each of the prior pdfs, the joint pdf (7.10) can be evaluated by integration of each individual realization forward in time using stochastic model equations. The prior pdfs do not need to be Gaussian distributed. The densities can be represented to a desired accuracy by using a sufficiently large number N, of realizations for each of them.

The dynamical model equation (7.1) can be rewritten as a stochastic model, similar to (4.33), as

$$d\boldsymbol{\psi} = \boldsymbol{G}(\boldsymbol{\psi}, \boldsymbol{\alpha})dt + \boldsymbol{h}(\boldsymbol{\psi}, \boldsymbol{\alpha})d\boldsymbol{q}, \qquad (9.1)$$

where we have now introduced the poorly known parameters $\boldsymbol{\alpha}$. Thus, to a small increment in time dt, is associated a random increment $d\boldsymbol{q}$, representing the model error, leading to an increment in the model state $d\boldsymbol{\psi}$. The model errors are described by the samples of $f(\boldsymbol{\psi}_i|\boldsymbol{\psi}_{i-1},\boldsymbol{\alpha})$.

As in Sect. 4.3 it is possible to derive Kolmogorov's equation for the evolution of the pdf in time. The use of the stochastic model (9.1) to integrate an ensemble of model states forward in time is equivalent to solving Kolmogorov's equation using a Monte Carlo method. It turns out that this is the most efficient way to solve this equation for high dimensional and nonlinear problems where analytical solutions don't exist and direct numerical integration becomes impossible due to the numerical cost. Further, using the Monte Carlo approach there are no approximations other than the use of a limited ensemble size. Thus, an ensemble representation of the prior pdfs and a stochastic ensemble integration results in a consistent ensemble representation of the joint pdf for the model evolution.

Combining the joint pdf for the model evolution (7.10) with the Bayesian update equation (7.12) we get

$$f(\boldsymbol{\psi}_1,\ldots,\boldsymbol{\psi}_k,\boldsymbol{\alpha},\boldsymbol{\psi}_0,\boldsymbol{\psi}_b|\boldsymbol{d})$$
$$\propto f(\boldsymbol{\psi}_1,\ldots,\boldsymbol{\psi}_k,\boldsymbol{\alpha},\boldsymbol{\psi}_0,\boldsymbol{\psi}_b)\prod_{j=1}^{m}f(\boldsymbol{d}_j|\boldsymbol{\psi}_{i(j)},\boldsymbol{\alpha}). \quad (9.2)$$

The computation of the Bayesian analysis (9.2) is complicated for arbitrary distributions and high dimensions. However, the use of importance sampling makes it possible to evaluate the mean and covariance of the posterior distribution in (9.2).

We adopt for simplicity a notation where $\boldsymbol{\psi}$ contains the model solution at all time instants and also includes the initial and boundary data, and the parameters. The expected value of a function of $h(\boldsymbol{\psi})$, given the posterior distribution in (9.2), then becomes

$$\begin{aligned} E[h(\boldsymbol{\psi})] &= \int h(\boldsymbol{\psi})f(\boldsymbol{\psi}|\boldsymbol{d})d\boldsymbol{\psi} \\ &= \frac{\int h(\boldsymbol{\psi})f(\boldsymbol{d}|\boldsymbol{\psi})f(\boldsymbol{\psi})d\boldsymbol{\psi}}{f(\boldsymbol{d})} \\ &= \frac{\int h(\boldsymbol{\psi})f(\boldsymbol{d}|\boldsymbol{\psi})f(\boldsymbol{\psi})d\boldsymbol{\psi}}{\int f(\boldsymbol{d}|\boldsymbol{\psi})f(\boldsymbol{\psi})d\boldsymbol{\psi}} \\ &\approx \frac{\sum_i h(\boldsymbol{\psi}_i)f(\boldsymbol{d}|\boldsymbol{\psi}_i)}{\sum_i f(\boldsymbol{d}|\boldsymbol{\psi}_i)}. \end{aligned} \quad (9.3)$$

The summation is over the ensemble members. Thus, we can evaluate expected values of functions of $\boldsymbol{\psi}$ using the ensemble representation for the model prediction. Using $h(\boldsymbol{\psi}) = \boldsymbol{\psi}$ results in the variance minimizing estimator which is the expected value for $\boldsymbol{\psi}$ of the posterior distribution in (9.2).

Further, defining $s(\boldsymbol{\psi}) = (\boldsymbol{\psi} - E[\boldsymbol{\psi}])(\boldsymbol{\psi} - E[\boldsymbol{\psi}])^{\mathrm{T}}$ results in the posterior error covariance.

In *van Leeuwen and Evensen* (1996) the formula (9.3) was examined for solving the inverse problem using a nonlinear ocean circulation model with 6400 unknowns. It was found that the weights for most of the ensemble members became negligible and only very few ensemble members contributed in the summation. Thus, it was concluded that a very large ensemble size would be needed to properly represent the full pdf for the posterior.

Another class of methods named particle filters solves the full Bayesian update equation using importance resampling techniques. They introduce a resampling step, which results in a new ensemble having the correct posterior distribution. Some resampling schemes are discussed in *Chen et al.* (2004) and in several of the articles in *Doucet et al.* (2001). Common for these is that they use schemes where ensemble members with low weights are rejected while multiple copies are generated of the ensemble members with large weights. This helps reducing the effect of degeneracy resulting from using an ensemble where only a few ensemble members have significant weights. There are several applications where these methods have worked well for low-dimensional systems, but common for these is the requirement of a very large number of ensemble members, a need for resampling of the posterior joint pdf, and extremely high computational cost for high-dimensional models.

Some other implementations of nonlinear filters have been based on either a kernel approximation, *Miller et al.* (1999), *Anderson and Anderson* (1999) and *Miller and Ehret* (2002); or a particle interpretation, *Pham* (2001), *van Leeuwen* (2003) and *Chen et al.* (2004), although more research is needed before these can be claimed to be practical for realistic high dimensional systems. See also the Sequential Monte Carlo Methods Particle Filtering webpage, `www-sigproc.eng.cam.ac.uk/smc`, for more information.

9.2 Linear ensemble analysis update

For the case with a linear dynamical model and Gaussian prior pdfs, the pdf for the model prediction in (7.10) will also be Gaussian. The variance minimizing analysis in this case also equals the MLH estimate.

We can evaluate the mean of the ensemble prediction $\overline{\boldsymbol{\psi}^{\mathrm{f}}}(\boldsymbol{x},t)$, as a function of space and time, and its associated ensemble error covariance $\boldsymbol{C}_{\psi\psi}^{\mathrm{f}}(\boldsymbol{x}_1,t_1,\boldsymbol{x}_2,t_2)$. We also have the measurements $\boldsymbol{d}$, with error covariance $\boldsymbol{C}_{\epsilon\epsilon}$. The linear variance minimizing analysis or MLH estimate is then, from (9.2), using (8.8), defined by the minimum of

$$\mathcal{J}[\boldsymbol{\psi}^{\mathrm{a}}] = \left(\boldsymbol{\psi}^{\mathrm{a}} - \overline{\boldsymbol{\psi}^{\mathrm{f}}}\right)^{\mathrm{T}} \bullet \left(\boldsymbol{C}_{\psi\psi}^{\mathrm{f}}\right)^{-1} \bullet \left(\boldsymbol{\psi}^{\mathrm{a}} - \overline{\boldsymbol{\psi}^{\mathrm{f}}}\right) \\ + \left(\boldsymbol{d} - \mathcal{M}[\boldsymbol{\psi}^{\mathrm{a}}]\right)^{\mathrm{T}} \boldsymbol{C}_{\epsilon\epsilon}^{-1} \left(\boldsymbol{d} - \mathcal{M}[\boldsymbol{\psi}^{\mathrm{a}}]\right). \tag{9.4}$$

This defines a Gauss-Markov interpolation in space and time and has the well-known minimizing solution and associated error covariance estimate given by

$$\psi^{\mathrm{a}} = \psi^{\mathrm{f}} + \mathcal{M}^{\mathrm{T}}[C^{\mathrm{f}}_{\psi\psi}]\left(\mathcal{M}^{\mathrm{T}}[\mathcal{M}[C^{\mathrm{f}}_{\psi\psi}]] + C_{\epsilon\epsilon}\right)^{-1}\left(d - \mathcal{M}[\psi^{\mathrm{f}}]\right), \quad (9.5)$$

$$C^{\mathrm{a}}_{\psi\psi} = C^{\mathrm{f}}_{\psi\psi} - \mathcal{M}^{\mathrm{T}}[C^{\mathrm{f}}_{\psi\psi}]\left(\mathcal{M}^{\mathrm{T}}[\mathcal{M}[C^{\mathrm{f}}_{\psi\psi}]] + C_{\epsilon\epsilon}\right)^{-1}\mathcal{M}[C^{\mathrm{f}}_{\psi\psi}]. \quad (9.6)$$

These equations should be compared with the analysis equations derived in Chap. 3 for the time-independent problem, in particular (3.26) which defines the problem and (3.39), (3.46) and (3.54), for the solution and error estimate. The derivation of (9.5) and (9.6) is identical to the one given for the time independent case.

From these equations it is also seen that if we define the representer functions as the measurements of the space-time error covariance for the model prediction

$$r = \mathcal{M}[C^{\mathrm{f}}_{\psi\psi}], \quad (9.7)$$

then the analysis equations (9.5) and (9.6) becomes just

$$\psi^{\mathrm{a}} = \psi^{\mathrm{f}} + r^{\mathrm{T}}\left(\mathcal{M}^{\mathrm{T}}[r] + C_{\epsilon\epsilon}\right)^{-1}\left(d - \mathcal{M}[\psi^{\mathrm{f}}]\right), \quad (9.8)$$

$$C^{\mathrm{a}}_{\psi\psi} = C^{\mathrm{f}}_{\psi\psi} - r^{\mathrm{T}}\left(\mathcal{M}^{\mathrm{T}}[r] + C_{\epsilon\epsilon}\right)^{-1}r. \quad (9.9)$$

Comparison of (9.8) with (5.60) illustrates the similarity between the representer method and Gauss-Markov interpolation in space and time. A more elaborate discussion is given by *McIntosh* (1990) and *Bennett* (1992, 2002). In *Bennett* (1992) is is actually shown that the representers equal measurements of the space time error covariance matrix. Thus, for linear dynamics and Gaussian priors, the representer method and (9.5) will provide the same result in the limit of an infinite ensemble size.

For a nonlinear dynamical model, the pdf for the model evolution will become non-Gaussian even if the prior pdfs are Gaussian. In this case (9.5) and (9.6) will provide only an approximate solution. Still these formulas may provide a useful solution if the prior pdf is nearly Gaussian. It should again be pointed out that only the update is linear and the updated ensemble will inherit some of the non-Gaussian contributions contained in the prior ensemble. Thus, the method is doing more than just resampling a Gaussian posterior pdf. The actual ensemble implementation of (9.5) is described below and results in the Ensemble Smoother method.

9.3 Ensemble representation of error statistics

The ensemble covariance is defined as

9.3 Ensemble representation of error statistics

$$\boldsymbol{C}_{\psi\psi} = \overline{(\boldsymbol{\psi} - \overline{\boldsymbol{\psi}})(\boldsymbol{\psi} - \overline{\boldsymbol{\psi}})^{\mathrm{T}}}. \tag{9.10}$$

The ensemble mean $\overline{\boldsymbol{\psi}}$, is regarded as the best-guess estimate, while the ensemble spread defines the error variance. The covariance is determined by the smoothness of the ensemble members. A covariance matrix can always be represented by an ensemble of model states and this representation is not unique.

As in *Evensen* (2003) we have defined the matrix holding the ensemble members $\boldsymbol{\psi}(\boldsymbol{x}, t_i) \in \Re^{n_\psi}$, at time t_i, where n_ψ is the number of variables in the state vector. Further, we augment the state vector with the poorly known parameters $\boldsymbol{\alpha}(\boldsymbol{x}) \in \Re^{n_\alpha}$, where n_α is the number of parameters in $\boldsymbol{\alpha}$, and write the matrix $\boldsymbol{A}(\boldsymbol{x}, t_i) \in \Re^{n \times N}$, with $n = n_\psi + n_\alpha$, holding the N ensemble members of $\boldsymbol{\psi}$ and $\boldsymbol{\alpha}$ at time t_i, as

$$\boldsymbol{A}_i = \boldsymbol{A}(\boldsymbol{x}, t_i) = \begin{pmatrix} \boldsymbol{\psi}^1(\boldsymbol{x}, t_i) & \boldsymbol{\psi}^2(\boldsymbol{x}, t_i) & \ldots & \boldsymbol{\psi}^N(\boldsymbol{x}, t_i) \\ \boldsymbol{\alpha}^1(\boldsymbol{x}, t_i) & \boldsymbol{\alpha}^2(\boldsymbol{x}, t_i) & \ldots & \boldsymbol{\alpha}^N(\boldsymbol{x}, t_i) \end{pmatrix}. \tag{9.11}$$

Note that we have used a time index on $\boldsymbol{\alpha}$ even though the parameters are supposed to be constant in time. This is to be able to distinguish between the estimates of $\boldsymbol{\alpha}$ at different times, which in the EnKF and EnKS change at each update with measurements.

The ensemble mean is stored in each column of $\overline{\boldsymbol{A}}(\boldsymbol{x}, t_i)$ which can be defined as

$$\overline{\boldsymbol{A}}(\boldsymbol{x}, t_i) = \boldsymbol{A}(\boldsymbol{x}, t_i) \boldsymbol{1}_N, \tag{9.12}$$

where $\boldsymbol{1}_N \in \Re^{N \times N}$ is the matrix where each element is equal to $1/N$. We can then define the ensemble perturbation matrix as

$$\boldsymbol{A}'(\boldsymbol{x}, t_i) = \boldsymbol{A}(\boldsymbol{x}, t_i) - \overline{\boldsymbol{A}}(\boldsymbol{x}, t_i) = \boldsymbol{A}(\boldsymbol{x}, t_i)(\boldsymbol{I} - \boldsymbol{1}_N). \tag{9.13}$$

The ensemble covariances $\boldsymbol{C}^{\mathrm{e}}_{\psi\psi}(\boldsymbol{x}_1, \boldsymbol{x}_2, t_i) \in \Re^{n \times n}$, can be defined as

$$\boldsymbol{C}^{\mathrm{e}}_{\psi\psi}(\boldsymbol{x}_1, \boldsymbol{x}_2, t_i) = \frac{\boldsymbol{A}'(\boldsymbol{x}_1, t_i)\bigl(\boldsymbol{A}'(\boldsymbol{x}_2, t_i)\bigr)^{\mathrm{T}}}{N-1}. \tag{9.14}$$

Now, given the ensemble matrices for the different instants in time $\boldsymbol{A}(\boldsymbol{x}, t_{i'})$, for $i' = 1, \ldots, i$, we can define the ensemble matrix for the joint state from t_0 to t_i as

$$\widetilde{\boldsymbol{A}}_i = \begin{pmatrix} \boldsymbol{A}(\boldsymbol{x}, t_0) \\ \vdots \\ \boldsymbol{A}(\boldsymbol{x}, t_i) \end{pmatrix}. \tag{9.15}$$

The space-time ensemble covariance between the model states at two arbitrary times t_1 and t_2 then becomes

$$\widetilde{\boldsymbol{C}}^{\mathrm{e}}_{\psi\psi}(\boldsymbol{x}_1, t_1, \boldsymbol{x}_2, t_2) = \frac{\widetilde{\boldsymbol{A}}'_i(\boldsymbol{x}_1, t_1)\bigl(\widetilde{\boldsymbol{A}}'_i(\boldsymbol{x}_2, t_2)\bigr)^{\mathrm{T}}}{N-1}. \tag{9.16}$$

9.4 Ensemble representation for measurements

At the data time $t_{i(j)}$, we have given a vector of measurements $\boldsymbol{d}_j \in \Re^{m_j}$, with m_j being the number of measurements at this time. We can define the N vectors of perturbed measurements as

$$\boldsymbol{d}_j^l = \boldsymbol{d}_j + \boldsymbol{\epsilon}_j^l, \quad l = 1, \ldots, N, \quad (9.17)$$

which can be stored in the columns of a matrix

$$\boldsymbol{D}_j = \left(\boldsymbol{d}_j^1, \boldsymbol{d}_j^2, \ldots, \boldsymbol{d}_j^N\right) \in \Re^{m_j \times N}. \quad (9.18)$$

The ensemble of measurement perturbations, with mean equal to zero, can be stored in the matrix

$$\boldsymbol{E}_j = \left(\boldsymbol{\epsilon}_j^1, \boldsymbol{\epsilon}_j^2, \ldots, \boldsymbol{\epsilon}_j^N\right) \in \Re^{m_j \times N}, \quad (9.19)$$

from which we can construct the ensemble representation of the measurement error covariance matrix

$$\boldsymbol{C}_{\epsilon\epsilon}^{\mathrm{e}}(t_{i(j)}) = \frac{\boldsymbol{E}_j \boldsymbol{E}_j^{\mathrm{T}}}{N-1}. \quad (9.20)$$

9.5 Ensemble Smoother (ES)

The ES was proposed by *van Leeuwen and Evensen* (1996) as a linear variance minimizing smoother analysis. It computes an approximate update of (9.2) using the linear update (9.5). In fact, it can be shown that if each individual ensemble member is updated independently using (9.5), using the perturbed observations from (9.18), then the updated ensemble will have the correct mean and covariance as defined by the analysis (9.5) and (9.6). It was shown in *Burgers et al.* (1998) that the perturbation of measurements is required to obtain the correct covariance.

The linear ES analysis equation then becomes for $\widetilde{\boldsymbol{A}}_k^{\mathrm{a}}$, as defined in (9.15),

$$\widetilde{\boldsymbol{A}}_k^{\mathrm{a}} = \widetilde{\boldsymbol{A}}_k + \mathcal{M}^{\mathrm{T}}[\widetilde{\boldsymbol{C}}_{\psi\psi}^{\mathrm{e}}] \left(\mathcal{M}^{\mathrm{T}}[\mathcal{M}[\widetilde{\boldsymbol{C}}_{\psi\psi}^{\mathrm{e}}]] + \boldsymbol{C}_{\epsilon\epsilon}^{\mathrm{e}}\right)^{-1} \left(\boldsymbol{D} - \mathcal{M}[\widetilde{\boldsymbol{A}}_k]\right), \quad (9.21)$$

where we have used

$$\boldsymbol{D} = \begin{pmatrix} \boldsymbol{D}_1 \\ \vdots \\ \boldsymbol{D}_m \end{pmatrix}, \quad \mathcal{M} = \begin{pmatrix} \mathcal{M}_1 \\ \vdots \\ \mathcal{M}_m \end{pmatrix}, \quad (9.22)$$

and

$$\boldsymbol{C}_{\epsilon\epsilon}^{\mathrm{e}} = \begin{pmatrix} \boldsymbol{C}_{\epsilon\epsilon}^{\mathrm{e}}(t_{i(1)}) & & \\ & \ddots & \\ & & \boldsymbol{C}_{\epsilon\epsilon}^{\mathrm{e}}(t_{i(m)}) \end{pmatrix}. \quad (9.23)$$

9.5 Ensemble Smoother (ES)

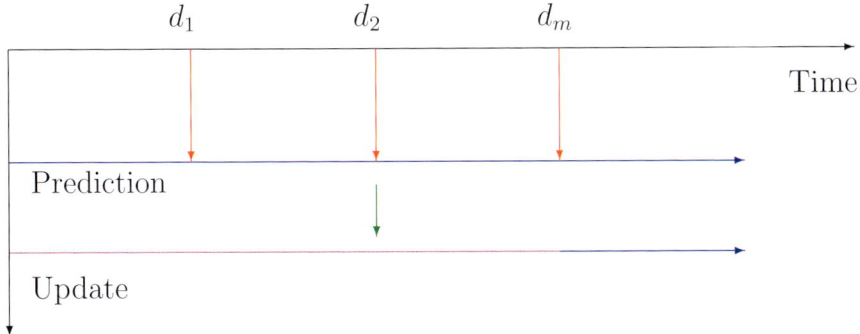

Fig. 9.1. Illustration of the update procedure used in the ES. The horizontal axis is time and the measurements are indicated at regular intervals. The vertical axis indicates the number of updates with measurements. The blue arrows represent the forward ensemble integration, while the red arrows are the introduction of measurements

The total number of measurements is $M = \sum_{j=1}^{m} m_j$. Thus, we have $\boldsymbol{D} \in \Re^{M \times N}$, $\mathcal{M} \in \Re^M$, and $\boldsymbol{C}_{\epsilon\epsilon}^{\mathrm{e}} \in \Re^{M \times M}$.

We now define the ensemble of innovation vectors as

$$\boldsymbol{D}' = \boldsymbol{D} - \mathcal{M}[\widetilde{\boldsymbol{A}}_k], \tag{9.24}$$

the measurements of the ensemble perturbations $\boldsymbol{S} \in \Re^{M \times N}$, as

$$\boldsymbol{S} = \mathcal{M}[\widetilde{\boldsymbol{A}}'_k], \tag{9.25}$$

and the matrix $\boldsymbol{C} \in \Re^{M \times M}$ as

$$\boldsymbol{C} = \boldsymbol{S}\boldsymbol{S}^\mathrm{T} + (N-1)\boldsymbol{C}_{\epsilon\epsilon}^{\mathrm{e}}. \tag{9.26}$$

Using (9.24–9.26) together with the definitions of the ensemble error covariance matrices in (9.16) and (9.20), the analysis (9.21) can be expressed as

$$\begin{aligned}
\widetilde{\boldsymbol{A}}_k^{\mathrm{a}} &= \widetilde{\boldsymbol{A}}_k + \widetilde{\boldsymbol{A}}'_k \mathcal{M}^\mathrm{T}[\widetilde{\boldsymbol{A}}'_k] \left(\mathcal{M}[\widetilde{\boldsymbol{A}}'_k]\mathcal{M}^\mathrm{T}[\widetilde{\boldsymbol{A}}'_k] + (N-1)\boldsymbol{C}_{\epsilon\epsilon}^{\mathrm{e}}\right)^{-1} \boldsymbol{D}' \\
&= \widetilde{\boldsymbol{A}}_k + \widetilde{\boldsymbol{A}}_k(\boldsymbol{I} - \boldsymbol{1}_N)\boldsymbol{S}^\mathrm{T}\boldsymbol{C}^{-1}\boldsymbol{D}' \\
&= \widetilde{\boldsymbol{A}}_k \left(\boldsymbol{I} + (\boldsymbol{I} - \boldsymbol{1}_N)\boldsymbol{S}^\mathrm{T}\boldsymbol{C}^{-1}\boldsymbol{D}'\right) \\
&= \widetilde{\boldsymbol{A}}_k \left(\boldsymbol{I} + \boldsymbol{S}^\mathrm{T}\boldsymbol{C}^{-1}\boldsymbol{D}'\right) \\
&= \widetilde{\boldsymbol{A}}_k \boldsymbol{X},
\end{aligned} \tag{9.27}$$

where we have used (9.13) and $\boldsymbol{1}_N \boldsymbol{S}^\mathrm{T} \equiv \boldsymbol{0}$. Thus, the updated ensemble can be considered as a combination of the forecast ensemble members.

Equation (9.27) converges towards the exact solution of the Bayesian formulation with increasing ensemble size if the assumption of Gaussian statistics is true. This requires that all priors are Gaussian and that a linear model is used. In this linear case it will also converge towards the representer solution.

The representer solution and the ES solution will differ in the case with nonlinear dynamics. Using the ES we should be concerned about the validity of the Gaussian approximation and the required ensemble size. When using the representer method we need to consider the convergence of the iteration, the validity of the tangent linear approximation, and whether the modal trajectory is a good estimator. Further, the computation of the posterior errors is not straight forward in the representer method.

In *Evensen and van Leeuwen* (2000) it was illustrated that the ES may have problems with nonlinear dynamical models. The method was examined with the nonlinear Lorenz model where it turned out that the Gaussian approximation for the pdf of the model evolution was too crude.

9.6 Ensemble Kalman Smoother (EnKS)

We will now present an alternative approach, by *Evensen and van Leeuwen* (2000), which solves the recursion (7.15–7.18) using an ensemble representation for the error statistics.

In (7.15), the joint pdf for the model prediction until $t_{i(1)}$ is

$$f(\boldsymbol{\psi}_1,\ldots,\boldsymbol{\psi}_{i(1)},\boldsymbol{\alpha},\boldsymbol{\psi}_0,\boldsymbol{\psi}_b) \propto \\ f(\boldsymbol{\alpha})f(\boldsymbol{\psi}_0)f(\boldsymbol{\psi}_b) \prod_{i=1}^{i(1)} f(\boldsymbol{\psi}_i|\boldsymbol{\psi}_{i-1},\boldsymbol{\alpha}). \quad (9.28)$$

Similar to the procedure used in the ES, this joint pdf can be evaluated using a large ensemble of realizations for each of the prior pdfs and integrating these forward in time using the stochastic model equations.

The stochastic integration results in an ensemble representation of the joint pdf for the model solution $\boldsymbol{\psi}_1,\ldots,\boldsymbol{\psi}_{i(1)}$, the initial condition $\boldsymbol{\psi}_0$, the boundary condition $\boldsymbol{\psi}_b$, and the poorly known parameters $\boldsymbol{\alpha}$.

The major problem is now the efficient computation of the joint pdf conditional on the measurements $\boldsymbol{d}_1$, given the ensemble representation of (9.28); i.e. we need to solve (7.15) rewritten as

$$f(\boldsymbol{\psi}_1,\ldots,\boldsymbol{\psi}_{i(1)},\boldsymbol{\alpha},\boldsymbol{\psi}_0,\boldsymbol{\psi}_b|\boldsymbol{d}_1) \propto \\ f(\boldsymbol{\psi}_1,\ldots,\boldsymbol{\psi}_{i(1)},\boldsymbol{\alpha},\boldsymbol{\psi}_0,\boldsymbol{\psi}_b)f(\boldsymbol{d}_1|\boldsymbol{\psi}_{i(1)},\boldsymbol{\alpha}), \quad (9.29)$$

which gives the update based on the first set of measurements at $t_{i(1)}$.

The EnKS is similar to the ES, except that it processes the measurements sequentially in time. Starting from the initial ensemble stored in $\boldsymbol{A}_0$, a forward stochastic integration of the ensemble until the first available data set, gives the ensemble prediction

$$\widetilde{\boldsymbol{A}}^{\text{f}}_{i(1)} = \begin{pmatrix} \boldsymbol{A}_0 \\ \boldsymbol{A}^{\text{f}}_1 \\ \vdots \\ \boldsymbol{A}^{\text{f}}_{i(1)} \end{pmatrix}. \tag{9.30}$$

Using the ES update (9.27) with (9.30) using the first set of measurements $\boldsymbol{d}_1$, which solves (9.29) under the assumption of a Gaussian pdf for the predicted ensemble, we get

$$\begin{aligned}
\widetilde{\boldsymbol{A}}^{\text{a}}_{i(1)} &= \widetilde{\boldsymbol{A}}^{\text{f}}_{i(1)} + \widetilde{\boldsymbol{A}}^{\text{f}'}_{i(1)} \mathcal{M}^{\text{T}}_1 [\widetilde{\boldsymbol{A}}^{\text{f}'}_{i(1)}] \\
&\quad \times \left(\mathcal{M}_1 [\widetilde{\boldsymbol{A}}^{\text{f}'}_{i(1)}] \mathcal{M}^{\text{T}}_1 [\widetilde{\boldsymbol{A}}^{\text{f}'}_{i(1)}] + (N-1) \boldsymbol{C}^{\text{e}}_{\epsilon\epsilon}(t_{i(1)}) \right)^{-1} \boldsymbol{D}'_1 \\
&= \widetilde{\boldsymbol{A}}^{\text{f}}_{i(1)} + \widetilde{\boldsymbol{A}}^{\text{f}}_{i(1)} (\boldsymbol{I} - \boldsymbol{1}_N) \boldsymbol{S}^{\text{T}}_1 \boldsymbol{C}^{-1}_1 \boldsymbol{D}'_1 \\
&= \widetilde{\boldsymbol{A}}^{\text{f}}_{i(1)} \left(\boldsymbol{I} + (\boldsymbol{I} - \boldsymbol{1}_N) \boldsymbol{S}^{\text{T}}_1 \boldsymbol{C}^{-1}_1 \boldsymbol{D}'_1 \right) \\
&= \widetilde{\boldsymbol{A}}^{\text{f}}_{i(1)} \left(\boldsymbol{I} + \boldsymbol{S}^{\text{T}}_1 \boldsymbol{C}^{-1}_1 \boldsymbol{D}'_1 \right) \\
&= \widetilde{\boldsymbol{A}}^{\text{f}}_{i(1)} \boldsymbol{X}_1.
\end{aligned} \tag{9.31}$$

Here we have used the definitions of innovation vectors,

$$\boldsymbol{D}'_j = \boldsymbol{D}_j - \mathcal{M}_j [\widetilde{\boldsymbol{A}}^{\text{f}}_{i(j)}], \tag{9.32}$$

the measurements of the ensemble perturbations $\boldsymbol{S}_j \in \Re^{m_j \times N}$,

$$\boldsymbol{S}_j = \mathcal{M}_j [\widetilde{\boldsymbol{A}}^{\text{f}'}_{i(j)}], \tag{9.33}$$

and the matrix $\boldsymbol{C}_j \in \Re^{m_j \times m_j}$,

$$\boldsymbol{C}_j = \boldsymbol{S}_j \boldsymbol{S}^{\text{T}}_j + (N-1) \boldsymbol{C}_{\epsilon\epsilon}(t_{i(j)}). \tag{9.34}$$

The update (9.31) is identical to the ES update in the case where the time interval covers $t \in [t_0, t_{i(1)}]$, and the data are all contained in $\boldsymbol{d}_1$. The EnKS provides an approximate ensemble representation for the joint pdf conditional on $\boldsymbol{d}_1$, in (9.29), and this serves as a prior for a continued ensemble integration until the next time when measurements are available, and then a new update is computed.

The general update equation for the measurements at the time $t_{i(j)}$, can be written

$$\begin{aligned}
f(\boldsymbol{\psi}_1, \ldots, \boldsymbol{\psi}_{i(j)}, \boldsymbol{\alpha}, \boldsymbol{\psi}_0, \boldsymbol{\psi}_b | \boldsymbol{d}_1, \ldots, \boldsymbol{d}_j) &\propto \\
f(\boldsymbol{\psi}_1, \ldots, \boldsymbol{\psi}_{i(j)}, \boldsymbol{\alpha}, \boldsymbol{\psi}_0, \boldsymbol{\psi}_b | \boldsymbol{d}_1, \ldots, \boldsymbol{d}_{j-1}) &f(\boldsymbol{d}_j | \boldsymbol{\psi}_{i(j)}, \boldsymbol{\alpha}).
\end{aligned} \tag{9.35}$$

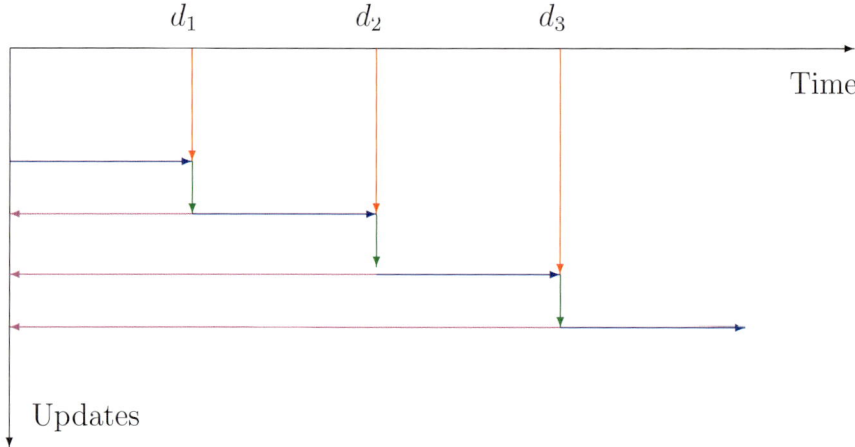

Fig. 9.2. Illustration of the update procedure used in the EnKS. The horizontal axis is time and the measurements are indicated at regular intervals. The vertical axis indicates the number of updates with measurements. The blue arrows represent the forward ensemble integration, the red arrows are the introduction of measurements, while the green arrows denote updates. Thus, the blue arrows indicate the EnKF solution as a function of time, which is updated every time measurements are available. The magenta arrows are the updates for the EnKS, which goes backward in time, and which is computed following the EnKF update every time measurements are available

Now, define the ensemble prediction matrix

$$\widetilde{\boldsymbol{A}}^{\mathrm{f}}_{i(j)} = \begin{pmatrix} \widetilde{\boldsymbol{A}}^{\mathrm{a}}_{i(j-1)} \\ \boldsymbol{A}^{\mathrm{f}}_{i(j-1)+1} \\ \vdots \\ \boldsymbol{A}^{\mathrm{f}}_{i(j)} \end{pmatrix}, \qquad (9.36)$$

where the ensemble prediction $\boldsymbol{A}^{\mathrm{f}}_{i(j-1)+1}, \ldots, \boldsymbol{A}^{\mathrm{f}}_{i(j)}$ is obtained by ensemble integration starting from the final analyzed result in $\widetilde{\boldsymbol{A}}^{\mathrm{a}}_{i(j-1)}$. We can then compute the EnKS update based on (9.35), using the measurements at time $t_{i(j)}$ as,

$$\widetilde{\boldsymbol{A}}^{\mathrm{a}}_{i(j)} = \widetilde{\boldsymbol{A}}^{\mathrm{f}}_{i(j)} \boldsymbol{X}_j, \qquad (9.37)$$

with $\boldsymbol{X}_j$ defined as

$$\boldsymbol{X}_j = \boldsymbol{I} + \boldsymbol{S}_j^{\mathrm{T}} \boldsymbol{C}_j^{-1} \boldsymbol{D}'_j. \qquad (9.38)$$

Here the predicted ensemble $\widetilde{\boldsymbol{A}}^{\mathrm{f}}_{i(j)}$ has been updated from all previous measurements $\boldsymbol{d}_1, \ldots, \boldsymbol{d}_{j-1}$. The update from measurements at time $t_{i(j)}$ adds the incremental information included in the measurements at the time $t_{i(j)}$.

Further, the combination X_j, is only dependent on the ensemble at the time $t_{i(j)}$, and then only at the measurement locations. Thus, the update can be characterized as a weakly nonlinear combination of the prior ensemble.

9.7 Ensemble Kalman Filter (EnKF)

The EnKF can be most easily characterized as a simplification of the EnKS where the analysis acts on the ensemble only at the measurement times. Thus, there is no information propagated backward in time like in the EnKS.

We now only consider the analysis step at time $t_{i(j)}$, and the analysis equation (9.37) is rewritten as

$$\boldsymbol{A}^{\text{a}}_{i(j)} = \boldsymbol{A}^{\text{f}}_{i(j)} \boldsymbol{X}_j, \tag{9.39}$$

where the ensembles at all prior times are discarded in the analysis.

9.7.1 EnKF with linear noise free model

Referring to the notation used in Fig. 7.1, let us examine the EnKF with a linear model with no model errors, i.e.

$$\boldsymbol{A}_{i+1} = \boldsymbol{F} \boldsymbol{A}_i. \tag{9.40}$$

It was shown in *Evensen* (2004) that, given the initial ensemble stored in $\boldsymbol{A}_0$, the ensemble forecast at time t_k, becomes

$$\boldsymbol{A}_k = \boldsymbol{F}^k \boldsymbol{A}_0. \tag{9.41}$$

If the EnKF is used to update the solution at every time t_j, where $j = 1, \ldots, J$, the ensemble solution at time t_k becomes

$$\boldsymbol{A}_k = \boldsymbol{F}^k \boldsymbol{A}_0 \prod_{j=1}^{J} \boldsymbol{X}_j, \tag{9.42}$$

where $\boldsymbol{X}_j$ is the matrix defined by (9.38) which when multiplied with the ensemble forecast matrix at time $t_{i(j)}$ produces the analysis ensemble at that time. Thus, starting with $\boldsymbol{A}_0$, the assimilation solution at time $t_{i(1)}$ is obtained by multiplication of $\boldsymbol{F}^{i(1)}$ with $\boldsymbol{A}_0$ to produce the forecast at time $t_{i(1)}$ followed by the multiplication of the forecast with $\boldsymbol{X}_1$.

Note that the expression $\boldsymbol{A}_0 \prod_{j=1}^{J} \boldsymbol{X}_j$ is the EnKS solution at time t_0. Thus, for the linear noise-free model, (9.42) can also be interpreted as a forward integration of the smoother solution from the initial time t_0, until t_k, where $\boldsymbol{A}_k$ is produced.

This means that for a linear model without model errors, the EnKF solution at all times is a combination of the initial ensemble members, and the

130   9 Ensemble methods

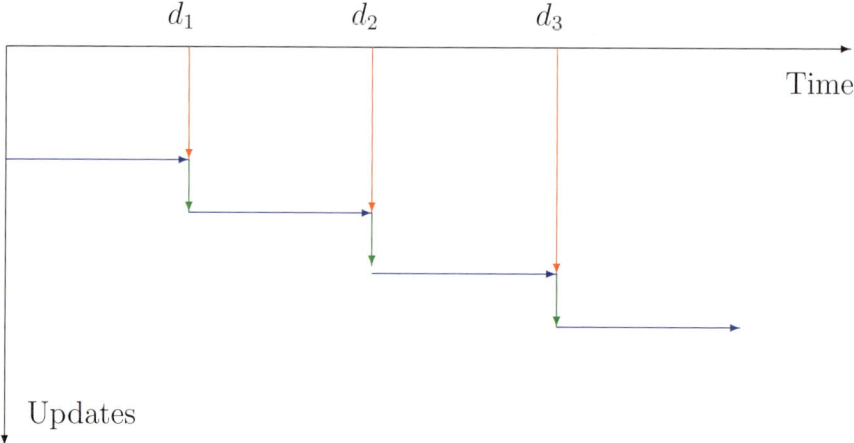

Fig. 9.3. Illustration of the update procedure used in the EnKF. The horizontal axis is time and the measurements are indicated at regular intervals. The vertical axis indicates the number of updates with measurements. The blue arrows represent the forward ensemble integration, the red arrows are the introduction of measurements, while the green arrows is the EnKF update algorithm. Thus, the blue arrows indicate the EnKF solution as a function of time, which is updated every time measurements are available

dimension of the affine space spanned by the initial ensemble does not change with time as long as the operators $\boldsymbol{F}$ and $\boldsymbol{X}_j$ are of full rank. Thus, the quality of the EnKF solution is dependent on the rank and conditioning of the initial ensemble matrix, $\boldsymbol{A}_0$.

9.7.2 EnKS using EnKF as a prior

The EnKS is a straight forward extension of the EnKF. As the EnKF uses the ensemble covariances in space to spread the information from the measurements, the EnKS uses the ensemble covariances in space and time to spread the information also backward in time.

Thus, we can write the analysis update at a time t_l from measurements available at a later time $t_{i(j)}$ as,

$$\boldsymbol{A}^{\mathrm{a}}(\boldsymbol{x}, t_l) = \boldsymbol{A}(\boldsymbol{x}, t_l) + \boldsymbol{A}'(\boldsymbol{x}, t_l)\boldsymbol{S}_j^{\mathrm{T}}\boldsymbol{C}_j^{-1}\boldsymbol{D}'_j, \qquad (9.43)$$

where $\boldsymbol{D}'_j$ from (9.32), $\boldsymbol{S}_j$ from (9.33), and $\boldsymbol{C}_j$ from (9.34) are evaluated using the ensemble and measurements at the time $t_{i(j)}$.

It is then seen that the update at the time t_l, uses exactly the same combination of ensemble members as was defined by $\boldsymbol{X}_j$ in (9.38) for the EnKF analysis at the time $t_{i(j)}$. Thus, we can write the EnKS analysis at a time $t_i \in [t_{i(j-1)}, t_{i(j)})$, as

$$\boldsymbol{A}_{\text{EnKS}}(\boldsymbol{x}, t_i) = \boldsymbol{A}_{\text{EnKF}}(\boldsymbol{x}, t_i) \prod_{l=j}^{J} \boldsymbol{X}_l. \tag{9.44}$$

It is then a simple exercise to compute the EnKS analysis as soon as the EnKF solution has been found. This requires only the storage of the coefficient matrices $\boldsymbol{X}_j$, for $j = 1, \ldots, J$, and the EnKF ensemble matrices for the previous times where we want to compute the EnKS analysis. Note that the EnKF ensemble matrices are large, but it is possible to store only specific variables at selected locations where the EnKS solution is needed. An illustration of the sequential processing of measurements is given in Fig. 9.2.

9.8 Example with the Lorenz equations

The example from *Evensen* (1997) was in *Evensen and van Leeuwen* (2000) used to intercompare the ES, EnKS and EnKF, and the results from this intercomparison are now presented. The chaotic Lorenz model by *Lorenz* (1963) is used. It was discussed in Chap. 6, and consists of a system of three coupled and nonlinear ordinary differential equations, (6.5–6.7) with initial conditions (6.8–6.10).

9.8.1 Description of experiments

For all the cases to be discussed the initial conditions for the reference case are given by $(x_0, y_0, z_0) = (1.508870, -1.531271, 25.46091)$ and the time interval is $t \in [0, 40]$. The observations and initial conditions are simulated by adding normal distributed noise with zero mean and variance equal to 2.0 to the reference solution. All of the variables x, y and z are measured. The initial conditions used are also assumed to have the same variance as the observations. These are the same values as were used in *Miller et al.* (1994) and *Evensen* (1997).

The model error covariance is defined to be diagonal with variances equal to 2.000, 12.13, and 12.31 for the three equations (6.5–6.7), respectively. These numbers define the error variance growth expected over one time unit in the model. The reference case is generated by integrating the model equations including the stochastic forcing corresponding to the specified model error variances. The stochastic forcing is included through a term like $\sqrt{\Delta t}\sqrt{\sigma^2} d\omega$ where σ^2 is the model error variance, and $d\omega$ is drawn from the distribution $\mathcal{N}(0, 1)$.

In the calculation of the ensemble statistics an ensemble of 1000 members is used. This is a fairly large ensemble but it is chosen to prevent the possibility of drawing erroneous conclusions due to the use of a too small ensemble. The same simulation was rerun with various ensemble sizes and the differences between the results were negligible even using 50 members of the ensemble.

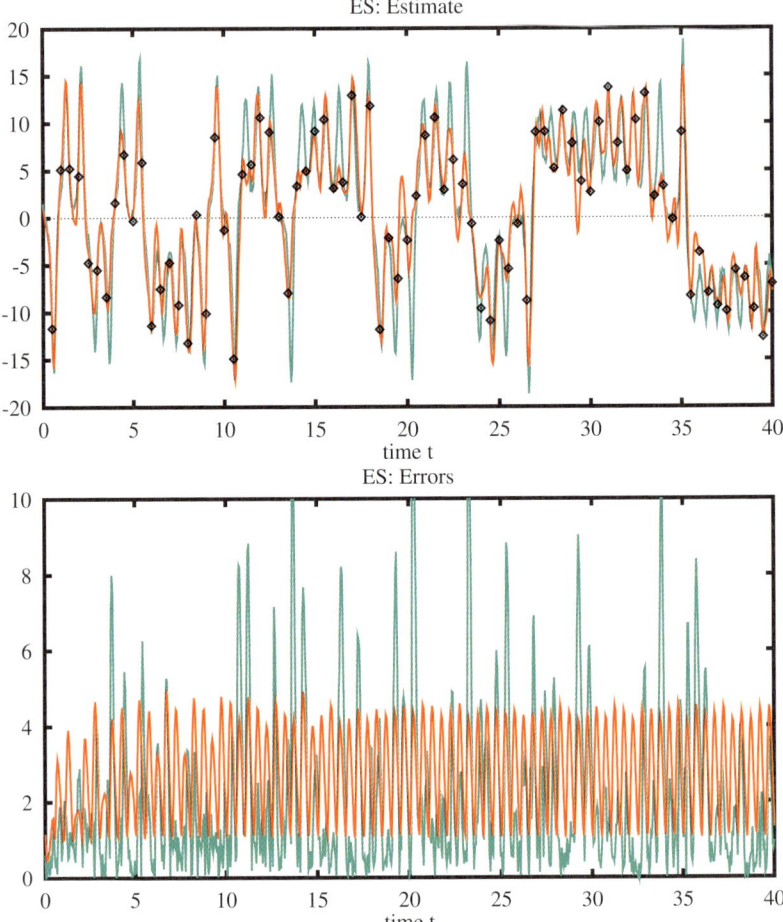

Fig. 9.4. Ensemble Smoother: The inverse estimate *(red line)* and reference solution *(blue line)* for x are shown in the upper plot. The lower plot shows the corresponding estimated standard deviations *(red line)* and the absolute value of the difference between the reference solution and the estimate, i.e. the real posterior errors *(blue line)*. Reproduced from *Evensen and van Leeuwen* (2000)

9.8.2 Assimilation Experiment

The three methods discussed above will now be examined and compared in an experiment where the distance between the measurements is $\Delta t_{\text{obs}} = 0.5$, which is similar to Experiment B in *Evensen* (1997).

In the upper plots in Figs. 9.4–9.7, the red line denotes the estimate and the blue line is the reference solution. In the lower plots the red line is the standard deviation estimated from ensemble statistics, while the blue line is the true residuals with respect to the reference solution.

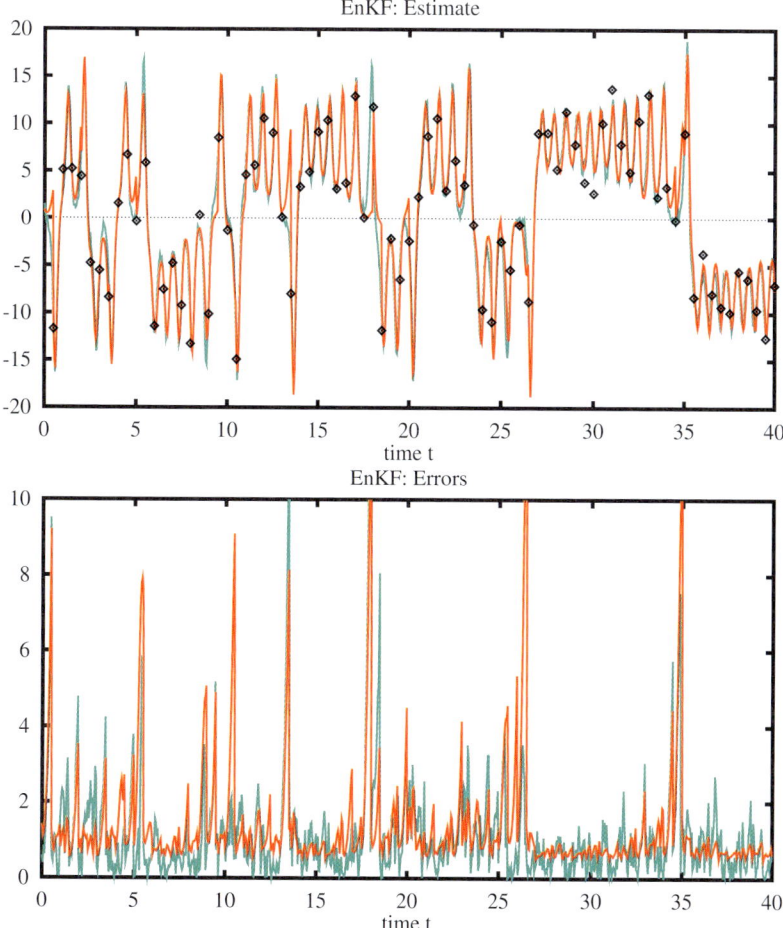

Fig. 9.5. Ensemble Kalman Filter: See explanation in Fig. 9.4. Reproduced from *Evensen and van Leeuwen* (2000)

Ensemble Smoother Solution

The ES solution for the x-component and the estimated error variance are given in Fig. 9.4. It was found that the ES performed rather poorly with the current data density. Note, however, that even if the fit to the reference trajectory is rather poor, it captures most of the transitions. The main problem is related to the estimate of the amplitudes in the reference solution. This is linked to the appearance of non-Gaussian contributions in the distribution for the model evolution, which can be expected in such a strongly nonlinear case.

Remember that the smoother solution consists of a first guess estimate, which is the mean of the freely evolving ensemble, plus a linear combination of

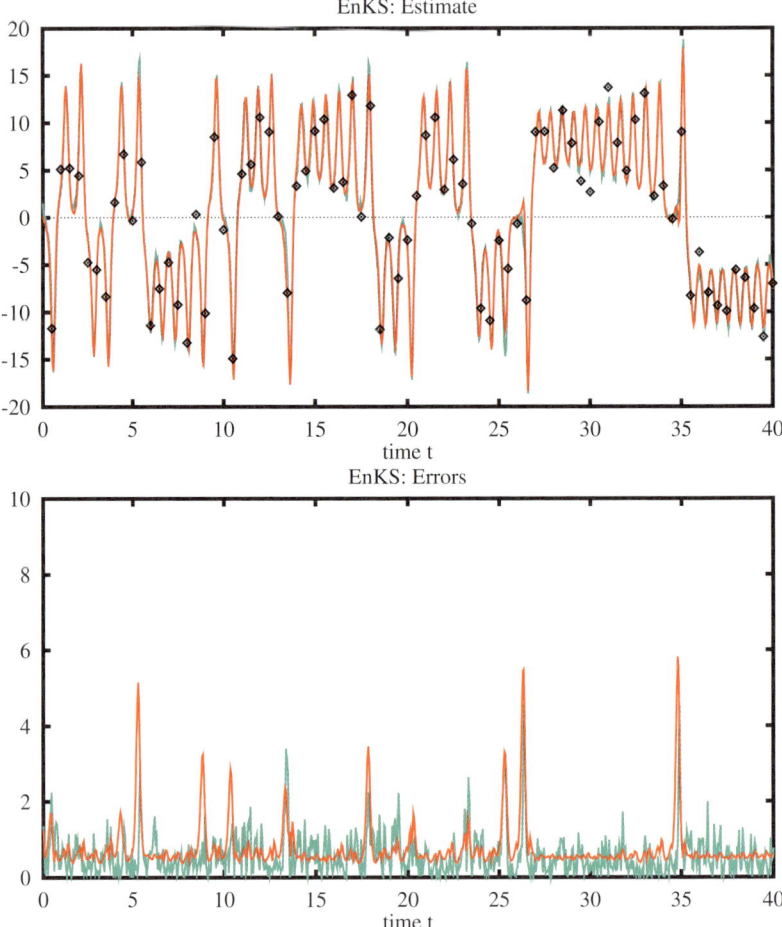

Fig. 9.6. Ensemble Kalman Smoother: See explanation in Fig. 9.4. Reproduced from *Evensen and van Leeuwen* (2000)

time-dependent influence functions or representers which are calculated from the ensemble statistics. Thus, the method becomes equivalent to a variance-minimizing objective analysis method where the time dimension is included.

In the ensemble smoother the posterior error variances can easily be calculated by performing an analysis for each of the ensemble members and then evaluating the variance of the new ensemble. Clearly, the error estimates are not large enough at the peaks where the smoother performs poorly. This is again a result of neglecting the non-Gaussian contribution from the probability distribution for the model evolution. Thus, the method assumes the distribution is Gaussian and believes it is doing well. Otherwise the error estimate looks reasonable with minima at the measurement locations and maxima in

9.8 Example with the Lorenz equations   135

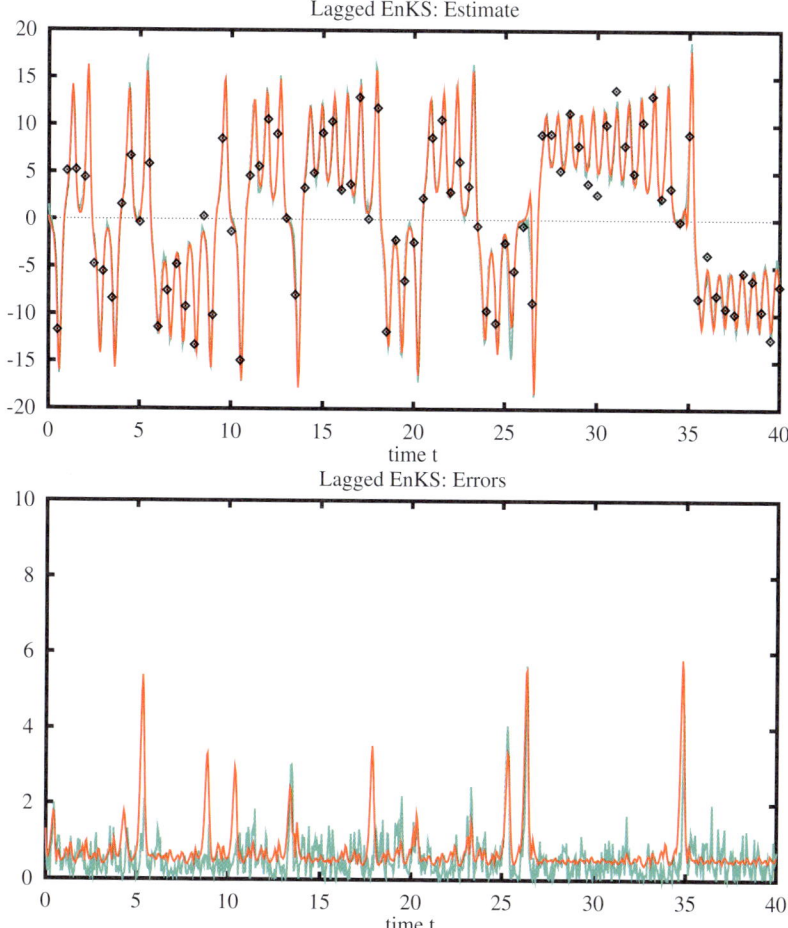

Fig. 9.7. Lagged Ensemble Kalman Smoother: See explanation in Fig. 9.4. Reproduced from *Evensen and van Leeuwen* (2000)

between the measurements. Note again that if a linear model is used the posterior density will be Gaussian and the ensemble smoother will, in the limit of an infinite ensemble size, provide the same solution as the Kalman smoother or the representer method.

Ensemble Kalman Filter Solution

The EnKF does a reasonably good job at tracking the reference solution with the lower data density, as can be seen in Fig. 9.5. One transition is missed near $t = 18$, and there are also a few other locations where the EnKF has problems, e.g. $t = 1, 5, 9, 10, 13, 17, 19, 23, 26,$ and 34. The error variance estimate is consistent, showing large peaks at the locations where the estimate obviously has problems tracking the reference solution. Note also the similarity between the absolute value of the residual between the reference solution and the estimate, and the estimated standard deviation. For all peaks in the residual there is a corresponding peak for the error variance estimate.

The error estimates show the same behaviour as was found by *Miller et al.* (1994) with very strong error growth when the model solution passes through the unstable regions of the state space, and otherwise weak error variance growth or even decay in the stable regions. Note for example the low error variance when $t \in [28, 34]$ corresponding to the oscillation of the solution around one of the attractors.

The probably surprising result is that the EnKF performs better than the ensemble smoother. This is at least surprising based on linear theory, where one has learned that the Kalman smoother solution at the end of the time interval is identical to the Kalman filter solution, and the additional information introduced by propagating the contribution of future measurements backward in time further reduces the error variance compared to the filter solution. Note again that if the model dynamics are linear, the EnKF will give the same solution as the Kalman filter, and the ensemble smoother will give the same result as the Kalman smoother, in the limit of an infinite ensemble size.

Ensemble Kalman Smoother Solution

In Fig. 9.6 the solution obtained by the EnKS is shown. Clearly, the estimate is an improvement upon the EnKF estimate. The solution is smoother in time and seems to provide a better fit to the reference trajectory. Looking in particular at the problematic locations in the EnKF solution, these are all recovered in the smoother estimate. Note, for example, the additional transitions in $t = 1, 5, 13$, and 34, in the EnKF solution which have now been eliminated in the smoother. The missed transition at $t = 17$ has also been recovered in the smoother solution.

The error estimates are reduced throughout the time interval. In particular the large peaks in the EnKF solution are now significantly reduced. As for the EnKF solution there are corresponding peaks in the error estimates for all the peaks in the residuals which proves that the EnKS error estimate is consistent with the true errors.

This is a very promising result. In fact the EnKS solution with $\Delta t_{\text{obs}} = 0.5$ seemed to do as well or better than the EnKF solution with $\Delta t_{\text{obs}} = 0.25$ (see *Evensen*, 1997).

In Fig. 9.7 the result from a lagged smoother is shown. In this case the measurement information in propagated backward in time only for a short time interval. This is motivated by the assumption that the impact of measurements is negligible outside an interval of length similar to the predictability time of the model. A time lag of 5 time units was used and the results are almost indistinguishable from the full smoother solution. Thus, a significant saving of storage and CPU should be possible for more realistic applications when using the lagged smoother.

9.9 Discussion

The similarity or connection between the EnKS and EnKF has been clarified. The EnKS is the optimal smoother solution for the linear problems with Gaussian statistics. The EnKF is a simplification which does not project information backward in time. After the final measurement time $t_{i(m)}$, the EnKF and EnKS state and parameter estimates are identical and the EnKF is therefore ideal for forecasting purposes.

The ensemble methods introduce an approximation by using only the mean and covariance of the prior joint pdf when computing the posterior ensemble in (9.35). Thus, it is effectively assumed that the prior joint pdf is Gaussian when computing the updates. This means that the EnKS and the EnKF will not give the correct answer if the prior joint pdf has non-Gaussian contributions. On the other hand the ensemble methods have proven to work well with a large number of different nonlinear dynamical models.

The ES method is similar to simple kriging or Gauss-Markov interpolation in space and time, using an ensemble representation for the space-time error covariance matrix. For a linear problem this will give exactly the same results as solving the problem with sequential processing of measurements, or minimizing the generalized inverse formulation (8.20). However, when the model is nonlinear, the long integration of the model, unconstrained by measurements, allows for the development of strongly non-Gaussian contributions in the prior density. In *Evensen and van Leeuwen* (2000) the EnKF, EnKS, and ES were compared using the highly nonlinear Lorenz equations, and it was demonstrated that the non-Gaussian contributions in the ES lead to results which were significantly worse than those obtained using the EnKF and EnKS. Further it was suggested that the sequential introduction of measurements, with Gaussian distributed errors, actually introduced "Gaussianity" to the ensemble representing the conditional joint density.

The derivation of the ensemble methods allowed for the estimation of poorly known model parameters. Examples involving parameter estimation using the EnKF and EnKS will be presented in the following chapters.

10
Statistical optimization

Optimization problems are often solved by minimizing a cost function in search of the global minimum. The solution then corresponds to the maximum likelihood estimate. Many solution methods, e.g. gradient methods, search only for the minimum of the cost function, and do not provide information about the uncertainty of the solution. The uncertainty can be estimated using statistical sampling based on the Metropolis or hybrid Monte Carlo methods from Chap. 6, or by examining the inverse of the Hessian of the cost function around the minimum value. We will now formulate an optimization problem in a Bayesian setting and show how it can be solved using the EnKS. This results in a statistical estimate of the solution and provides error estimates. Several examples are used to illustrate the difference between the exact Bayesian solution and the approximate EnKS solution. Furthermore, the examples illustrate properties of the EnKS when used with non-Gaussian distributions and nonlinear measurement operators.

10.1 Definition of the minimization problem

The EnKS can be used to solve time independent optimization problems. A typical problem could involve a set of parameters $\boldsymbol{\alpha}(\boldsymbol{x}) \in \Re^{n_\alpha}$, which is input to a function or model which outputs a vector of fields $\boldsymbol{\psi}(\boldsymbol{x}) \in \Re^{n_\psi}$, on the spatial domain $\mathcal{D}$. In addition we have available some observations of the true field $\boldsymbol{\psi}^{\mathrm{t}}(\boldsymbol{x})$. The problem is then to find the set of input parameters $\boldsymbol{\alpha}$, which gives the best possible correspondence between the simulated fields and the observations. Such optimization problems are usually solved by first defining an appropriate cost function and then solving for the minimum. However, if the functional mapping is nonlinear, the cost function is likely to contain local minima and the global minimum may be hard to find. Furthermore, traditional methods do not allow the functional mapping to contain errors nor do they provide any information about the uncertainties of the solution.

10.1.1 Parameters

We start by defining a set of first-guess parameters $\boldsymbol{\alpha}^{\rm f}(\boldsymbol{x}) \in \Re^{n_\alpha}$, which can be either constants or functions of the spatial coordinate, and we assume that they contain stochastic errors $\boldsymbol{\alpha}'(\boldsymbol{x}) \in \Re^{n_\alpha}$, with mean equal to zero and known covariance $\boldsymbol{C}_{\alpha\alpha}(\boldsymbol{x}_1, \boldsymbol{x}_2) \in \Re^{n_\alpha \times n_\alpha}$. This is represented in the following equation

$$\boldsymbol{\alpha}(\boldsymbol{x}) = \boldsymbol{\alpha}^{\rm f}(\boldsymbol{x}) + \boldsymbol{\alpha}'(\boldsymbol{x}), \tag{10.1}$$

which states that the estimated value of $\boldsymbol{\alpha}$ should be close to the prior $\boldsymbol{\alpha}^{\rm f}$, but allowed to deviate from it according to the uncertainty represented by the stochastic error term.

10.1.2 Model

We then define our function or model which connects the simulated realization $\boldsymbol{\psi}(\boldsymbol{x})$, to the parameters $\boldsymbol{\alpha}(\boldsymbol{x})$, as

$$\boldsymbol{\psi}(\boldsymbol{x}) - \boldsymbol{G}(\boldsymbol{\alpha}) + \boldsymbol{q}(\boldsymbol{x}), \tag{10.2}$$

where $\boldsymbol{G}(\boldsymbol{\alpha}) \in \Re^{n_\psi}$ is the nonlinear model operator and $\boldsymbol{q}(\boldsymbol{x}) \in \Re^{n_\psi}$ is an additive stochastic term representing the errors in the model. We assume that the model errors have a Gaussian distribution with mean equal to zero and known covariance $\boldsymbol{C}_{qq}(\boldsymbol{x}_1, \boldsymbol{x}_2) \in \Re^{n_\psi \times n_\psi}$. Thus, for any realization $\boldsymbol{\alpha}_j$, we can simulate a realization $\boldsymbol{\psi}_j(\boldsymbol{x})$. The case with non-additive model errors, e.g. $\boldsymbol{G}(\boldsymbol{\alpha}, \boldsymbol{q})$, can be treated using an approach which is similar to the one used for estimation of time correlated model errors in Chap. 12.

10.1.3 Measurements

The M measurements of the true mapping are stored in the data vector $\boldsymbol{d} \in \Re^M$. We assume that the measurements can be related to a simulated realization through the measurement functional

$$\mathcal{M}\big[\boldsymbol{\psi}(\boldsymbol{x})\big] = \boldsymbol{d} + \boldsymbol{\epsilon}, \tag{10.3}$$

where $\boldsymbol{\epsilon} \in \Re^M$ represents random measurement errors. Here $\mathcal{M}\big[\boldsymbol{\psi}(\boldsymbol{x})\big] \in \Re^M$ just projects the functional mapping $\boldsymbol{\psi}(\boldsymbol{x})$, onto the measurements. It will typically be similar to (7.6) but excluding the time variable in this case. Thus, given a field $\boldsymbol{\psi}(\boldsymbol{x})$, we can find the prediction of the measurement of the field by evaluating $\mathcal{M}\big[\boldsymbol{\psi}(\boldsymbol{x})\big]$. Also for the random measurement errors $\boldsymbol{\epsilon}$, we assume Gaussian statistics with zero mean and known covariance $\boldsymbol{C}_{\epsilon\epsilon} \in \Re^{M \times M}$.

10.1.4 Cost function

A cost function can be defined as

$$\begin{aligned}\mathcal{J}[\boldsymbol{\alpha}, \boldsymbol{\psi}] = &\iint_{\mathcal{D}} (\boldsymbol{\alpha} - \boldsymbol{\alpha}^{\text{f}})_1^{\text{T}} \boldsymbol{W}_{\alpha\alpha}(\boldsymbol{x}_1, \boldsymbol{x}_2)(\boldsymbol{\alpha} - \boldsymbol{\alpha}^{\text{f}})_2 d\boldsymbol{x}_1 d\boldsymbol{x}_2 \\ &+ \iint_{\mathcal{D}} (\boldsymbol{\psi} - \boldsymbol{G}(\boldsymbol{\alpha}))_1 \boldsymbol{W}_{qq}(\boldsymbol{x}_1, \boldsymbol{x}_2)(\boldsymbol{\psi} - \boldsymbol{G}(\boldsymbol{\alpha}))_2 d\boldsymbol{x}_1 d\boldsymbol{x}_2 \\ &+ \Big(\boldsymbol{d} - \mathcal{M}[\boldsymbol{\psi}]\Big)^{\text{T}} \boldsymbol{W}_{\epsilon\epsilon} \Big(\boldsymbol{d} - \mathcal{M}[\boldsymbol{\psi}]\Big). \end{aligned} \quad (10.4)$$

This is a fairly general cost function which measures the errors in the first-guess parameters, the model and the measurements, in a weighted least squares sense. The subscripts, 1 and 2, denote functions of $\boldsymbol{x}_1$ and $\boldsymbol{x}_2$, respectively. It is natural to assume that the weights $\boldsymbol{W}_{\alpha\alpha}$ and $\boldsymbol{W}_{\epsilon\epsilon}$ are inverses of the error covariances, $\boldsymbol{C}_{\alpha\alpha}$ and $\boldsymbol{C}_{\epsilon\epsilon}$, as before, see Chap. 8. For the weight, $\boldsymbol{W}_{qq}(\boldsymbol{x}_1, \boldsymbol{x}_2)$, we define

$$\int_{\mathcal{D}} \boldsymbol{W}_{qq}(\boldsymbol{x}_1, \boldsymbol{x}_2) \boldsymbol{C}_{qq}(\boldsymbol{x}_2, \boldsymbol{x}_3) d\boldsymbol{x}_2 = \delta(\boldsymbol{x}_1 - \boldsymbol{x}_3) \boldsymbol{I}, \quad (10.5)$$

with $\delta(\boldsymbol{x}_1 - \boldsymbol{x}_2)$ being the Dirac delta function and $\boldsymbol{I} \in \Re^{n_\psi \times n_\psi}$ the diagonal identity matrix.

If the model is assumed to be perfect we can rewrite the cost function as

$$\begin{aligned}\mathcal{J}[\boldsymbol{\alpha}] = &\iint_{\mathcal{D}} (\boldsymbol{\alpha} - \boldsymbol{\alpha}^{\text{f}})_1^{\text{T}} \boldsymbol{W}_{\alpha\alpha}(\boldsymbol{x}_1, \boldsymbol{x}_2)(\boldsymbol{\alpha} - \boldsymbol{\alpha}^{\text{f}})_2 d\boldsymbol{x}_1 d\boldsymbol{x}_2 \\ &+ \Big(\boldsymbol{d} - \mathcal{M}[\boldsymbol{G}(\boldsymbol{\alpha})]\Big)^{\text{T}} \boldsymbol{W}_{\epsilon\epsilon} \Big(\boldsymbol{d} - \mathcal{M}[\boldsymbol{G}(\boldsymbol{\alpha})]\Big). \end{aligned} \quad (10.6)$$

This is the standard cost function which is minimized in many applications.

10.2 Bayesian formalism

In a Bayesian formalism we can derive the cost function by assuming that we have given the pdf for the parameters $\boldsymbol{\alpha}$ as $f(\boldsymbol{\alpha})$, and the pdf for the model as $f(\boldsymbol{\psi}|\boldsymbol{\alpha})$. Furthermore, we have the likelihood for the measurements $\boldsymbol{d}$, given as

$$f(\boldsymbol{d}|\boldsymbol{\alpha}, \boldsymbol{\psi}) = f(\boldsymbol{d}|\boldsymbol{\psi}), \quad (10.7)$$

since the measurements, in this case, are assumed to be independent of $\boldsymbol{\alpha}$.

Bayes' theorem states that

$$\begin{aligned}f(\boldsymbol{\alpha}, \boldsymbol{\psi}|\boldsymbol{d}) &\propto f(\boldsymbol{d}|\boldsymbol{\alpha}, \boldsymbol{\psi}) f(\boldsymbol{\alpha}, \boldsymbol{\psi}) \\ &= f(\boldsymbol{d}|\boldsymbol{\psi}) f(\boldsymbol{\psi}|\boldsymbol{\alpha}) f(\boldsymbol{\alpha}).\end{aligned} \quad (10.8)$$

If we assume Gaussian statistics for all the errors we get

$$f(\boldsymbol{\alpha}) \propto \exp\left(-\frac{1}{2}\iint_{\mathcal{D}}(\boldsymbol{\alpha}-\boldsymbol{\alpha}^{\mathrm{f}})_1^{\mathrm{T}}\boldsymbol{W}_{\alpha\alpha}(\boldsymbol{x}_1,\boldsymbol{x}_2)(\boldsymbol{\alpha}-\boldsymbol{\alpha}^{\mathrm{f}})_2 d\boldsymbol{x}_1 d\boldsymbol{x}_2\right), \quad (10.9)$$

$$\begin{aligned}f(\boldsymbol{\psi}|\boldsymbol{\alpha}) \propto \exp\Big(&-\frac{1}{2}\iint_{\mathcal{D}}(\boldsymbol{\psi}-\boldsymbol{G}(\boldsymbol{\alpha}))_1\\&\times \boldsymbol{W}_{qq}(\boldsymbol{x}_1,\boldsymbol{x}_2)(\boldsymbol{\psi}-\boldsymbol{G}(\boldsymbol{\alpha}))_2 d\boldsymbol{x}_1 d\boldsymbol{x}_2\Big),\end{aligned} \quad (10.10)$$

and

$$f(\boldsymbol{d}|\boldsymbol{\psi}) \propto \exp\left(-\frac{1}{2}\big(\boldsymbol{d}-\mathcal{M}[\boldsymbol{\psi}]\big)^{\mathrm{T}}\boldsymbol{W}_{\epsilon\epsilon}\big(\boldsymbol{d}-\mathcal{M}[\boldsymbol{\psi}]\big)\right). \quad (10.11)$$

Insertion of these into (10.8) gives

$$f(\boldsymbol{\alpha},\boldsymbol{\psi}|\boldsymbol{d}) \propto \exp\left(-\frac{1}{2}\mathcal{J}[\boldsymbol{\alpha},\boldsymbol{\psi}]\right). \quad (10.12)$$

Maximization of (10.12), which results in the maximum likelihood solution, is equivalent to minimization of the cost function as defined in (10.4).

Standard minimization of the cost function (10.4) using gradient methods may be difficult since this requires derivatives of $G(\boldsymbol{\alpha})$ and $\mathcal{M}[\boldsymbol{\psi}]$ and if the model operator in sufficiently nonlinear these methods are likely to get trapped in local minima. Furthermore, the dimension of the problem becomes high since we need to minimize with respect to both $\boldsymbol{\alpha}$ and $\boldsymbol{\psi}(\boldsymbol{x})$ simultaneously.

10.3 Solution by ensemble methods

The EnKS does not minimize the cost function directly. Rather it takes the pdfs and likelihood functions as a starting point, and represents these using large ensembles of realizations. To illustrate, we could start by sampling N realizations $\boldsymbol{\alpha}_j^{\mathrm{f}}$, from $f(\boldsymbol{\alpha})$ as defined in (10.9). We then compute the N realizations $\boldsymbol{\psi}_j^{\mathrm{f}}$, by evaluating the stochastic model (10.2) for the N parameter sets $\boldsymbol{\alpha}_j^{\mathrm{f}}$. The simulated realizations are then measured to generate an ensemble of predicted measurements. Thus, we have,

$$\boldsymbol{\alpha}_j^{\mathrm{f}} = \boldsymbol{\alpha}^{\mathrm{f}} + \boldsymbol{\alpha}_j', \quad (10.13)$$
$$\boldsymbol{\psi}_j^{\mathrm{f}}(\boldsymbol{x}) = G(\boldsymbol{\alpha}_j^{\mathrm{f}}) + \boldsymbol{q}_j(\boldsymbol{x}), \quad (10.14)$$
$$\widehat{\boldsymbol{d}}_j = \mathcal{M}[\boldsymbol{\psi}_j^{\mathrm{f}}], \quad (10.15)$$

where $\widehat{\boldsymbol{d}}_j$ is the prediction of the measurements given $\boldsymbol{\alpha}_j$. Note that in (10.15) it would also be possible to introduce a stochastic error term to take into account representation errors in the measurement operator.

It is also possible to combine these equations and write

10.3 Solution by ensemble methods

$$\widehat{d}_j = \mathcal{M}\left[G(\boldsymbol{\alpha}^{\mathrm{f}} + \boldsymbol{\alpha}'_j) + \boldsymbol{q}_j(\boldsymbol{x})\right], \tag{10.16}$$

where only $\boldsymbol{\alpha}$ is used as a state vector, but we will retain the form (10.13–10.15). The state vector which originally consisted of only $\boldsymbol{\alpha}$ can then be extended to include both the functional mapping and the predicted measurements, i.e. we define the realizations

$$\boldsymbol{\Psi}_j^{\mathrm{f}} = \begin{pmatrix} \boldsymbol{\alpha}_j^{\mathrm{f}} \\ \psi_j^{\mathrm{f}}(\boldsymbol{x}) \\ \widehat{d}_j \end{pmatrix}. \tag{10.17}$$

From the N realizations $\boldsymbol{\Psi}_j^{\mathrm{f}}$, it is possible to compute the symmetrical ensemble covariance

$$\boldsymbol{C}_{\Psi\Psi}^{\mathrm{f}} = \begin{pmatrix} \boldsymbol{C}_{\alpha\alpha}^{\mathrm{f}}(\boldsymbol{x}_1,\boldsymbol{x}_2) & \boldsymbol{C}_{\alpha\psi}^{\mathrm{f}}(\boldsymbol{x}_1,\boldsymbol{x}_2) & \boldsymbol{C}_{\alpha d}^{\mathrm{f}}(\boldsymbol{x}_1) \\ \boldsymbol{C}_{\psi\alpha}^{\mathrm{f}}(\boldsymbol{x}_1,\boldsymbol{x}_2) & \boldsymbol{C}_{\psi\psi}^{\mathrm{f}}(\boldsymbol{x}_1,\boldsymbol{x}_2) & \boldsymbol{C}_{\psi d}^{\mathrm{f}}(\boldsymbol{x}_1) \\ \boldsymbol{C}_{d\alpha}^{\mathrm{f}}(\boldsymbol{x}_2) & \boldsymbol{C}_{d\psi}^{\mathrm{f}}(\boldsymbol{x}_2) & \boldsymbol{C}_{dd}^{\mathrm{f}} \end{pmatrix}. \tag{10.18}$$

Thus, we have defined the first-guess covariance matrices between the components of the state vector; $\boldsymbol{C}_{\alpha\alpha}^{\mathrm{f}} \in \Re^{n_\alpha \times n_\alpha}$, $\boldsymbol{C}_{\psi\psi}^{\mathrm{f}} \in \Re^{n_\psi \times n_\psi}$, $\boldsymbol{C}_{dd}^{\mathrm{f}} \in \Re^{M \times M}$, $\boldsymbol{C}_{\alpha\psi}^{\mathrm{f}} \in \Re^{n_\alpha \times n_\psi}$, $\boldsymbol{C}_{\alpha d}^{\mathrm{f}} \in \Re^{n_\alpha \times M}$ and $\boldsymbol{C}_{d\psi}^{\mathrm{f}} \in \Re^{M \times n_\psi}$.

We can now define the cost function

$$\mathcal{J}[\boldsymbol{\Psi}] = (\boldsymbol{\Psi} - \boldsymbol{\Psi}^{\mathrm{f}})^{\mathrm{T}} \boldsymbol{W}_{\Psi\Psi}(\boldsymbol{\Psi} - \boldsymbol{\Psi}^{\mathrm{f}}) \\ + (\boldsymbol{d} - \boldsymbol{M}\boldsymbol{\Psi}^{\mathrm{f}})^{\mathrm{T}} \boldsymbol{W}_{\epsilon\epsilon}(\boldsymbol{d} - \boldsymbol{M}\boldsymbol{\Psi}^{\mathrm{f}}). \tag{10.19}$$

Note that $\boldsymbol{W}_{\epsilon\epsilon}$ is the inverse of the error covariance matrix of the measurement errors $\boldsymbol{C}_{\epsilon\epsilon}$, while $\boldsymbol{C}_{dd}$ is the ensemble covariance matrix of the model predicted measurements. We have defined $\boldsymbol{M}$ as a matrix operator which extracts the predicted measurements from $\boldsymbol{\Psi}$, i.e.

$$\boldsymbol{M} = \begin{pmatrix} \boldsymbol{0}^{n_\alpha \times n_\alpha} & \boldsymbol{0}^{n_\alpha \times n_\psi} & \boldsymbol{0}^{n_\alpha \times M} \\ \boldsymbol{0}^{n_\psi \times n_\alpha} & \boldsymbol{0}^{n_\psi \times n_\psi} & \boldsymbol{0}^{n_\psi \times M} \\ \boldsymbol{0}^{M \times n_\alpha} & \boldsymbol{0}^{M \times n_\psi} & \boldsymbol{M}^{M \times M} \end{pmatrix}. \tag{10.20}$$

The first-guess estimate is computed as the mean of the first-guess ensemble and we write, with the overline denoting ensemble average,

$$\overline{\boldsymbol{\Psi}^{\mathrm{f}}} = \begin{pmatrix} \overline{\boldsymbol{\alpha}^{\mathrm{f}}} \\ \overline{\psi^{\mathrm{f}}} \\ \overline{\widehat{d}} \end{pmatrix}, \tag{10.21}$$

where $\overline{\boldsymbol{\alpha}^{\mathrm{f}}} = \boldsymbol{\alpha}^{\mathrm{f}}$. We have defined the inverse of the covariance $\boldsymbol{C}_{\Psi\Psi}$ as $\boldsymbol{W}_{\Psi\Psi}$, using the now-familiar definitions for the inverses of covariances which are functions of the spatial coordinate.

10.3.1 Variance minimizing solution

From the theory outlined in Chaps. 3 and 9, it is easy to show that the variance minimizing solution $\boldsymbol{\Psi}^{\mathrm{a}}$, of (10.19) becomes

$$\boldsymbol{\Psi}^{\mathrm{a}} = \boldsymbol{\Psi}^{\mathrm{f}} + \boldsymbol{C}_{\Psi\Psi}\boldsymbol{M}^{\mathrm{T}}\left(\boldsymbol{M}\boldsymbol{C}_{\Psi\Psi}\boldsymbol{M}^{\mathrm{T}} + \boldsymbol{C}_{\epsilon\epsilon}\right)^{-1}(\boldsymbol{d} - \boldsymbol{M}\boldsymbol{\Psi}^{\mathrm{f}}). \tag{10.22}$$

This can be written in simpler form as

$$\begin{pmatrix} \boldsymbol{\alpha}^{\mathrm{a}} \\ \psi^{\mathrm{a}} \\ \widehat{\boldsymbol{d}}^{\mathrm{a}} \end{pmatrix} = \begin{pmatrix} \boldsymbol{\alpha}^{\mathrm{f}} \\ \psi^{\mathrm{f}} \\ \widehat{\boldsymbol{d}} \end{pmatrix} + \begin{pmatrix} \boldsymbol{C}_{\alpha d} \\ \boldsymbol{C}_{\psi d} \\ \boldsymbol{C}_{dd} \end{pmatrix}(\boldsymbol{C}_{dd} + \boldsymbol{C}_{\epsilon\epsilon})^{-1}\left(\boldsymbol{d} - \mathcal{M}[G(\boldsymbol{\alpha}^{\mathrm{f}})]\right), \tag{10.23}$$

or if only $\boldsymbol{\alpha}$ is solved for we write

$$\boldsymbol{\alpha}^{\mathrm{a}} = \boldsymbol{\alpha}^{\mathrm{f}} + \boldsymbol{C}_{\alpha d}\left(\boldsymbol{C}_{dd} + \boldsymbol{C}_{\epsilon\epsilon}\right)^{-1}\left(\boldsymbol{d} - \mathcal{M}[G(\boldsymbol{\alpha}^{\mathrm{f}})]\right). \tag{10.24}$$

10.3.2 EnKS solution

The EnKS solves (10.23) using an ensemble representation for $\boldsymbol{\Psi}$; i.e. given an ensemble of realizations $\boldsymbol{\alpha}_j^{\mathrm{f}}$, for the parameters we compute the corresponding ensembles of realizations, $\psi_j^{\mathrm{f}}(\boldsymbol{x})$ and $\widehat{\boldsymbol{d}}_j$, using the defined prior error statistics for the stochastic terms. The covariances in $\boldsymbol{C}_{\Psi\Psi}$ are all evaluated directly from the ensemble of realizations $\boldsymbol{\Psi}_j$.

The EnKS can be used to update the whole ensemble, $\boldsymbol{\Psi}_j$ with $j = 1, N$, not just the mean, and the result is a full ensemble of parameters $\boldsymbol{\alpha}_j^{\mathrm{a}}$, consistent with the priors and data. Further, the spread of the ensemble of parameters also determines the uncertainty of the estimated parameters.

The actual procedure is similar to the one used in Chap. 9. We store the ensemble members in the matrix $\boldsymbol{A}$, defined as

$$\boldsymbol{A} = (\boldsymbol{\Psi}_1, \boldsymbol{\Psi}_2, \ldots, \boldsymbol{\Psi}_N). \tag{10.25}$$

Then the ensemble mean is stored in each column of $\overline{\boldsymbol{A}}$ which can be defined as

$$\overline{\boldsymbol{A}} = \boldsymbol{A}\boldsymbol{1}_N, \tag{10.26}$$

where $\boldsymbol{1}_N \in \Re^{N \times N}$ is the matrix where each element is equal to $1/N$. We can then define the ensemble perturbation matrix as

$$\boldsymbol{A}' = \boldsymbol{A} - \overline{\boldsymbol{A}} = \boldsymbol{A}(\boldsymbol{I} - \boldsymbol{1}_N). \tag{10.27}$$

The first-guess ensemble-covariance representation of $\boldsymbol{C}_{\Psi\Psi}^{\mathrm{f}}$ in (10.18), can be defined as

$$\boldsymbol{C}_{\Psi\Psi}^{\mathrm{e}} = \frac{\boldsymbol{A}'\boldsymbol{A}'^{\mathrm{T}}}{N - 1}. \tag{10.28}$$

Example	$F(x)$	x_{prior}	y_{prior}	σ_x	σ_y	σ_q
1a	$y = x$	1.0	−1.0	1.0	0.3	1.0
1b	$y = x$	1.0	−1.0	1.0	0.3	0.1
2	$y = x^2$	1.0	−1.0	1.0	0.3	1.0
3	$y = x^2(x^2 - 2)$	1.0	−1.0	1.0	0.3	1.0
4a	$y = \cos(x)$	1.0	−1.0	1.0	0.3	1.0
4b	$y = \cos(x)$	1.0	−1.0	1.0	0.3	0.1
4c	$y = \cos(x)$	1.0	−1.0	4.0	0.3	1.0

Table 10.1. Parameters used in the different examples. Here x_{prior} is the first-guess of x, while y_{prior} is the "observation" of y. The standard deviations for the errors in the priors and the model are σ_x, σ_y and σ_q

We then define N vectors of perturbed measurements as

$$d_j = d + \epsilon_j, \quad j = 1, \ldots, N, \qquad (10.29)$$

which can be stored in the columns of a matrix

$$D = (d_1, d_2, \ldots, d_N) \in \Re^{M \times N}. \qquad (10.30)$$

The ensemble of measurement perturbations, with mean equal to zero, can be stored in the matrix

$$E = (\epsilon_1, \epsilon_2, \ldots, \epsilon_N) \in \Re^{M \times N}, \qquad (10.31)$$

from which we can construct the ensemble representation of the measurement error covariance matrix

$$C_{\epsilon\epsilon}^{\text{e}} = \frac{EE^{\text{T}}}{N-1}. \qquad (10.32)$$

We can then write

$$A^{\text{a}} = A^{\text{f}} + A'^{\text{f}}(MA'^{\text{f}})^{\text{T}} \left(MA'^{\text{f}}(MA'^{\text{f}})^{\text{T}} + EE^{\text{T}} \right) \left(D - MA^{\text{f}} \right), \qquad (10.33)$$

which is the equation solved in the EnKS. This equation has the nice property that the covariance of A^{a} is the correct expected covariance of the analyzed estimate.

10.4 Examples

A simple example is now used to illustrate the difference between standard minimization problems and statistical estimation. We start by defining a simple scalar model or mapping $y = F(x)$, where x now takes the role of the poorly known parameter α, and y takes the role of the observed variable ψ. The standard cost function for this problem becomes

$$J[x] = (x - x_0)^2/\sigma_x^2 + (d - F(x))^2/\sigma_y^2. \tag{10.34}$$

When using a Bayesian approach, we can evaluate the product of the Gaussian pdf for the prior and the pdf for the model evolution, assuming Gaussian model errors, i.e.

$$f(x, y) = f(y|x)f(x) \propto \exp\left(-\frac{1}{2}\frac{(x - x_0)^2}{\sigma_x^2} - \frac{1}{2}\frac{(y - F(x))^2}{\sigma_q^2}\right). \tag{10.35}$$

The joint conditional pdf becomes

$$f(x, y|d) \propto \exp\left(-\frac{1}{2}\frac{(x - x_0)^2}{\sigma_x^2} - \frac{1}{2}\frac{(y - F(x))^2}{\sigma_q^2} - \frac{1}{2}\frac{(d - y)^2}{\sigma_y^2}\right). \tag{10.36}$$

Figs. 10.1–10.7 display the resulting cost functions and pdfs for several mappings as defined in Table 10.1, and using different input parameters. The joint pdf with its marginal pdfs, modes and mean are shown in the upper left plot. The upper right plot shows the similar pdf but as estimated from a large ensemble of realizations. The lower left plot shows the joint pdf conditional on the measurement and the lower right plot is the corresponding pdf as computed from the samples conditioned on the data using the EnKS.

In Cases 1a and 1b, shown in Figs. 10.1 and 10.2, we assume the linear model $F(x) = x$. In these cases the cost function becomes quadratic, and the marginal pdfs are all Gaussian as would be expected. This case in particular illustrates the impact of model errors. In Case 1a the joint pdf for the prediction in the upper plots shows a large uncertainty while in Case 1b, it is narrow and nearly aligned along the line $y = x$. In Case 1b the most likely solution is found close to the line $y = x$ and consistent with the prior for y, i.e. the pdf for the measurement of y. It is also consistent with the minimum of the cost function. In Case 1a, a completely different solution is found which reflects that the model prediction has a great uncertainty and this leads to a situation where the measurement of y has less impact on the estimate of x. The apparent tilt of the predicted joint pdf in Case 1a is expected. The reason is that, given a value for x, the model uncertainty introduces an uncertainty in the y value (which is symmetrical in the y-direction about a point on the $y = x$ line). In Cases 1a and 1b the maximum likelihood estimate from the joint pdf is identical to the maximum likelihood estimate from the marginal pdfs as well as the estimated mean. This will be true only in the case with a linear model and Gaussian priors. It is also clear that the EnKS in this case produces a consistent result, as is expected.

In Case 2 we introduce a nonlinearity using the function $F(x) = x^2$. Still the problem has only one global minimum and no local minima. In this case we see from Fig. 10.3 that both the joint pdf and marginal pdfs become non-Gaussian. We can also differentiate between the maximum likelihood estimate from the joint and marginal pdfs as well as the mean. Thus, here we will have to choose which estimator to use. From the two lower plots it is also clear that

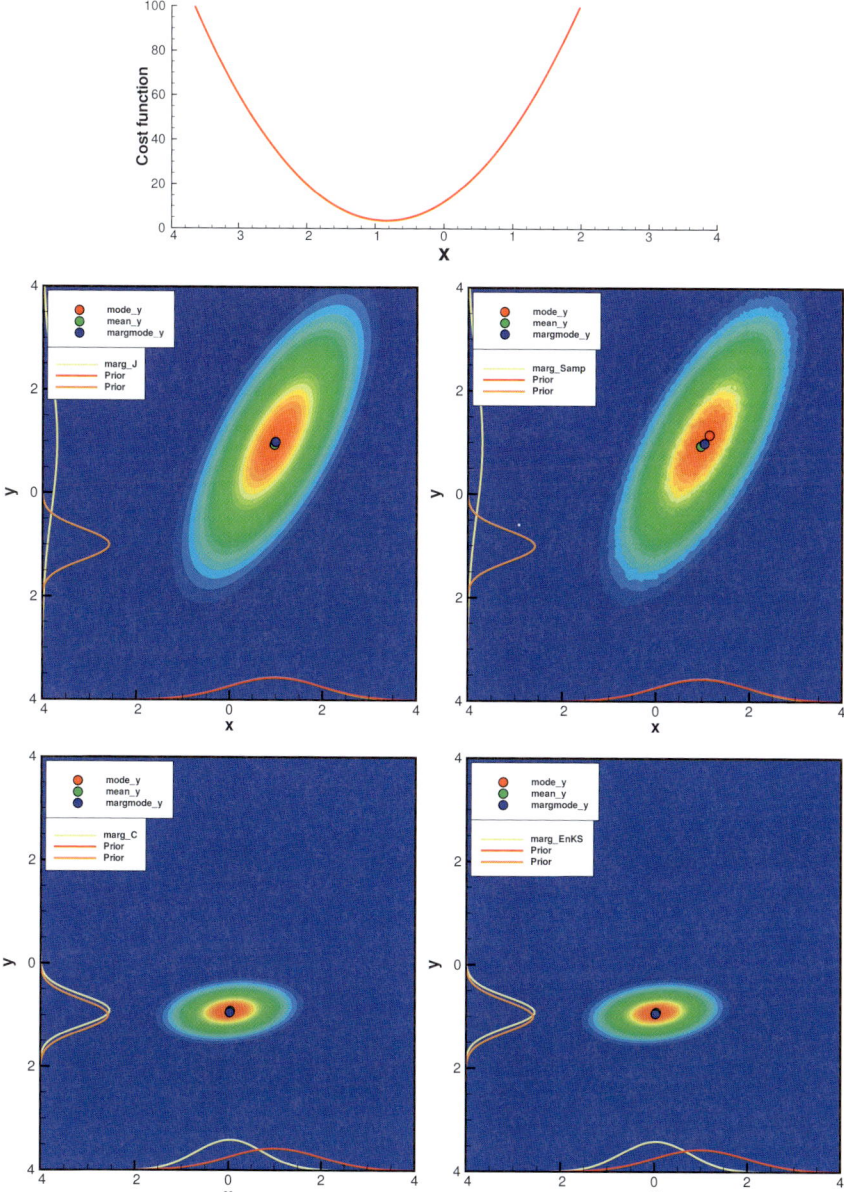

Fig. 10.1. Case 1a: Joint and conditional pdfs using the linear function $F(x) = x$

148   10 Statistical optimization

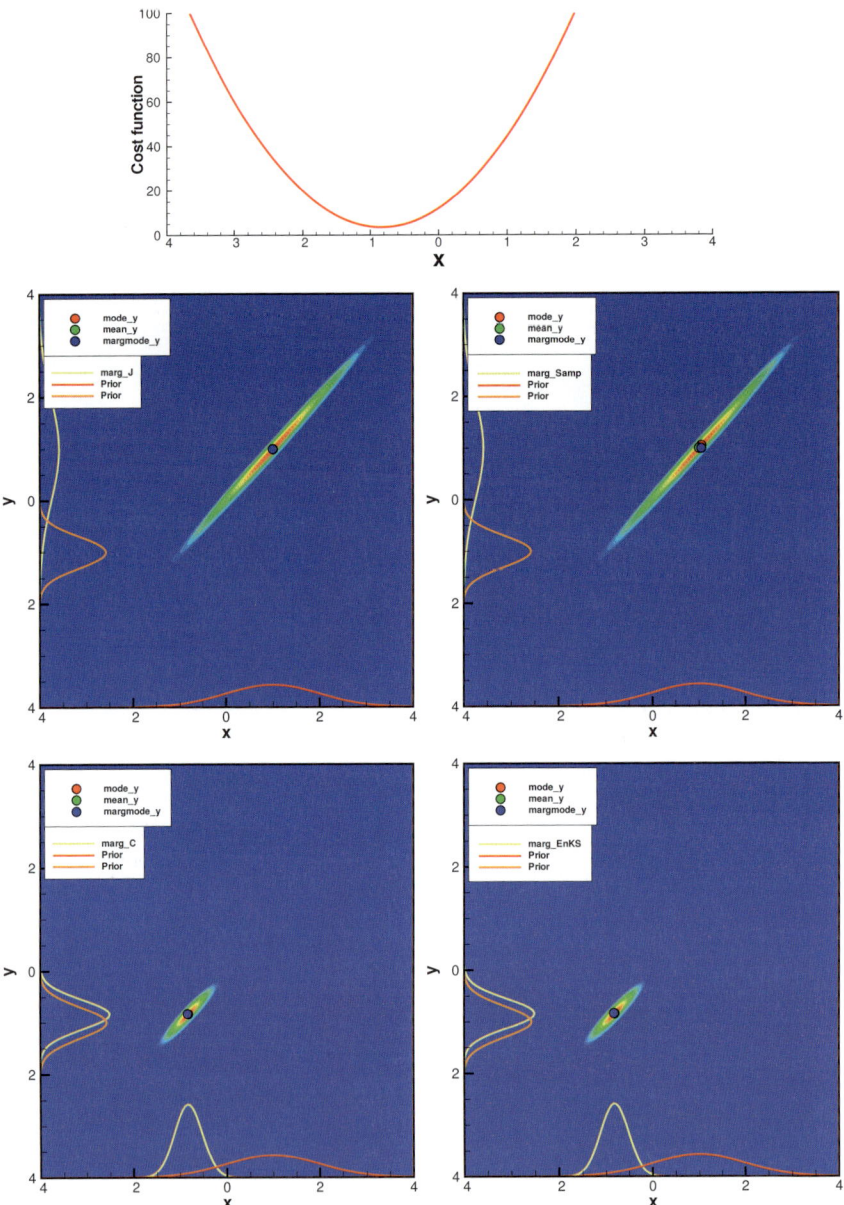

Fig. 10.2. Case 1b: Same as Fig. 10.1 but with more accurate model

10.4 Examples 149

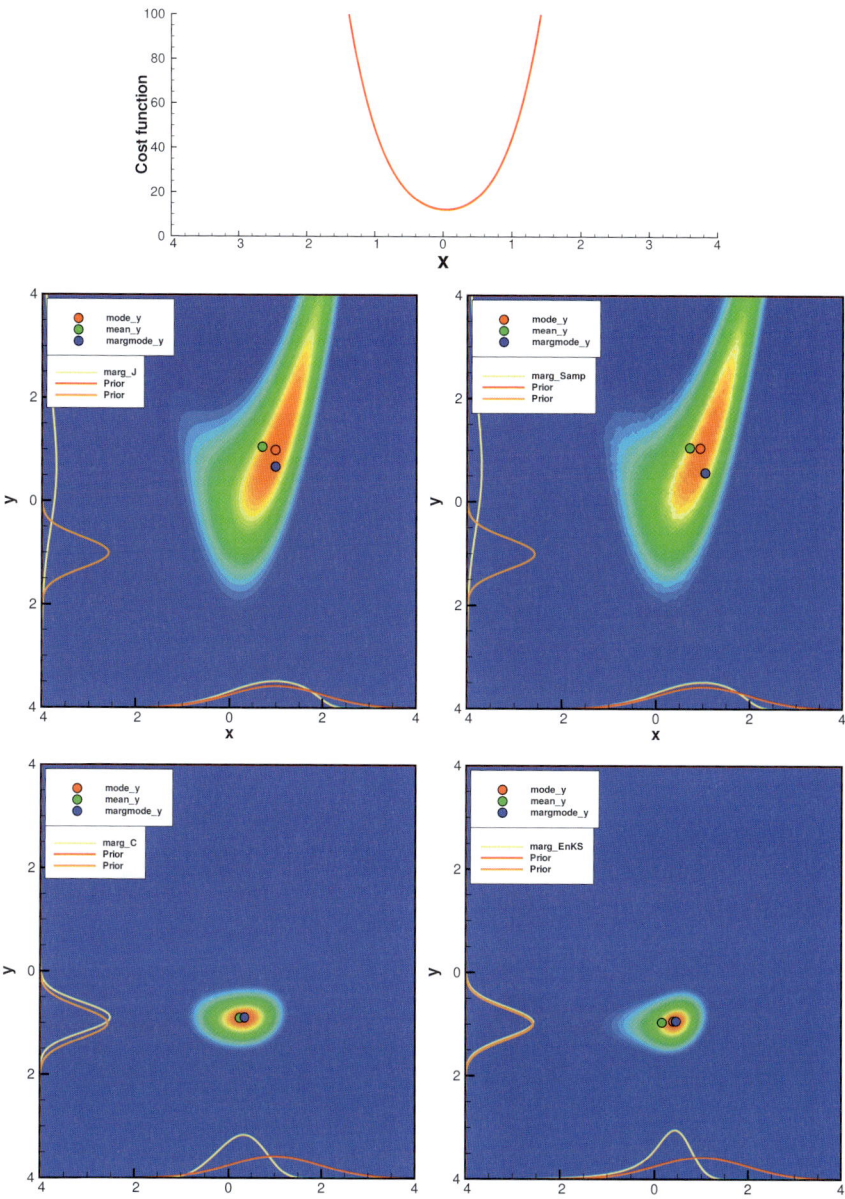

Fig. 10.3. Case 2: Joint and conditional pdfs using the quadratic function $F(x) = x^2$

150   10 Statistical optimization

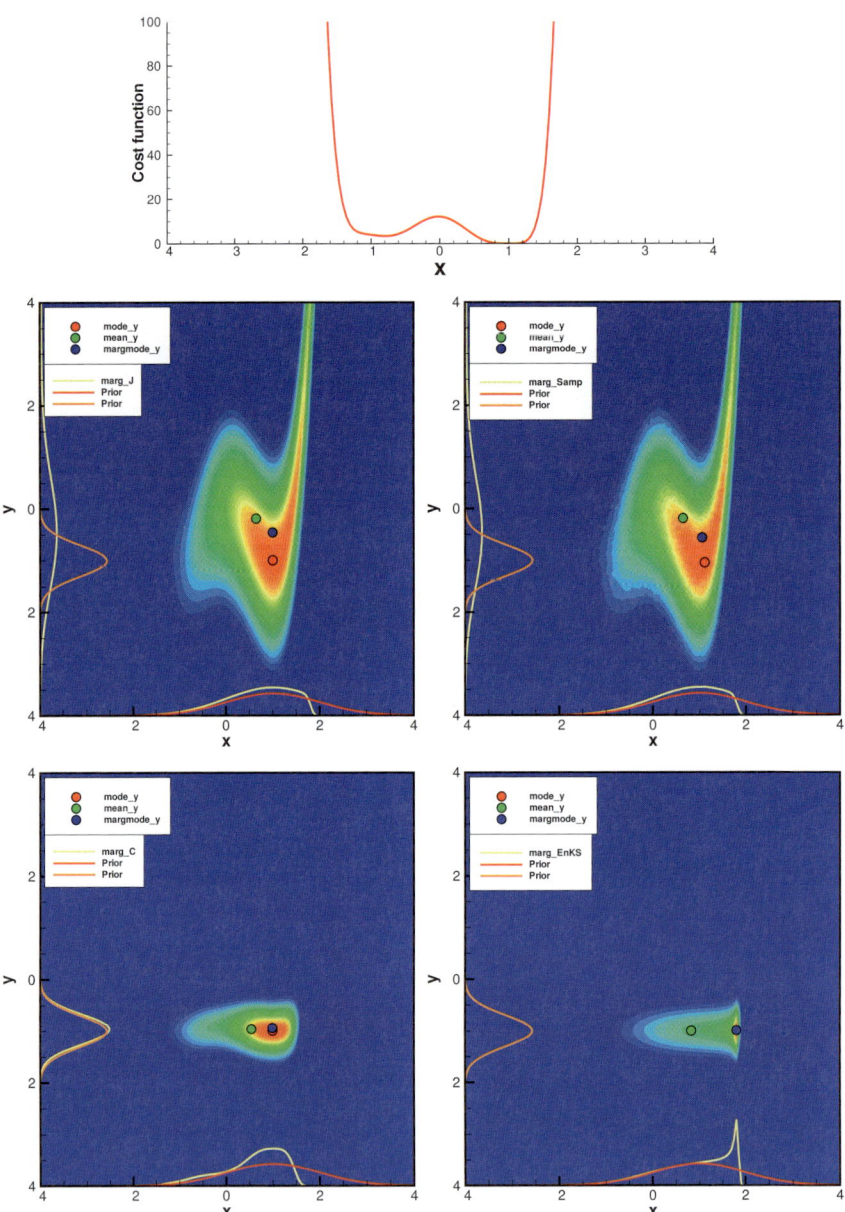

Fig. 10.4. Case 3: Joint and conditional pdfs using the nonlinear function $F(x) = x^2(x^2 - 2)$

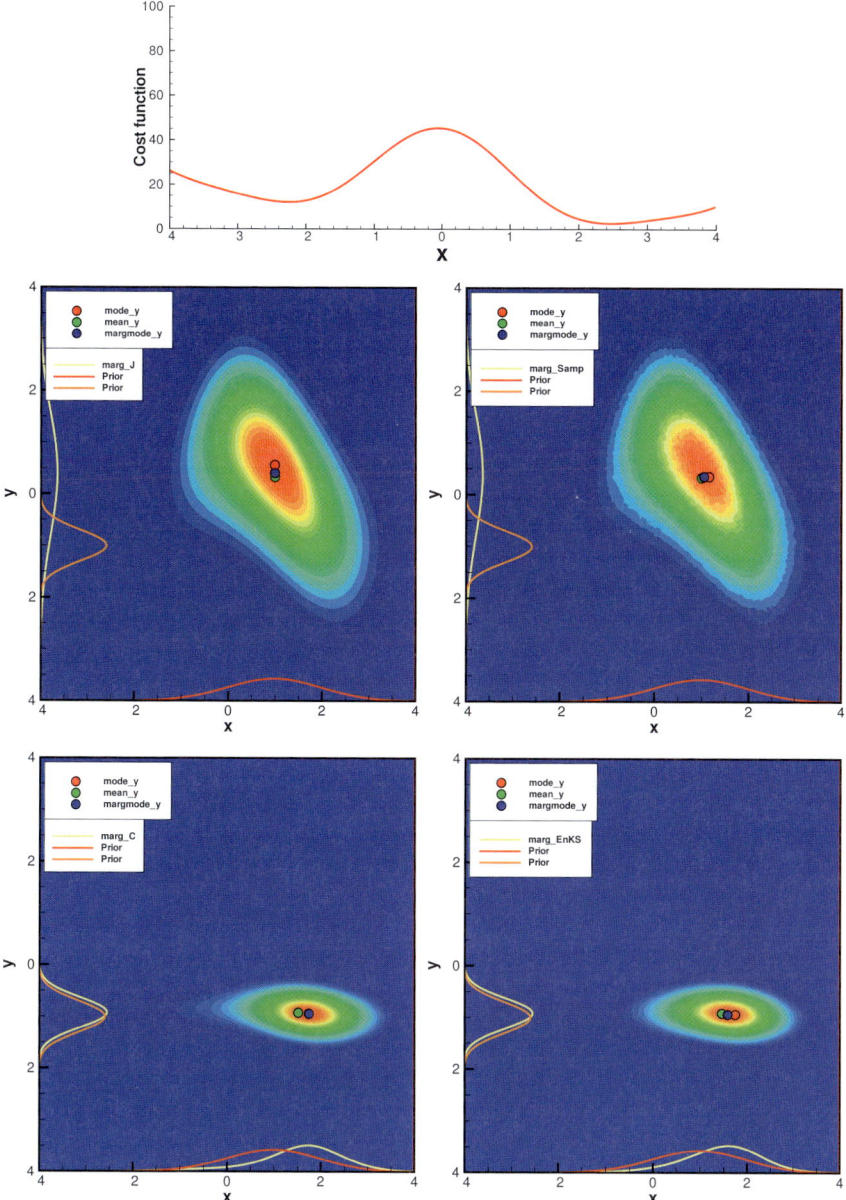

Fig. 10.5. Case 4a: Joint and conditional pdfs using the nonlinear function $F(x) = \cos(x)$

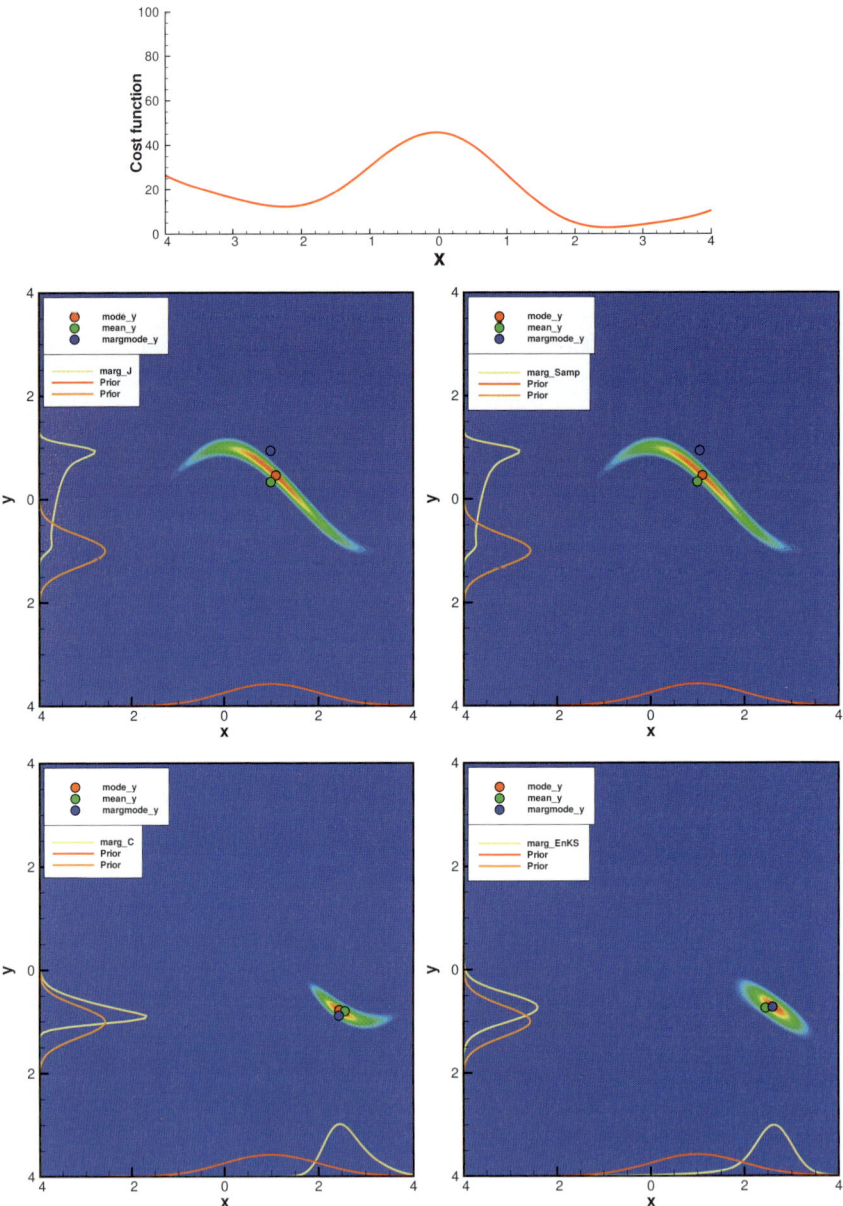

Fig. 10.6. Case 4b: Same as Fig. 10.5 but with high accuracy of the model

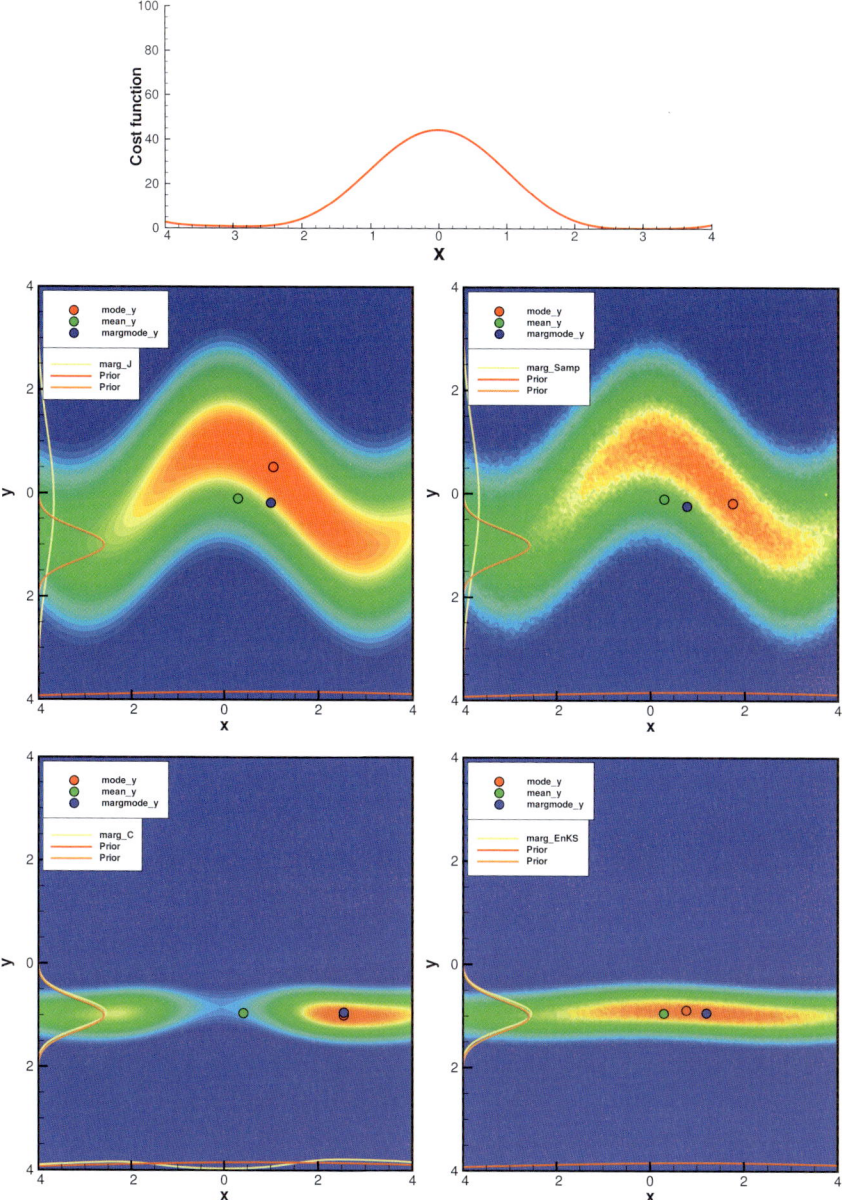

Fig. 10.7. Case 4c: Same as Fig. 10.5 but with weak penalty on first-guess which results in a bimodal pdf

the joint pdf estimated from the EnKS differs slightly from the analytical pdf. Thus, in this case the EnKS will give a slightly different estimate than an exact Bayesian solver, and this is due to the approximate linear update used.

In Case 3, shown in Fig. 10.4, we consider the function $F(x) = x^2(x^2 - 2)$, which leads to a cost function with both a local and global minimum. It is interesting to note that the introduction of model errors in this case leads to a predicted joint pdf which is unimodal. Thus, a unique solution is found and the EnKS solution contains some of the same characteristics as the exact analytical solution.

In Cases 4a–4c, shown in Figs. 10.5–10.7, we use the function $F(x) = \cos(x)$, and examine again the impact of the prior statistics for the model errors as well as errors in the initial guess. In Case 4a we set both the standard deviation for the model error and for the prior of x to one. Again the cost function contains an additional local minimum while the Bayesian approach leads to unimodal pdfs. The EnKS solution is fairly consistent. In Case 4b the model is very accurate, and again we converge towards a solution where the Bayesian estimate is close to the global minimum of the cost function. Note also that it is the rather accurate prior pdf which ensures that the joint pdf is unimodal. This is clearly illustrated in Case 4c where a low accuracy on the prior for x is used. In this case the joint conditional pdf has a bimodal structure and the mean falls between the peaks in the pdf and is not useful as an estimator. On the other hand, both the modes of the conditional joint and marginal pdfs provide realistic and similar estimates. The EnKS has a problem in this case and is not capable of reproducing the bimodal structure. It also provides a solution which has a fairly low probability.

10.5 Discussion

This chapter has considered the use of the EnKS as an optimization or parameter estimation method for nonlinear mappings. There is a clear analogy between this problem and the analysis step used in traditional data assimilation problems; e.g. if we consider the variable x, to be an initial state, and y to be a prediction by the nonlinear model, then this becomes analogous to the standard EnKS analysis step where the observation of y is assimilated. Alternatively, if we consider x to be the prediction at a certain time, and y to be a nonlinear measurement at the same time, related to x through an equation like (10.16) with $\boldsymbol{\alpha}$ replaced with x, then these examples resembles the EnKF update step using a nonlinear measurement functional.

Thus, it is clear that the EnKF and EnKS can handle certain levels of nonlinearity in both the model prediction and measurement functional. Even if the prior ensemble is non-Gaussian the ensemble methods will in many cases provide an updated ensemble having a realistic pdf. When the prior ensemble is non-Gaussian, the analyzed ensemble will inherit some of the non-Gaussian structures. On the other hand, it is also possible to make the EnKS and EnKF fail completely; e.g. if the weight on the prior is low and a multimodal pdf develops, this may result in non-physical solutions.

From the analytical (left columns) and ensemble representation (right columns) of the joint pdfs in Figs 10.1–10.7, it is clear that the unconditioned joint and marginal pdfs are indistinguishable in all the cases. This illustrates that the stochastic ensemble integration which solves Kolmogorov's equation (4.34) gives the same result as the multiplication of the prior pdf with the transition density, as is expected. Note also that, while Kolmogorov's equation provides only the marginal densities, the ensemble integration allows for computation of the joint pdf if we track ensemble members in time; i.e we can evaluate the joint density from the pairs of points (x^l, y^l) where $l = 1, \ldots, N$.

11

Sampling strategies for the EnKF

The purpose of this Chapter is to present some algorithms for generating ensemble members, and model and measurement perturbations. There is a number of simulation methods available for generation of random realizations with different kinds of statistical properties, and we refer to the text books by *Lantuéjoul* (2002) and *Chilés* (1999) for further information. It is also shown that by selecting the initial ensemble, the model noise and the measurement perturbations wisely, it is possible to achieve a significant improvement in the EnKF results, without increasing the size of the ensemble.

11.1 Introduction

The ensemble methods use Monte Carlo sampling for generation of the initial ensemble, the model noise and the measurement perturbations. When defining an ensemble of realizations we need to specify the statistical properties of the distribution we are sampling from. In particular we need to ensure that the smoothness properties of the realizations are realistic for the physical variables they represent. The smoothness of a realization can be described by a covariance function or even better by a quantity named the variogram. For a field where the smoothness is independent of position, the variogram becomes

$$\gamma(\boldsymbol{h}) = C(\boldsymbol{0}) - C(\boldsymbol{h}), \tag{11.1}$$

where $C(\boldsymbol{h})$ is the covariance of points located a distance $|\boldsymbol{h}|$ apart. It is easy to show that $\gamma(\boldsymbol{0}) = 0$, $\gamma(\boldsymbol{h}) \geq 0$ and $-\gamma(\boldsymbol{h}) = \gamma(\boldsymbol{h})$. An extensive discussion of the variogram and its use in geostatistics is given in *Wackernagel* (1998).

Typical variograms are shown in Fig. 11.1 for a field with exponential, spherical and Gaussian covariance functions. The exponential covariance function is defined as

$$C_{\mathrm{exp}}(\boldsymbol{h}) \propto \exp\left(-\frac{|\boldsymbol{h}|}{a}\right) \tag{11.2}$$

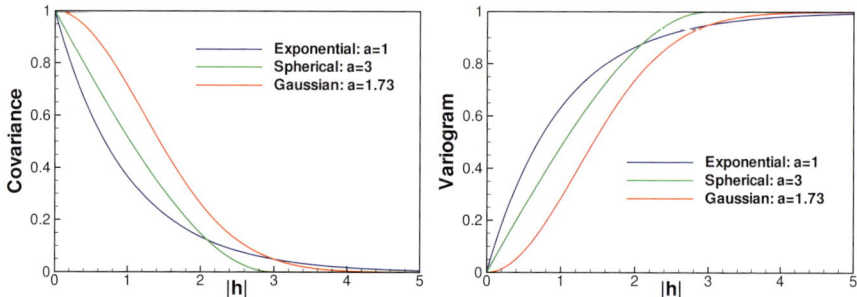

Fig. 11.1. The left plot shows exponential, spherical and Gaussian covariance functions and the right plot shows the corresponding variograms

with a being a de-correlation length. Note that the exponential correlation function is continuous but not differentiable at the origin. The spherical correlation function is given by

$$C_{\text{sphere}}(\boldsymbol{h}) = \begin{cases} 1 - 1.5|\boldsymbol{h}|/a + 0.5|\boldsymbol{h}|^3/a^3 & \text{for } 0 \leq |\boldsymbol{h}| \leq a \\ 0 & \text{for } |\boldsymbol{h}| > a, \end{cases} \quad (11.3)$$

where again a defines the de-correlation length. A Gaussian correlation function is given by

$$C_{\text{gauss}}(\boldsymbol{h}) \propto \exp\left(-\frac{|\boldsymbol{h}|^2}{a^2}\right). \quad (11.4)$$

We can define the range of the covariance functions as the distance where the covariance has a significant value. For the spherical covariance function the range is equal to a, while for the exponential and Gaussian it is common to define the ranges as $3a$ and $\sqrt{3}a$.

From the behaviour of the variograms when $|\boldsymbol{h}|$ approaches zero, it is clear that the Gaussian variogram corresponds to realizations which are rather smooth, while the exponential variogram corresponds to fields with more noisy behaviour. The spherical covariance functions corresponds to realizations with smoothness located somewhere between the exponential and Gaussian.

When simulating random fields, we need to know the statistical properties of the fields we are sampling to ensure that the realizations are physically acceptable for the process or variable they are meant to represent.

11.2 Simulation of realizations

The problem is now to simulate N realizations $\psi_i(\boldsymbol{x})$ for $i = 1\ldots N$, which has zero mean and covariance given by $C_{\psi\psi}(\boldsymbol{x}_1, \boldsymbol{x}_2)$. The following procedure can be used to compute smooth random fields with mean equal to zero, variance equal to one, and a specified covariance which determines the smoothness of the fields. The algorithm is an extension of the one presented in the Appendix of *Evensen* (1994b). We have used a Gaussian covariance function which makes sense in ocean simulations where smooth realizations are used. The method has some resemblance with the spectral method described by *Lantuéjoul* (2002) but uses a fast Fourier transform and exploits that the covariance matrix is diagonal in the Fourier space.

11.2.1 Inverse Fourier transform

Let $\psi = \psi(x,y)$ be a continuous field, which may be described by its Fourier transform

$$\psi(x,y) = \int_{-\infty}^{\infty}\int_{-\infty}^{\infty} \widehat{\psi}(\boldsymbol{k}) e^{i\boldsymbol{k}\cdot\boldsymbol{x}} d\boldsymbol{k}. \tag{11.5}$$

We are using an $n_x \times n_y$ grid. Further, we define $\boldsymbol{k} = (\kappa_l, \lambda_p)$, where l and p are integer indices and κ_l and λ_p are wave numbers in the x and y directions, respectively. We now get a discrete version of (11.5),

$$\psi(x_n, y_m) = \sum_{l,p} \widehat{\psi}(\kappa_l, \lambda_p) e^{i(\kappa_l x_n + \lambda_p y_m)} \Delta \boldsymbol{k}, \tag{11.6}$$

where $x_n = n\Delta x$ and $y_m = m\Delta y$. For the wave numbers, we have

$$\kappa_l = \frac{2\pi l}{x_{n_x}} = \frac{2\pi l}{n_x \Delta x}, \tag{11.7}$$

$$\lambda_p = \frac{2\pi p}{y_{n_y}} = \frac{2\pi p}{n_y \Delta y}, \tag{11.8}$$

$$\Delta \boldsymbol{k} = \Delta\kappa\Delta\lambda = \frac{(2\pi)^2}{n_x n_y \Delta x \Delta y}. \tag{11.9}$$

11.2.2 Definition of Fourier spectrum

In *Evensen* (1994a) the following Gaussian form were used for the Fourier coefficients,

$$\widehat{\psi}(\kappa_l, \lambda_p) = \frac{c}{\Delta \boldsymbol{k}} e^{-(\kappa_l^2 + \lambda_p^2)/r^2} e^{2\pi i \phi_{l,p}}, \tag{11.10}$$

where $\phi_{l,p} \in [0,1]$ is a uniformly distributed random number which introduces a random phase shift. With increasing l and p the wave numbers κ_l and λ_p will give an exponentially decreasing contribution, and large wave numbers corresponding to small scales are penalized. This choice of Fourier coefficients leads to isotropic covariances for the simulated fields, i.e. the smoothness is the same in all directions.

Here we have used the property that the Fourier transform of the Gaussian function (11.4) also becomes a Gaussian function. Clearly we can define other Fourier coefficients, e.g. corresponding to the exponential or spherical covariances if this is what we want to simulate.

A further extension of this algorithm to account for asymmetrical and rotated covariance functions is straight-forward. Defining de-correlation lengths for the principal directions in the Fourier space as r_1 and r_2, and a rotation angle as θ, we can define

$$a_{11} = \frac{\cos^2(\theta)}{r_1^2} + \frac{\sin^2(\theta)}{r_2^2}, \tag{11.11}$$

$$a_{22} = \frac{\sin^2(\theta)}{r_1^2} + \frac{\cos^2(\theta)}{r_2^2}, \tag{11.12}$$

$$a_{12} = \left(\frac{1}{r_2^2} - \frac{1}{r_1^2}\right)\cos(\theta)\sin(\theta), \tag{11.13}$$

and the Fourier coefficients as

$$\widehat{\psi}(\kappa_l, \lambda_p) = \frac{c}{\Delta k} e^{-\left(a_{11}\kappa_l^2 + 2a_{12}\kappa_l\lambda_p + a_{22}\lambda_p^2\right)} e^{2\pi i \phi_{l,p}}. \tag{11.14}$$

This is a Fourier spectrum which has different scales in the two principal directions and the principal direction is rotated an angle θ. With $r_1 = r_2 = r$ this formula reduces to (11.10).

Now, (11.14) may be inserted into (11.6), and we get

$$\psi(x_n, y_m) = c\sqrt{\Delta k} \sum_{l,p} e^{-\left(a_{11}\kappa_l^2 + 2a_{12}\kappa_l\lambda_p + a_{22}\lambda_p^2\right)} e^{2\pi i \phi_{l,p}} e^{i(\kappa_l x_n + \lambda_p y_m)}, \tag{11.15}$$

for the inverse Fourier transform which defines the random fields.

It should be noted that we want (11.15) to produce real fields only. Thus, when the summation over l, p is performed, all the imaginary contributions must add up to zero. This is satisfied whenever

$$\widehat{\psi}(\kappa_l, \lambda_p) = \widehat{\psi}^*(\kappa_{-l}, \lambda_{-p}), \tag{11.16}$$

where the asterisk denotes complex conjugate, and in addition

$$\operatorname{Im} \widehat{\psi}(\kappa_0, \lambda_0) = 0. \tag{11.17}$$

11.2.3 Specification of covariance and variance

The formula (11.15) can be used to generate an ensemble of random fields with a covariance determined by the parameters c, r_1 and r_2.

An expression for the covariance is given by

$$\overline{\psi(x_1, y_1)\psi(x_2, y_2)} = (\Delta k)^2 \sum_{l,p,r,s} \overline{\widehat{\psi}(\kappa_l, \lambda_p)\widehat{\psi}(\kappa_r, \lambda_s)} e^{i(\kappa_l x_1 + \lambda_p y_1 + \kappa_r x_2 + \lambda_s y_2)} \tag{11.18}$$

By using (11.16), and by noting that the summation runs over both positive and negative r and s, we may insert the complex conjugate instead, i.e.

$$\begin{aligned}\overline{\psi(x_1,y_1)\psi(x_2,y_2)} &= (\Delta k)^2 \sum_{l,p,r,s} \overline{\widehat{\psi}(\kappa_l,\lambda_p)\widehat{\psi}^*(\kappa_r,\lambda_s)} e^{i(\kappa_l x_1 - \kappa_r x_2 + \lambda_p y_1 - \lambda_s y_2)} \\ &= c^2 \sum_{l,p,r,s} e^{-\left(a_{11}(\kappa_l^2+\kappa_r^2)+2a_{12}(\kappa_l\lambda_p+\kappa_r\lambda_s)+a_{22}(\lambda_p^2+\lambda_s^2)\right)} \\ &\quad \overline{e^{2\pi i(\phi_{l,p}-\phi_{r,s})}} e^{i(\kappa_l x_1 - \kappa_r x_2 + \lambda_p y_1 - \lambda_s y_2)}.\end{aligned} \qquad (11.19)$$

We assume that the fields are uncorrelated in wave space. Thus, there is only a distance dependence for the covariance, and the statistical properties of the simulated fields will be independent of the position. We can then set $l=r$ and $p=s$, and the above expression becomes

$$\overline{\psi(x_1,y_1)\psi(x_2,y_2)} = c^2 \sum_{l,p} e^{-2\left(a_{11}\kappa_l^2 + 2a_{12}\kappa_l\lambda_p + a_{22}\lambda_p^2\right)} e^{i(\kappa_l(x_1-x_2)+\lambda_p(y_1-y_2))}. \qquad (11.20)$$

The variance at the location (x,y), should be equal to 1, and from this equation we then get

$$\overline{\psi(x,y)\psi(x,y)} = 1 = c^2 \sum_{l,p} e^{-2\left(a_{11}\kappa_l^2 + 2a_{12}\kappa_l\lambda_p + a_{22}\lambda_p^2\right)}. \qquad (11.21)$$

This equation is invariant with respect to θ and can therefore be expressed with $\theta = 0$ as

$$1 = c^2 \sum_{l,p} e^{-2\left(\kappa_l^2/r_1^2 + \lambda_p^2/r_2^2\right)}, \qquad (11.22)$$

and we can solve it for c.

Further, we define de-correlation lengths r_x and r_y for the spatial fields in the two principal directions, and we require the covariance along the principal directions corresponding to distances r_x and r_y both to be equal to e^{-1}. Thus, in (11.20) we set $\theta = 0$ and evaluate $\overline{\psi(x_1+r_x,y_1)\psi(x_1,y_1)}$ and $\overline{\psi(x_1,y_1+r_y)\psi(x_1,y_1)}$ which should both equal e^{-1}, to get

$$e^{-1} = c^2 \sum_{l,p} e^{-2(\kappa_l^2/r_1^2 + \lambda_p^2/r_2^2)} \cos(\kappa_l r_x), \qquad (11.23)$$

$$e^{-1} = c^2 \sum_{l,p} e^{-2(\kappa_l^2/r_1^2 + \lambda_p^2/r_2^2)} \cos(\lambda_p r_y). \qquad (11.24)$$

By inserting for c^2 from (11.22), we get

$$e^{-1} = \sum_{l,p} e^{-2(\kappa_l^2/r_1^2 + \lambda_p^2/r_2^2)} \cos(\kappa_l r_x) / \sum_{l,p} e^{-2(\kappa_l^2/r_1^2 + \lambda_p^2/r_2^2)}, \quad (11.25)$$

$$e^{-1} = \sum_{l,p} e^{-2(\kappa_l^2/r_1^2 + \lambda_p^2/r_2^2)} \cos(\lambda_p r_y) / \sum_{l,p} e^{-2(\kappa_l^2/r_1^2 + \lambda_p^2/r_2^2)}. \quad (11.26)$$

This is a system of two nonlinear equations which can be solved for r_1 and r_2. Thereafter we can compute c from (11.22). The formula (11.15) can then be used to simulate an ensemble of random fields with variance 1 and covariance determined by the de-correlation lengths r_x and r_y and the rotation angle θ.

Using that the denominator appearing in (11.25) and (11.26) is always positive and larger than zero, we can write the two conditions as

$$F_1 = \sum_{l,p} e^{-2(\kappa_l^2/r_1^2 + \lambda_p^2/r_2^2)} (\cos(\kappa_l r_x) - e^{-1}) = 0, \quad (11.27)$$

$$F_2 = \sum_{l,p} e^{-2(\kappa_l^2/r_1^2 + \lambda_p^2/r_2^2)} (\cos(\lambda_p r_y) - e^{-1}) = 0. \quad (11.28)$$

These are easily solved using a Newton method, where we also need the derivatives

$$\frac{\partial F_1}{\partial r_1} = \sum_{l,p} e^{-2(\kappa_l^2/r_1^2 + \lambda_p^2/r_2^2)} \frac{4\kappa_l^2}{r_1^3} (\cos(\kappa_l r_x) - e^{-1}), \quad (11.29)$$

$$\frac{\partial F_1}{\partial r_2} = \sum_{l,p} e^{-2(\kappa_l^2/r_1^2 + \lambda_p^2/r_2^2)} \frac{4\lambda_p^2}{r_2^3} (\cos(\kappa_l r_x) - e^{-1}), \quad (11.30)$$

$$\frac{\partial F_2}{\partial r_1} = \sum_{l,p} e^{-2(\kappa_l^2/r_1^2 + \lambda_p^2/r_2^2)} \frac{4\kappa_l^2}{r_1^3} (\cos(\lambda_p r_y) - e^{-1}), \quad (11.31)$$

$$\frac{\partial F_2}{\partial r_2} = \sum_{l,p} e^{-2(\kappa_l^2/r_1^2 + \lambda_p^2/r_2^2)} \frac{4\lambda_p^2}{r_2^3} (\cos(\lambda_p r_y) - e^{-1}). \quad (11.32)$$

An efficient approach for finding the inverse transform in (11.15) is to apply a two-dimensional fast Fourier transform (FFT). The inverse FFT is calculated on a grid which is a few characteristic lengths larger than the computational domain to ensure non-periodic fields (*Evensen*, 1994b).

To summarize, we are now able to simulate two-dimensional pseudo random fields with variance equal to one and a prescribed anisotropic covariance.

11.3 Simulating correlated fields

A simple formula can be used to introduce correlations between the simulated realizations. This is useful in ocean and atmospheric models where there can be vertical correlations between levels or layers in the model. As an example, a simulated temperature field at two nearby depths will be correlated if there is strong vertical mixing such as in the ocean mixed layer. Another example relates to the simulation of model errors where we will expect there to be a finite time correlation.

The following equation can be used for simulating correlated realizations:

$$\psi_k(\boldsymbol{x}) = \rho \psi_{k-1}(\boldsymbol{x}) + \sqrt{1-\rho^2} w_k(\boldsymbol{x}). \quad (11.33)$$

Here we assume that $w_k(\boldsymbol{x})$ is a random realization sampled from a distribution with zero mean and variance equal to one, while $\psi_{k-1}(\boldsymbol{x})$ is the previous realization, to which $\psi_k(\boldsymbol{x})$ should be correlated. The $w_k(\boldsymbol{x})$ fields are typically generated by an algorithm similar to the one described in the previous section. Thus, starting with $\psi_1(\boldsymbol{x}) = w_1(\boldsymbol{x})$ the formula (11.33) can be used to sequentially simulating the correlated fields.

The coefficient $\rho \in [0, 1)$ determines the correlation of the stochastic forcing, e.g. $\rho = 0$ generates a white sequence, while $\rho = 1$ will remove the stochastic forcing and we obtain a random field identical to initial guess $\psi_0(\boldsymbol{x}) = w_0(\boldsymbol{x})$. More generally the covariance between $\psi_i(\boldsymbol{x})$ and $\psi_j(\boldsymbol{x})$ becomes

$$\overline{\psi_i(\boldsymbol{x})\psi_j(\boldsymbol{x})} = \rho^{|i-j|}. \quad (11.34)$$

The variance of the simulated fields will be equal to one.

11.4 Improved sampling scheme

Sampling errors can, according to the central limit theorem, be reduced by an increase of the ensemble size.

Based on the works by *Pham* (2001) and *Nerger et al.* (2005) it should be possible to introduce some improvements in the EnKF, by using a more clever sampling for the initial ensemble, the model noise and the measurement perturbations. We will now examine a sampling scheme, which effectively produces results similar to those obtained in the SEIK filter by *Pham* (2001). The scheme does not add significantly to the computational cost of the EnKF and may lead to a significant improvement in the results.

We start by defining an error covariance matrix $\boldsymbol{C}_{\psi\psi}$. We can assume this to be the initial error covariance matrix for the discrete model state. Given $\boldsymbol{C}_{\psi\psi}$ we can compute the eigenvalue decomposition

$$\boldsymbol{C}_{\psi\psi} = \boldsymbol{Z}\boldsymbol{\Lambda}\boldsymbol{Z}^\mathrm{T}, \quad (11.35)$$

where the matrices $\boldsymbol{Z}$ and $\boldsymbol{\Lambda}$ contain the eigenvectors and eigenvalues of $\boldsymbol{C}_{\psi\psi}$.

In the SEIK filter by *Pham* (2001) an algorithm was used where the initial ensemble was sampled from the first dominant eigenvectors of $\boldsymbol{C}_{\psi\psi}$. This introduces a maximum rank and conditioning of the ensemble matrix, and it also ensures that the ensemble provides a best possible representation of the error covariance matrix for a given ensemble size.

Now, one can approximate the error covariance matrix with its ensemble representation $C^e_{\psi\psi} \simeq C_{\psi\psi}$,

$$C^e_{\psi\psi} = \frac{1}{N-1} A'(A')^T \qquad (11.36)$$

$$= \frac{1}{N-1} U\Sigma V^T V \Sigma U^T \qquad (11.37)$$

$$= \frac{1}{N-1} U\Sigma^2 U^T. \qquad (11.38)$$

This is similar to the definition (9.14) but excluding the time dimension and using a discrete representation ψ, of the state. Here, A' contains the ensemble perturbations, and is defined as a discrete version of the formula (9.13), while U, Σ and V^T result from a singular value decomposition[1], and contain the singular vectors and singular values of A'. In the limit when the ensemble size goes to infinity the n singular vectors in U will converge towards the n eigenvectors in Z and the square of the singular values Σ^2, divided by $N-1$, will converge towards the eigenvalues Λ.

This shows that there are two strategies for defining an accurate ensemble approximation $C^e_{\psi\psi}$, of $C_{\psi\psi}$.

1. We can increase the ensemble size N, by sampling additional model states and adding these to the ensemble. As long as the addition of new ensemble members increases the space spanned by the overall ensemble, this will result in an ensemble covariance $C^e_{\psi\psi}$, which is a more accurate representation of $C_{\psi\psi}$.
2. Alternatively we can improve the rank/conditioning of the ensemble by ensuring that the first N singular vectors in U are similar to the N first eigenvectors in Z. Thus, the absolute error in the representation $C^e_{\psi\psi}$ of $C_{\psi\psi}$ will be smaller for ensembles generated with such an improved sampling than for Monte Carlo ensembles of a given moderate ensemble size. In other words, we want to generate A such that $\text{rank}(A) = N$ and the condition number, defined as the ratio between the largest and smallest singular values, $\kappa_2(A) = \sigma_1(A)/\sigma_N(A)$, is minimal. If the ensemble members stored in the columns of A are nearly dependent then $\kappa_2(A)$ is large.

This first approach is the standard Monte Carlo method used in the traditional EnKF where the convergence is slow. The second approach has a

---

[1] The singular value decomposition of a rectangular matrix $A \in \Re^{m\times n}$ is $A = U\Sigma V^T$ where $U \in \Re^{m\times m}$ and $V \in \Re^{n\times n}$ are orthogonal matrices and $\Sigma \in \Re^{m\times n}$ contains the $p = \min(m,n)$ singular values $\sigma_1 \geq \sigma_2 \geq \cdots \geq \sigma_p \geq 0$ on the diagonal. Further, $U^T A V = \Sigma$. Note that numerical algorithms for computing the SVD when $m > n$ often offers to compute only the first p singular vectors in U since the remaining singular vectors (columns in U) are normally not needed. However, for the expression $UU^T = I$ to be true the full U must be used.

flavour of quasi-random sampling which ensures much better convergence with increasing sample size. That is, we chose ensemble members which have less linear dependence and therefore span a larger space.

For most applications the size of $C_{\psi\psi}$ is too large to allow for the direct computation of eigenvectors. An alternative algorithm for generating an N-member ensemble with better conditioning, is to first generate a "start ensemble" which is larger than N, and then to resample N members along the first N dominant singular vectors of this larger start ensemble.

The algorithm goes as follows: First sample a large ensemble of model states with, e.g. β times N members, and store the ensemble perturbations in $\hat{A}' \in \Re^{n \times \beta N}$. Then perform the following steps:

1. Compute the SVD, $\hat{A}' = \hat{U}\hat{\Sigma}\hat{V}^{\mathrm{T}}$, where the columns of $\hat{U}$ are the singular vectors and the diagonal of $\hat{\Sigma}$ contains the singular values σ_i (note that with a multivariate state it may be necessary to scale the variables in $\hat{A}'$ first).
2. Retain only the first $N \times N$ quadrant of $\hat{\Sigma}$ which is stored in $\Sigma \in \Re^{N \times N}$, i.e. we set $\sigma_i = 0, \forall i > N$.
3. Scale the nonzero singular values with $\sqrt{\beta}$, which is done to obtain the correct variance in the new ensemble.
4. Generate a random orthogonal matrix[2] $V_1^{\mathrm{T}} \in \Re^{N \times N}$.
5. Generate an N-member ensemble using only the first N singular vectors in $\hat{U}$, stored in U, the nonzero singular values stored in Σ and the orthogonal matrix V_1^{T}.

Thus, we are using the formula

$$A' = U\frac{1}{\sqrt{\beta}}\Sigma V_1^{\mathrm{T}}. \qquad (11.39)$$

When the size of the start ensemble approaches infinity, the singular vectors will converge towards the eigenvectors of $C_{\psi\psi}$. It is of course assumed that the ensemble perturbations are sampled with the correct covariance as given by $C_{\psi\psi}$. As long as the initial ensemble is chosen large enough this algorithm will provide an ensemble which is similar to what is used in the SEIK filter, and the SVD algorithm has a lower computational cost than the explicit eigenvalue decomposition of $C_{\psi\psi}$ when n is large.

Before the ensemble perturbation matrix A' is used, it is important to ensure that the mean is zero and the variance takes a value as specified. This can be done by subtracting an eventual ensemble mean and then rescaling the ensemble members to obtain the correct variance. As will be seen below, this has a positive impact on the quality of the results. Note that the removal

---

[2] A random orthogonal matrix $V_1^{\mathrm{T}} \in \Re^{N \times N}$, is most effectively generated by a QR factorization of a matrix where the elements are independent normal distributed numbers with zero mean and variance equal to one.

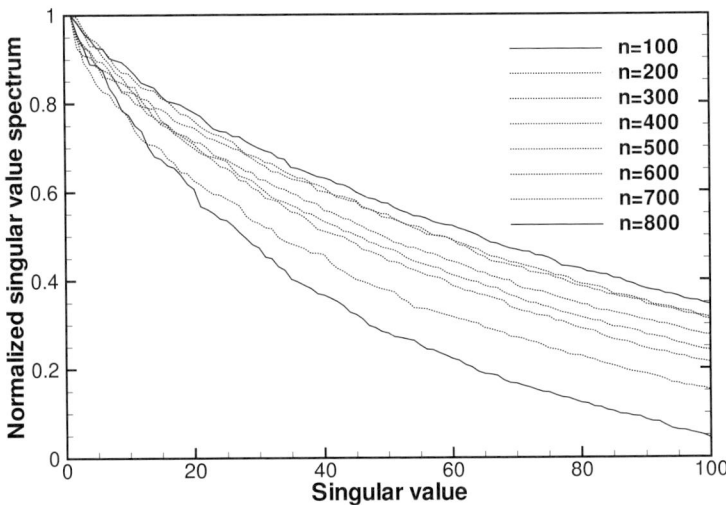

Fig. 11.2. The plot shows the normalized singular values of ensembles which are generated using start ensembles of different sizes, with the lower line corresponding to the start ensemble of 100 members. Clearly, the condition of the ensemble improves when a larger start ensemble is used

of the mean of the ensemble leaves the maximum possible rank of $\boldsymbol{A}'$ to be $N - 1$.

As an example, a 100-member ensemble has been generated using start ensembles of $100, 200, \ldots, 800$ members. The size of the one-dimensional model state is 1001 and the characteristic length scale of the solution is 4 grid cells. The singular values (normalized to the first singular value) for the resulting ensemble is plotted in Fig. 11.2 for the different sizes of start ensemble. Clearly, there is a benefit of using this sampling strategy. The ratio between singular values 100 and 1, is 0.21 when standard sampling is used. With increasing size of the start ensemble the conditioning improves until it reaches 0.59 for 800 members in the start ensemble.

We now assume a linear model operator defined by the full rank matrix $\boldsymbol{G}$. With zero model noise the ensemble at a later time t_k, can be written as

$$\boldsymbol{A}_k = \boldsymbol{G}^k \boldsymbol{A}_0. \tag{11.40}$$

Thus, the rank introduced in the initial ensemble will be preserved as long as $\boldsymbol{G}$ is full rank, and $\boldsymbol{A}_k$ will span the same space as $\boldsymbol{A}_0$.

With system noise the time evolution of the ensemble becomes

$$\boldsymbol{A}_k = \boldsymbol{G}^k \boldsymbol{A}_0 + \sum_{i=1}^{k} \boldsymbol{G}^{k-i} \boldsymbol{Q}_i, \tag{11.41}$$

where $\boldsymbol{Q}_i$ denote the ensemble of model noise used at time t_i. Thus, the rank and conditioning of the ensemble will also depend on the rank and conditioning of the model noise introduced.

For a nonlinear model operator, $\boldsymbol{G}(\boldsymbol{\psi}, \boldsymbol{q})$, where $\boldsymbol{q}$ is the model noise, the evolution of the ensemble can be written as

$$\boldsymbol{A}_k = \boldsymbol{G}_k\left(\ldots \boldsymbol{G}_2\left(\boldsymbol{G}_1\left(\boldsymbol{A}_0, \boldsymbol{Q}_1\right), \boldsymbol{Q}_2\right)\ldots, \boldsymbol{Q}_k\right). \quad (11.42)$$

Using a nonlinear model there is no guarantee that the nonlinear transformations will preserve the rank of $\boldsymbol{A}$ and the introduction of wisely sampled model noise may be crucial to maintain an ensemble with good rank properties during the simulation. Thus, the same procedure as used when generating the initial ensemble could be used when simulating the system noise. This will ensure that a maximum rank is introduced into the ensemble, and this may also counteract any rank reduction introduced by the model operator.

The EnKS and EnKF analysis algorithms in (9.37) and (9.39) with $\boldsymbol{X}_j$ defined in (9.38), use perturbed measurements through $\boldsymbol{D}'_j$. The improved sampling procedure could then be used when generating the measurement perturbations. This will lead to a better conditioning of the ensemble of perturbations and the ensemble covariance $\boldsymbol{C}^{\text{e}}_{\epsilon\epsilon}$ will get closer to $\boldsymbol{C}_{\epsilon\epsilon}$. The impact of improved sampling of measurement perturbations is significant and will be demonstrated in the examples below.

11.5 Experiments

The impact of ensemble size and improved sampling will now be discussed in some detail using the one-dimensional linear advection model from Sect. 4.1.3. The solution of this model is exactly known, and this allows us to run realistic experiments with zero model errors to examine the impact of the sampling schemes used.

In most of the following experiments an ensemble size of 100 members is used. A larger start ensemble is used in many of the experiments to generate ensemble members and/or measurement perturbations which provide a better representation of the error covariance matrix. Otherwise, the experiments differ in the sampling of measurement perturbations and the analysis scheme used. In Fig. 4.1 an example was shown from one of the experiments.

11.5.1 Overview of experiments

Several experiments have been carried out as listed in Table 11.1. For each of the experiments, 50 EnKF simulations were performed to allow for a statistical comparison. In each simulation, the only difference is the random seed used. Thus, every simulation will have a different and random true state, first guess, initial ensemble, set of measurements and measurement perturbations.

11 Sampling strategies for the EnKF

Experiment	N	Sample fix	β_{ini}	β_{mes}	Residual	Std. dev.
A	100	F	1	1	0.762	0.074
B	100	T	1	1	0.759	0.053
C	100	T	2	1	0.715	0.065
D	100	T	4	1	0.683	0.062
E	100	T	6	1	0.679	0.071
H	100	T	6	30	0.627	0.053
I	100	T	1	30	0.706	0.060
B150	150	T	1	1	0.681	0.053
B200	200	T	1	1	0.651	0.061
B250	250	T	1	1	0.626	0.058

Table 11.1. Summary of experiments. The first column is the experiment name, in the second column N is the ensemble size used, then "Sample fix" is true or false and indicates if the sample mean and variance is corrected, β_{ini} is a number which defines the size of the start ensemble used for generating the initial ensemble as $\beta_{\text{ini}} N$, similarly β_{mes} denote the size of the start ensemble used for generating the measurement perturbations, followed by the analysis algorithm used. The two last columns contain the average RMS errors of the 50 simulations in each experiment and the standard deviation of these

The standard version of the EnKF analysis scheme is used with a full rank matrix $\boldsymbol{C} = \boldsymbol{S}\boldsymbol{S}^{\text{T}} + (N-1)\boldsymbol{C}_{\epsilon\epsilon}$ which is factorized by computing the eigenvalue decomposition $\boldsymbol{Z}\boldsymbol{\Lambda}\boldsymbol{Z}^{\text{T}} = \boldsymbol{C}$, to get

$$\boldsymbol{C}^{-1} = \boldsymbol{Z}\boldsymbol{\Lambda}^{-1}\boldsymbol{Z}^{\text{T}}, \tag{11.43}$$

where all matrices are of dimension $m \times m$. Thus, we solve the standard EnKF analysis solving (9.39) with the definition (9.38), where measurements are perturbed, i.e. at each assimilation time we compute

$$\boldsymbol{A}^{\text{a}} = \boldsymbol{A}^{\text{f}}\left(\boldsymbol{I} + \boldsymbol{S}^{\text{T}}\boldsymbol{Z}\boldsymbol{\Lambda}^{-1}\boldsymbol{Z}^{\text{T}}(\boldsymbol{D} - \mathcal{M}[\boldsymbol{A}^{\text{f}}])\right), \tag{11.44}$$

where we have dropped the update index j.

In all the experiments the residuals were computed as the Root Mean Square (RMS) errors of the difference between the estimate and the true solution taken over the complete space and time domain. For each of the experiments we have plotted the mean and standard deviation of the residuals in Fig. 11.3.

It is also of interest to examine how well the predicted errors represent the actual residuals (RMS as a function of time). In the summary Figs. 11.4 and 11.5 we have plotted the average of the predicted errors from the 50 simulations as the thick full line. The thin full lines indicate the one standard deviation spread of the predicted errors from the 50 simulations. The average of the RMS errors from the 50 simulations is plotted as the thick dotted line, with the associated one standard deviation spread shown by the dotted thin lines.

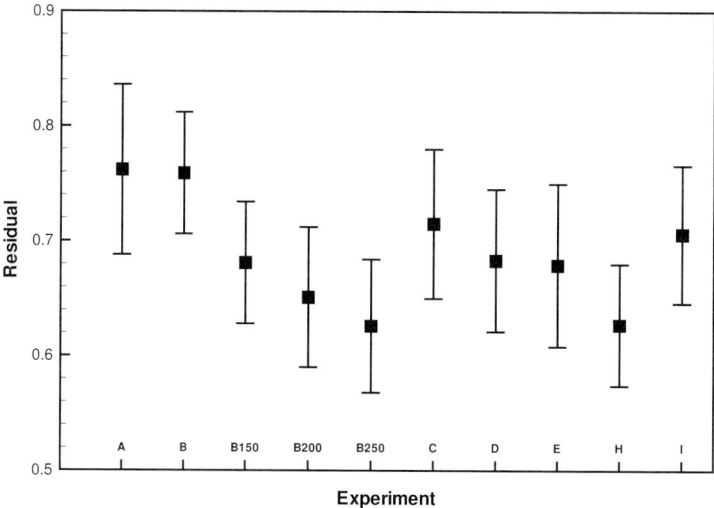

Fig. 11.3. Mean and standard deviation of the residuals from each of the experiments

The Table 11.2 gives the probabilities that the average residuals from the experiments are equal, as computed from the Student's t-test. Probabilities lower than, say 0.5, indicate statistically that the distributions from two experiments are significantly different.

The further details of the different experiments are described below.

Exp. A is the pure Monte Carlo case using a start ensemble of 100 members where all random variables are sampled "randomly". Thus, the mean and variance of the initial ensemble and the measurement perturbations will fluctuate within the accuracy that can be expected using a 100 member sample size.

Exp. B is similar to *Exp. A* except that the sampled ensemble perturbations are corrected to have mean zero and the correct specified variance. This is done by subtracting an eventual mean from the random sample and then dividing the members by the square root of the ensemble variance. As will be seen below, this leads to a small improvement in the assimilation results and this correction is therefore used in all the following experiments. This experiment is used as a reference case in the further discussion which illustrates the performance of the standard EnKF algorithm.

Exps. B150, B200 and *B250* are similar to *Exp. B* but using respectively ensemble sizes of 150, 200 and 250 members.

Exps. C, D and *E* are similar to *Exp. B* except that the start ensembles used to generate the initial 100 member ensemble contain respectively 200, 400

Exp	B	B150	B200	B250	C	D	E	H	I
A	**0.86**	0	0	0	0	0	0	0	0
B		0	0	0	0	0	0	0	0
B150			0.01	0	0.01	**0.86**	**0.86**	0	0.03
B200				0.04	0	0.01	0.04	0.04	0
B250					0	0	0	**0.91**	0
C						0.01	0.01	0	0.48
D							**0.75**	0	0.06
E								0	0.04
H									0

Table 11.2. Statistical probability that two experiments provide an equal mean for the residuals as computed using the Student's t-test. A probability close to one indicates that it is likely that the two experiments provide distributions of residuals with similar mean. The t-test numbers higher than 0.5 are printed in bold

and 600 members. *Exp. E* is used as a reference case illustrating the impact of the improved initial sampling algorithm.

Exp. H examines the combined impact of improved sampling of both measurement perturbations and the initial ensemble. The results should be compared with those of *Exp. E* where we now examine the additional impact improved sampling of measurement perturbations.

Exp. I should be compared with *Exps. H* and *B*. It uses improved sampling of measurement perturbations but standard sampling for the initial conditions. Thus, comparing it with results from *Exp. B* gives the impact of improved sampling of measurement perturbations.

11.5.2 Impact from ensemble size

We now compare the experiments *Exps. B, B150, B200* and *B250* to evaluate the impact of ensemble size on the performance of the EnKF. From Fig. 11.3 it is seen that the residuals, as expected, are decreasing when the ensemble size is increased. In practical applications we are naturally limited by the number of ensemble members we can afford to run. However, from the central limit theorem we have that the accuracy in the EnKF estimate will improve proportionally to the square root of the ensemble size. In most published applications of the EnKF a typical ensemble size has been around 100 members. This is clearly much less than effective dimension of the solution space of many dynamical models, but in these cases a so-called localization or local analysis computation is often used. This effectively increases the dimension of the space where the solution is searched for and will be discussed in more detail in the Appendix.

When comparing the time evolution of the residuals and the estimated standard deviations for the *Exps. B, B150* and *B250* in Fig. 11.4, we observe that the residuals show a larger spread between the simulations than the estimated standard deviations. The estimated standard deviations are internally consistent between the simulations performed in each of the experiments. The residuals are also generally larger than the ensemble standard deviations, although there is a significant improvement observed due to the increase in ensemble size.

11.5.3 Impact of improved sampling for the initial ensemble

Using the procedure outlined in Sect. 11.4 several experiments have been performed using start ensembles of 100–600 members to examine the impact of using an initial ensemble with better properties. The standard *Exp. B* is used as a reference while in the *Exps. C, D* and *E*, larger start ensembles of respectively 200, 400 and 600 members are used to generate the initial 100 member ensemble.

In Fig. 11.3 it is seen that just doubling the size of the start ensemble to 200 members (*Exp. C*) has a significant positive effect on the results, and using a start ensemble of 400 members (*Exp. D*) leads to a further improvement. The use of an even larger start ensemble of 600 members (*Exp. E*) does not provide a statistically significant improvement over *Exp. D* in this particular application, with a rather small state space.

The time evolutions of the residuals and the estimated standard deviations for the *Exps. B* and *E* in Figs. 11.4 and 11.5, show the same trend as was found for the *Exps. B, B150* and *B250* above, where residuals show a larger spread between the simulations than the estimated standard deviations and the residuals are larger than the ensemble standard deviations. Some improvement is seen when going from *Exp. B* to *Exp. E* due to the improved sampling of the initial ensemble.

It was also found when comparing *Exps. B150* and *B200* with *Exp. E* that an ensemble size between 150 and 200 is needed in the standard EnKF to get similar improvement as was obtained with improved sampling of a 100 member initial ensemble, using a start ensemble of 600.

These experiments clearly show that the improved sampling is justified for the initial ensemble. It is computed once and the additional computational cost is marginal. Thus, the improved sampling could be utilized to apply the filter algorithm with a smaller ensemble size and less computing time than required in the normal EnKF algorithm while still obtaining a comparable residual.

11.5.4 Improved sampling of measurement perturbations.

The *Exps. H* and *I* uses the improved sampling of measurement perturbations with a large start ensemble of perturbations of 30 times the ensemble size. The impact of this improved sampling is illustrated by comparing the *Exp. I* with

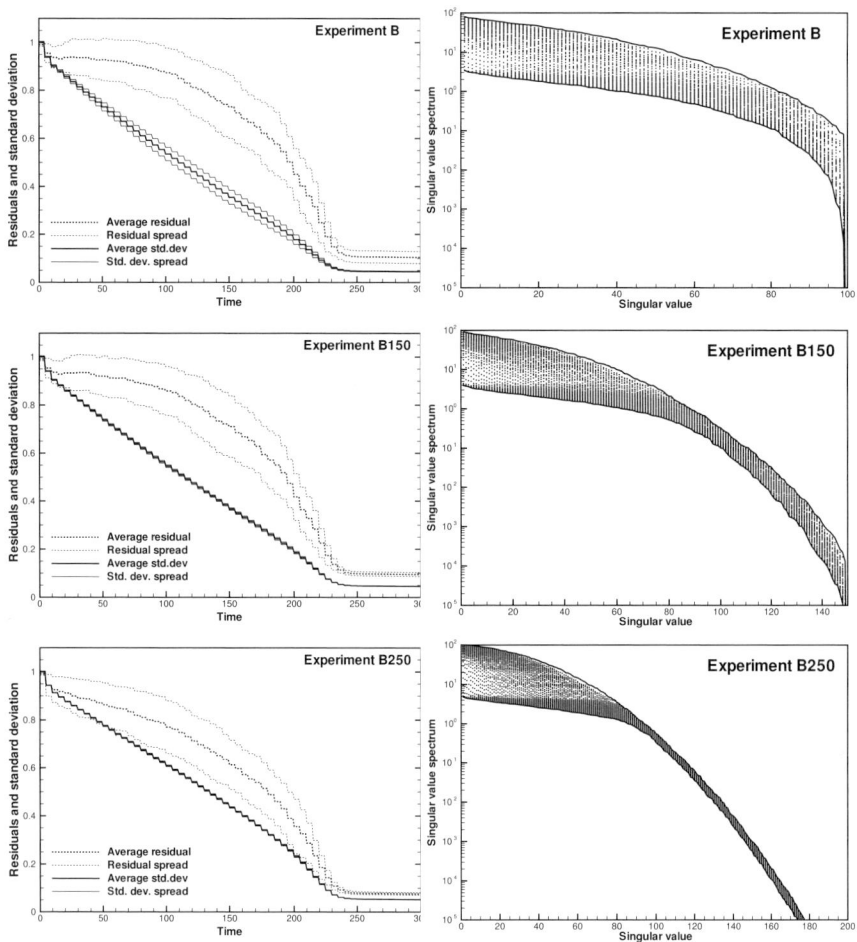

Fig. 11.4. RMS residuals and ensemble singular value spectra for some of the experiments. The left column shows the time evolution for RMS residuals *(dashed lines)* and estimated standard deviations *(full lines)*. The thick lines show the means over the 50 simulations and the thin lines show the means plus/minus one standard deviation. The right column shows the time evolution of the ensemble singular value spectra for the experiments

Exps. B, and then *Exp. H* with *Exp. E*, in Fig. 11.3. There is clearly a significant positive impact resulting from the improved sampling of measurement perturbations.

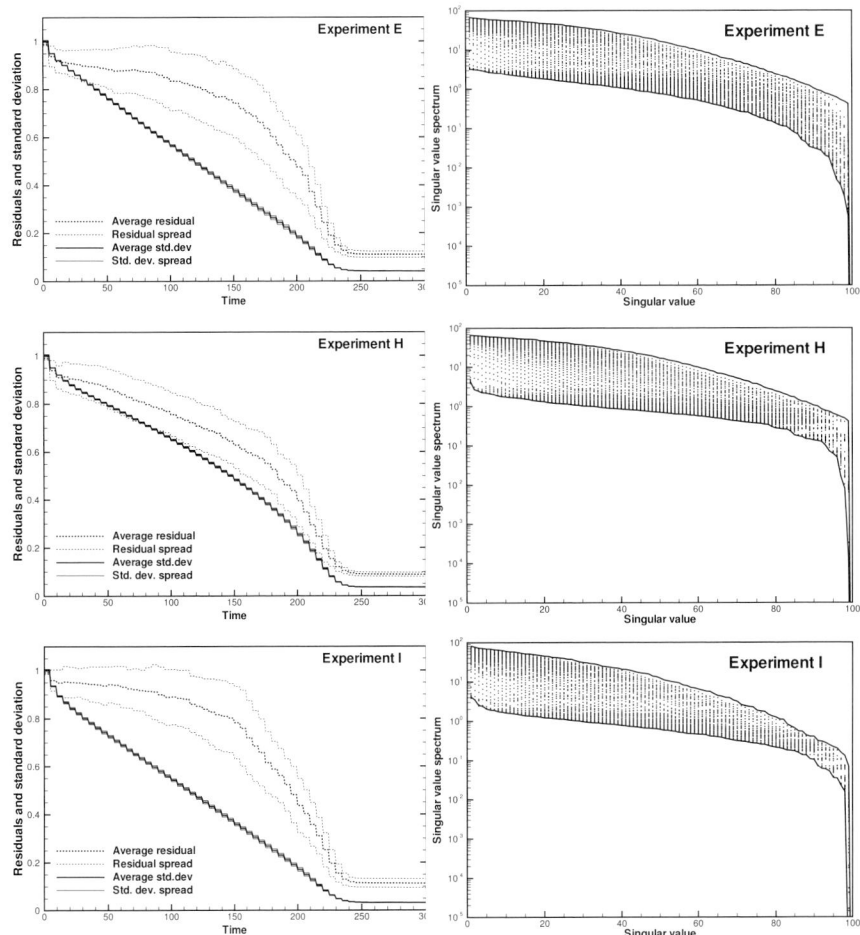

Fig. 11.5. See explanation in Fig. 11.4

11.5.5 Evolution of ensemble singular spectra

Finally, it is of interest to examine how the rank and conditioning of the ensemble evolves in time and is impacted by the computation of the analysis. In Figs. 11.4 and 11.5 we have plotted the singular values for the ensemble at each analysis time for some of the experiments. The initial singular spectrum of the ensemble is plotted as the upper thick line. Then the dotted lines indicate the reduction of the ensemble variance introduced at each analysis update, until the end of the experiment where the singular spectrum is given by the lower thick line.

It is clear from *Exps. B* and *E* that the conditioning of the initial ensemble improves when the new sampling scheme is used. Furthermore, it is seen from

Exps. B, B150 and *B250* that increasing the ensemble size does not add much to the representation of variance in the error subspace. This can be expected with the simple low dimensional model state considered here.

11.5.6 Summary

The previous experiments have quantified the impact of using an improved sampling scheme in the EnKF. The improved sampling attempts to generate ensembles with full rank and a conditioning which is better than what can be obtained using random sampling. The improved sampling has been used for the generation of the initial ensemble as well as for the sampling of measurement noise.

In the experiments discussed here it was possible to obtain a significant improvement in the results from the standard EnKF analysis scheme if improved sampling is used both for the initial ensemble and the measurement perturbations. It is expected that similar improvements can be obtained in general since the improved sampling provides a better representation of the ensemble error covariances and of the state space where the solution is searched for.

It is important to point out that these results may not be directly transferable to other more complex dynamical models. In the cases discussed here the dimension of the state vector (1001 grid cells) is small compared to typical applications with ocean and atmospheric models. Thus, although we expect that the use of improved sampling schemes always will lead to an improvement in the results, it is not possible to quantify this improvement in general.

We have not examined fully, the potential impact a nonlinear model will have on the ensemble evolution. The use of nonlinear models will change the basis from that of the initial ensemble, and may even reduce the rank of the ensemble. This suggests that the improved sampling should be used for the model noise as well, to help maintain the conditioning of the ensemble during the forward integration.

From these experiments we can give the recommendation that the use of high order sampling should always be used for both the initial ensemble and the sampling of measurement perturbations. The experiments have shown that there is a potential for either a significant reduction of the computing time or an improvement of the EnKF results, using the improved sampling schemes.

12
Model errors

We will now discuss the use of model errors in the ensemble and representer methods. A particular focus will be on the impact of time-correlated model errors. A simple scalar equation is used to illustrate the use of the ensemble and the representer methods for combined parameter and state estimation in this case.

12.1 Simulation of model errors

In the previous chapter we learned how to introduce a correlation between the random fields. We will now study in more detail how this can be used to simulate time correlated model errors, and how we can introduce the correct variance in each realization to properly represent the magnitude of the actual model error.

Again we assume that $w_k(\boldsymbol{x})$ is a sequence of white noise drawn from a distribution of smooth pseudo random fields with mean equal to zero and variance equal to one.

Equation (11.33) ensures that $q_k(\boldsymbol{x})$ is drawn from a distribution with variance equal to one as long as the variance of the distribution for $q_{k-1}(\boldsymbol{x})$ equals one. Thus, this equation will produce a sequence of time correlated pseudo random fields with mean equal to zero and variance equal to one. The covariance in time between $q_i(\boldsymbol{x})$ and $q_j(\boldsymbol{x})$ is given by the formula (11.34).

12.1.1 Determination of ρ.

The factor ρ in (11.33) should be related to the time step used and a specified time de-correlation length τ. Equation (11.33), when excluding the stochastic term, resembles a difference approximation to

$$\frac{\partial q}{\partial t} = -\frac{1}{\tau} q, \qquad (12.1)$$

which states that q is damped with a ratio e^{-1}, over a time period $t = \tau$. A numerical approximation becomes

$$q_k = \left(1 - \frac{\Delta t}{\tau}\right) q_{k-1}, \qquad (12.2)$$

where Δt is the time step. Thus, we define ρ as

$$\rho = 1 - \frac{\Delta t}{\tau}, \qquad (12.3)$$

where $\tau \geq \Delta t$.

12.1.2 Physical model.

A discrete stochastic model is now defined as

$$\psi_k(\boldsymbol{x}) = G(\psi_{k-1}(\boldsymbol{x})) + \sqrt{\Delta t}\sigma c q_k(\boldsymbol{x}), \qquad (12.4)$$

where σ is the standard deviation of the model error and c is a factor to be determined. The choice of the stochastic term is explained next.

12.1.3 Variance growth due to the stochastic forcing.

To explain the choice of the stochastic term in (12.4) we will use a simple random walk model for illustration, i.e.

$$\psi_k(\boldsymbol{x}) = \psi_{k-1}(\boldsymbol{x}) + \sqrt{\Delta t}\sigma c q_k(\boldsymbol{x}). \qquad (12.5)$$

This equation can be rewritten as

$$\psi_k(\boldsymbol{x}) = \psi_0(\boldsymbol{x}) + \sqrt{\Delta t}\sigma c \sum_{i=0}^{k-1} q_{i+1}(\boldsymbol{x}). \qquad (12.6)$$

The variance can be found by squaring (12.6) and taking the ensemble average, i.e.

$$\overline{\psi_s(\boldsymbol{x})\psi_s(\boldsymbol{x})} = \overline{\psi_0(\boldsymbol{x})\psi_0(\boldsymbol{x})} + \Delta t\sigma^2 c^2 \overline{\left(\sum_{k=0}^{s-1} q_{k+1}(\boldsymbol{x})\right)\left(\sum_{k=0}^{s-1} q_{k+1}(\boldsymbol{x})\right)} \qquad (12.7)$$

$$= \overline{\psi_0(\boldsymbol{x})\psi_0(\boldsymbol{x})} + \Delta t\sigma^2 c^2 \sum_{j=0}^{s-1}\sum_{i=0}^{s-1} \overline{q_{i+1}(\boldsymbol{x})q_{j+1}(\boldsymbol{x})} \qquad (12.8)$$

$$= \overline{\psi_0(\boldsymbol{x})\psi_0(\boldsymbol{x})} + \Delta t\sigma^2 c^2 \sum_{j=0}^{s-1}\sum_{i=0}^{s-1} \rho^{|i-j|} \qquad (12.9)$$

$$= \overline{\psi_0(\boldsymbol{x})\psi_0(\boldsymbol{x})} + \Delta t\sigma^2 c^2 \left(-s + 2\sum_{i=0}^{s-1}(s-i)\rho^i\right) \qquad (12.10)$$

$$= \overline{\psi_0(\boldsymbol{x})\psi_0(\boldsymbol{x})} + \Delta t\sigma^2 c^2 \frac{s - 2\rho - s\rho^2 + 2\rho^{s+1}}{(1-\rho)^2}, \qquad (12.11)$$

12.1 Simulation of model errors

where (11.34) has been used and s denote the number of time steps. The double sum in (12.9) is just summing elements in a matrix and is replaced by a single sum operating on diagonals of constant values. The summation in (12.10) has an explicit solution (*Gradshteyn and Ryzhik*, 1979, formula 0.113).

We now define the number n such that $n\Delta t = 1$, thus n is the number of time steps over one time unit. It is clear from (12.11), that if $c = 1$, then the increase in variance over s time steps is equal to

$$\frac{s\sigma^2}{n} \frac{1 - \rho^2 - 2\rho/s + 2\rho^{s+1}/s}{(1-\rho)^2}. \tag{12.12}$$

Thus, with $\rho = 0$ the increase in variance is just $s\sigma^2/n$ as would be expected. However, with coloured noise the increase in variance may become significantly higher, dependent on the value of ρ.

In cases where we know the exact statistics of the stochastic noise process, although these cases are rare, this additional variance increase is realistic. On the other hand, in many cases we may have an estimate of the expected variance increase σ^2 over a time unit, and we may anticipate that the noise process is coloured. In that case we will need to use the scaling factor c to obtain a noise process which results in a realistic variance increase per time unit.

The two equations (11.33) and (12.4) provide the standard framework for introducing stochastic model errors when using the EnKF. The formula (12.11) provides the mean for scaling the perturbations in (12.4) when changing ρ and/or the number of time steps per time unit to ensure that the ensemble variance growth over a time unit equals σ^2.

It is natural to assume that the increase in variance over s time steps should be equal to $s\sigma^2/n$, e.g. if $s = n$ this corresponds to integration over one time unit and the increase in variance becomes σ^2. We then have the formula

$$\frac{s\sigma^2}{n} = c^2 \frac{s\sigma^2}{n} \frac{1 - \rho^2 - 2\rho/s + 2\rho^{s+1}/s}{(1-\rho)^2}, \tag{12.13}$$

which we can solve for c to get

$$c^2 = \frac{(1-\rho)^2}{1 - \rho^2 - 2\rho/s + 2\rho^{s+1}/s}. \tag{12.14}$$

If the sequence of model noise $q_k(\boldsymbol{x})$ is white in time, i.e. $\rho = 0$, we get $c \equiv 1$ as is expected. Thus, when (12.5) is iterated, $c = 1$ leads to the correct increase in ensemble variance given by σ^2 per time unit. The formula (12.14) is identical to the one proposed by *Evensen* (2003) but it was given for $s = n$ and integration over one time unit.

For red model noise, with $\rho \in (0, 1)$, the formula (12.14) still gives the correct answer, i.e. if the model is integrated s time steps, the variance at time step s has increased by $s\sigma^2/n$. However, a problem with this approach is that the variance increase is not linear, and if the integration is continued for

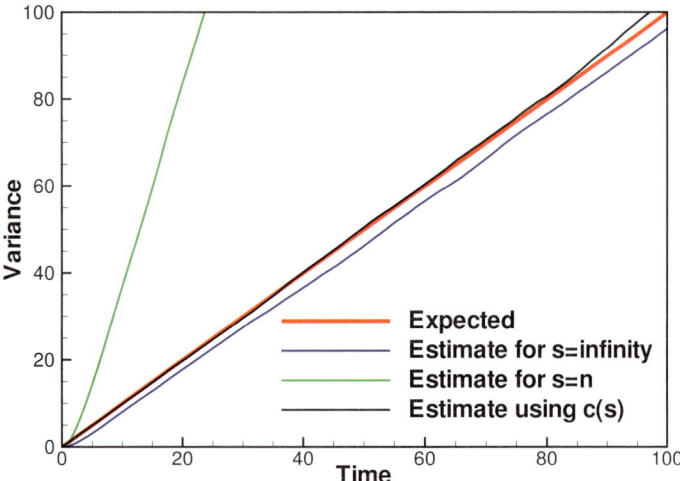

Fig. 12.1. The plot shows the variance using the different definitions for c. The expected variance is plotted as the red curve. The estimated variance from the Brownian motion (12.5) using the definitions (12.15) *(blue curve)*, and (12.14) with $s = n$ *(green curve)*. The formula (12.19) results in the black curve

more than s time steps the variance will increase too fast. This is seen from the green curve in Fig. 12.1 where c is evaluated for $s = n$, as in *Evensen* (2003), but the integration continues for a much longer time interval. The reason for the too large variance increase is that we have neglected correlations in time exceeding one time unit in the continued integration.

A better value for c, which can be used for long time integrations, is obtained by considering the limiting behaviour of the formula (12.11) when s becomes large. The solution for c when the number of time steps, s, goes to infinity in the formula (12.14) becomes

$$c^2 = \frac{1-\rho}{1+\rho}. \tag{12.15}$$

A plot of the estimated variance increase as a function of time for the Brownian motion process given by (12.5), is shown as the blue curve in Fig. 12.1. It is clear that the formula (12.15) gives a too weak variance increase initially but after an integration for a time period similar to the range of the exponential time correlation function used, the correct linear variance increase is obtained.

We can chose any value for s when evaluating the formula (12.14) for c. Thus, we can always obtain the correct variance at a certain time step, e.g. at a time when we are going to update the solution with measurements, but we would need to switch to the limiting value for c from (12.15) for the continued integration.

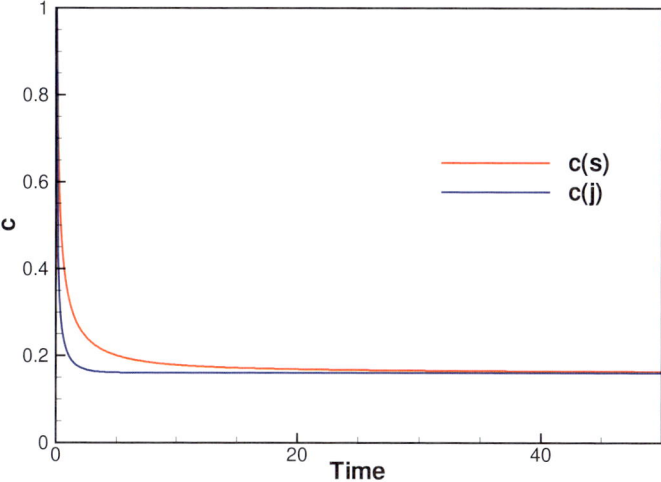

Fig. 12.2. The plot shows the value of c as a function of time, computed from (12.14) *(red curve)* and from (12.19) *(blue curve)*

It is possible to do better than this. We can use a formula with $c_i = c(i)$ being a function of the time step i, and require that the variance has the correct value at all time steps. We then need to introduce c_i inside the summation in (12.6) and we get

$$\psi_k(\boldsymbol{x}) = \psi_0(\boldsymbol{x}) + \sqrt{\Delta t}\sigma \sum_{i=1}^{k} c_i q_i(\boldsymbol{x}), \tag{12.16}$$

where we for simplicity also changed the summation index. As before we can write

$$\overline{\psi_s(\boldsymbol{x})\psi_s(\boldsymbol{x})} = \overline{\psi_0(\boldsymbol{x})\psi_0(\boldsymbol{x})} + \Delta t\sigma^2 \sum_{j=1}^{s}\sum_{i=1}^{s} c_i c_j \rho^{|i-j|}. \tag{12.17}$$

Now, assuming the increase in variance over s time steps to be equal to $s\sigma^2/n$ we get

$$\frac{s\sigma^2}{n} = \frac{\sigma^2}{n} \sum_{j=1}^{s}\sum_{i=1}^{s} c_i c_j \rho^{|i-j|}, \tag{12.18}$$

which can be rewritten as

$$s = \sum_{j=1}^{s-1}\sum_{i=1}^{s-1} c_i c_j \rho^{|i-j|} + \left(2\sum_{i=1}^{s-1} c_i \rho^{|s-i|}\right) c_s + c_s^2. \tag{12.19}$$

This is a recursion of second order scalar equations for c_s. Using that $c_1 = 1$ we can solve it recursively in each time step for $c_s, s \in (2,\infty)$, and we have resolved the issue with the too low initial variance increase. It is also clear that after a few time steps, exceeding the range of the time correlations specified, we approach the limiting value (12.15) for c. In Fig. 12.2 we have plotted c from (12.13) as a function of s as the red curve, and c from (12.19) as a function of time as the blue curve. Note that there is one sequence of positive and one of negative solutions for c_s which only differ in the sign and we can pick either.

12.1.4 Updating model noise using measurements

From the previous discussion is should be clear that when red model noise is used, correlations will develop between the red noise and the model variables. Thus, during the analysis step it is also possible to consistently update the model noise as well as the model state. This was illustrated in an example by *Reichle et al.* (2002). We now introduce a new state vector which consists of $\psi(\boldsymbol{x})$ augmented with $q(\boldsymbol{x})$. Equations (11.33) and (12.4) can then be written as

$$\begin{pmatrix} q_k(\boldsymbol{x}) \\ \psi_k(\boldsymbol{x}) \end{pmatrix} = \begin{pmatrix} \rho q_{k-1}(\boldsymbol{x}) \\ G(\psi_{k-1}(\boldsymbol{x})) + \sqrt{\Delta t} \sigma c q_k(\boldsymbol{x}) \end{pmatrix} + \begin{pmatrix} \sqrt{1-\rho^2} \boldsymbol{w}_{k-1} \\ 0 \end{pmatrix}. \tag{12.20}$$

During the analysis we can now compute covariances between the observed model variable and the model noise vector $q(\boldsymbol{x})$, which is updated together with the state vector. This will lead to a correction of the mean of $q(\boldsymbol{x})$ as well as a reduction of the variance in the model noise ensemble. Note that this procedure estimates the actual error in the model for each ensemble member, given the prescribed model error statistics.

The form of (11.33) ensures that, over time, $q_k(\boldsymbol{x})$ will approach a distribution with mean equal to zero and variance equal to one, as long as we don't update $q_k(\boldsymbol{x})$ in the analysis scheme. In the case when $q_k(\boldsymbol{x})$ is updated it will be relaxed back towards a process with zero mean and variance equal to one.

12.2 Scalar model

We now define a simple scalar equation containing a poorly known parameter α, which has a first guess value $\alpha_0 = 0$, while the true value is $\alpha = 1$. We also have a set of measurements of the true solution, which in this case becomes a constant $\psi(t) = 3$. Similarly to the system of equations (7.1–7.5) we now allow the model equation, the initial condition, the first guess parameter and the measurements to contain errors and write,

$$\frac{\partial \psi}{\partial t} = 1 - \alpha + q, \tag{12.21}$$

$$\psi(0) = 3 + a, \tag{12.22}$$

$$\alpha = 0 + \alpha', \tag{12.23}$$

$$\mathcal{M}[\psi] = \boldsymbol{d} + \boldsymbol{\epsilon}. \tag{12.24}$$

The model is defined on the interval $t \in [0, 50]$, thus using the notation from Chap. 7, $t_0 = 0$ and $t_k = 50$. We have used $G(\psi, \alpha) = 1 - \alpha$, so the model operator is linear and independent of ψ. There are nine measurements of ψ taken at the discrete times $t_{i(j)} = 5j$, for $j = 1, \ldots, 9$, and the measurement functional for measurement number j becomes just

$$\mathcal{M}_j[\psi] = \int_0^{50} \psi(t') \delta(t' - t_{i(j)}) dt' = \psi(t_{i(j)}). \tag{12.25}$$

It should be noted that the simple form used for $G(\psi, \alpha)$, will result in a linear inverse problem even though α is included as a variable to be estimated. This will not be the case in general, since linear models containing, e.g. a product of ψ and α, will lead to nonlinear inverse problems when the parameter α, is considered as a variable to be estimated.

12.3 Variational inverse problem

The formulation of the variational inverse problem and the representer method is now derived for the simple linear combined parameter and state estimation problem, using the methodology explained in Chap. 8.

12.3.1 Prior statistics

We have to make assumptions about the statistical properties of the error terms added to (12.21–12.24). It is common to assume simple statistical forms for the priors, i.e. the error terms all have zero mean and the statistics is described by a covariance. Further, we assume that the different error terms are uncorrelated.

For the model errors q, we assume an exponential correlation in time

$$C_{qq}(t_1, t_2) = \sigma^2 \exp(-|t_2 - t_1|/\tau), \tag{12.26}$$

with σ^2 being the model error variance and τ the correlation length in time. The weight W_{qq} is defined from

$$W_{qq}(t_1, t_2) \bullet C_{qq}(t_2, t_3) = \delta(t_1 - t_3), \tag{12.27}$$

where the bullet denote integration in t_2.

The error in the initial condition a, is determined by the variance C_{aa} with inverse $W_{aa} = 1/C_{aa}$, and similarly the error in α is given by $C_{\alpha\alpha}$ with inverse $W_{\alpha\alpha} = 1/C_{\alpha\alpha}$. For the measurements the errors are described by the measurement error covariance matrix $\boldsymbol{C}_{\epsilon\epsilon}$ with matrix inverse $\boldsymbol{W}_{\epsilon\epsilon}$.

12.3.2 Penalty function

The generalized inverse formulation (8.20) for the problem stated above becomes

$$\mathcal{J}[\psi,\alpha] = \left(\frac{\partial \psi}{\partial t} - 1 + \alpha\right)_{t_1} \bullet W_{qq}(t_1,t_2) \bullet \left(\frac{\partial \psi}{\partial t} - 1 + \alpha\right)_{t_2}$$
$$+ (\psi_0 - 3)W_{aa}(\psi_0 - 3) \tag{12.28}$$
$$+ (\alpha - 0)W_{\alpha\alpha}(\alpha - 0)$$
$$+ (\boldsymbol{d} - \mathcal{M}[\psi])^\mathrm{T} \boldsymbol{W}_{\epsilon\epsilon}(\boldsymbol{d} - \mathcal{M}[\psi]).$$

12.3.3 Euler–Lagrange equations

By setting the variational derivative of $\mathcal{J}[\psi,\alpha]$ with respect to α equal to zero, noting that ψ depends on α, we get the Euler–Lagrange equations

$$\frac{\partial \psi}{\partial t} = 1 - \alpha + C_{qq} \bullet \lambda, \tag{12.29}$$

$$\psi(0) = 3 + C_{aa}\lambda(0), \tag{12.30}$$

$$\frac{\partial \lambda}{\partial t} = -\mathcal{M}[\delta]\boldsymbol{W}_{\epsilon\epsilon}(\boldsymbol{d} - \mathcal{M}[\psi]), \tag{12.31}$$

$$\lambda(50) = 0, \tag{12.32}$$

$$\alpha = \alpha_0 - W_{\alpha\alpha}\int_0^{50} \lambda\, dt. \tag{12.33}$$

This is a coupled two point boundary problem in time, where the forward model (12.29) depends on the adjoint variable λ, and the adjoint backward model (12.31) depends on ψ. Note that the simple form of $G(\psi,\alpha)$ leads to an adjoint model (12.31) where the term $g_\psi \lambda$ in (8.42) vanishes.

If the true value of α is known, we eliminate the last equation (12.33) and are left with a linear inverse problem where the solution is defined by (12.29–12.32). This is still a coupled two point boundary value problem in time but a direct solution can be obtained using the representer method.

12.3.4 Iteration of parameter

We define an iteration of α to get a sequence of linear iterates for the Euler–Lagrange equations. Thus, we write

$$\alpha_l = \alpha_{l-1} - \gamma\left(\alpha_{l-1} - \alpha_0 + W_{\alpha\alpha}\int_0^{50}\lambda_{l-1}\,dt\right), \tag{12.34}$$

where the expression in the parentheses is the gradient of the penalty function with respect to the parameter α, and γ is a step length in a gradient descent method. Thus, for each iterate α_l, we need to solve

$$\frac{\partial \psi_l}{\partial t} = 1 - \alpha_l + C_{qq} \bullet \lambda_l, \qquad (12.35)$$

$$\psi_l(0) = 3 + C_{aa}\lambda_l(0), \qquad (12.36)$$

$$\frac{\partial \lambda_l}{\partial t} = -\mathcal{M}[\delta]\boldsymbol{W}_{\epsilon\epsilon}\left(\boldsymbol{d} - \mathcal{M}[\psi_l]\right), \qquad (12.37)$$

$$\lambda_l(50) = 0. \qquad (12.38)$$

12.3.5 Solution by representer expansions

Assume a solution of the form

$$\psi(t) = \psi_{\mathrm{F}}(t) + \boldsymbol{b}^{\mathrm{T}}\boldsymbol{r}(t), \qquad (12.39)$$

$$\lambda(t) = \lambda_{\mathrm{F}}(t) + \boldsymbol{b}^{\mathrm{T}}\boldsymbol{s}(t), \qquad (12.40)$$

i.e. the solution is a first guess solution $\psi_{\mathrm{F}}(t)$ and $\lambda_{\mathrm{F}}(t)$ plus a linear combination $\boldsymbol{b}$ of influence functions or representers $\boldsymbol{r}(t)$, and their adjoints $\boldsymbol{s}(t)$. There is one representer and associated adjoint for each measurement. We have now dropped the l-index for the iteration of the parameter α.

We insert (12.39) and (12.40) into (12.35–12.38). When assuming that $\boldsymbol{b}$ is undetermined and arbitrary we get a system of equations for the first guess solution,

$$\frac{\partial \psi_{\mathrm{F}}}{\partial t} = 1 - \alpha + C_{qq} \bullet \lambda_{\mathrm{F}}, \qquad (12.41)$$

$$\psi_{\mathrm{F}}(0) = 3 + C_{aa}\lambda_{\mathrm{F}}(0), \qquad (12.42)$$

$$\frac{\partial \lambda_{\mathrm{F}}}{\partial t} = 0, \qquad (12.43)$$

$$\lambda_{\mathrm{F}}(50) = 0. \qquad (12.44)$$

These equations have the solution $\lambda_{\mathrm{F}}(t) = 0$, and ψ_{F} is just the solution of the original dynamical model with no information from the measurements included.

By choosing the coefficients $\boldsymbol{b}$ to satisfy (5.60), we find the following set of equations for the representers and their adjoints

$$\frac{\partial \boldsymbol{r}}{\partial t} = C_{qq} \bullet \boldsymbol{s}, \qquad (12.45)$$

$$\boldsymbol{r}(0) = C_{aa}\boldsymbol{s}(0), \qquad (12.46)$$

$$\frac{\partial \boldsymbol{s}}{\partial t} = -\mathcal{M}[\delta], \qquad (12.47)$$

$$\boldsymbol{s}(50) = 0. \qquad (12.48)$$

These equations are now decoupled, i.e. (12.47) can be integrated backward in time from the final conditions (12.48) to find $\boldsymbol{s}$. Thereafter the system in (12.45) can be integrated forward in time from the initial conditions (12.46).

12 Model errors

The symmetric positive definite representer matrix $\mathcal{M}^{\mathrm{T}}[\boldsymbol{r}]$, can be constructed by measuring the representers as soon as they have been solved for. Knowing ψ_{F}, $\boldsymbol{b}$ which is found from (5.60) and $\boldsymbol{r}$, we can construct the optimal minimizing solution of the linear iterate from (12.39), given a value for α.

12.3.6 Variance growth due to model errors

In the previous sections we found that the variance growth of a stochastic model increased when the noise process representing the model errors became coloured. This also has implications for the representer method. If we want to compare solutions using the representer method and the ensemble methods with coloured noise, we also need to introduce a correction factor in the representer method.

We start by noting that the representer corresponding to a particular direct measurement equals the space-time covariance function for the corresponding measurement location, and its value at the measurement location is equal to the prior variance at that location.

On discrete form we can write the model error covariance as the matrix

$$\boldsymbol{C}(t_i, t_j) = \sigma^2 c_{\text{rep}} \exp(-|i-j|\Delta t/\tau), \tag{12.49}$$

for i and j taking values from 0 to the number of time steps and we have introduced the factor c_{rep} in the definition of the model error covariance.

Thus, we write the solution of (12.45) for each component, j of $\boldsymbol{r}$, at the corresponding measurement location $t_{i(j)}$, in discrete form as

$$r_j(t_{i(j)}) = r_j(t_{i(j)-1}) + \Delta t \sum_{i=0}^{i(j)} C(t_{i(j)}, t_i) s_j(t_i) \Delta t. \tag{12.50}$$

Note that the summation in the convolution for measurement j can be stopped at $i = i(j)$ since s_j is zero for $t_i > t_{i(j)}$. From this equation we can write

$$r_j(t_{i(j)}) = r_j(0) + \frac{\sigma^2 c_{\text{rep}}}{n} \sum_{k=1}^{i(j)} \sum_{i=0}^{i(j)} \exp(-|k-j|\Delta t/\tau) s_j(t_i) \Delta t, \tag{12.51}$$

where we have used that $n = 1/\Delta t$ is the number of time steps over one time unit.

Thus, as in the previous chapter we can now determine c_{rep} so that for each representer it will have the correct variance at the measurement location $t_{i(j)}$,

$$\frac{i(j)\sigma^2}{n} = \frac{\sigma^2 c_{\text{rep}}}{n} \sum_{k=1}^{i(j)} \sum_{i=0}^{i(j)} \exp(-|k-j|\Delta t/\tau) s_j(t_i) \Delta t, \tag{12.52}$$

which we can solve for c_{rep} to get

$$c_{\text{rep}} = i(j) / \sum_{k=1}^{i(j)} \sum_{i=0}^{i(j)} \exp(-|k-j|\Delta t/\tau) s_j(t_i) \Delta t. \qquad (12.53)$$

Note that we will get a slightly different value of c for measurements at different time locations, and probably a limiting value should be used, since a different value for c_{rep} at different time locations will lead to an unsymmetrical representer matrix.

12.4 Formulation as a stochastic model

For the ensemble methods we write the dynamical model (12.21) on stochastic form similarly to (12.20), i.e.

$$\begin{pmatrix} q_i \\ \psi_i \end{pmatrix} = \begin{pmatrix} \rho q_{i-1} \\ \psi_{i-1} + (1-\alpha)\Delta t + \sqrt{\Delta t}\sigma c_i q_i \end{pmatrix} + \begin{pmatrix} \sqrt{1-\rho^2}\, w_{i-1} \\ 0 \end{pmatrix}, \qquad (12.54)$$

where w_i is a white noise process with zero mean and unit variance, $\rho \in [0,1)$ determines the time correlation and c_i is the factor from (12.19) which is used to tune the variance increase in time during the stochastic integration.

12.5 Examples

We will now discuss some examples where the system (12.21–12.24) is solved using the representer method, the EnKF and the EnKS. We will discuss the standard state estimation case where the parameter α is known, the state estimation case when an erroneous value is used for α and the model therefore is biased, and the case where we estimate both the model state and the parameter. We will consider both the case with white and coloured model noise. The examples are similar to, but not identical to the ones from *Evensen* (2003).

In all the cases we have used an initial and measurement variance equal to nine and the model error variance is equal to one. In the cases with time correlated model noise the time scale of the correlation is $\tau = 2$. The number of ensemble members is 1000 and the time step is $\Delta t = 0.1$. In the cases with parameter estimation the parameter error variance is set to four.

The results from the experiments are presented in the Figs. 12.3–12.7. The measurements are plotted as bullets, the representer solution is plotted as the red line, the EnKF solution is given by the blue line and the EnKS solution is plotted as the green line. In addition we have included the EnKF and the EnKS solution plus and minus the estimated standard deviation as the blue and green dashed lines. Note that for the representer solution there is no error estimate, but if there where it would be identical to the EnKS error estimate in the limit of an infinite ensemble size.

12.5.1 Case A0

We first consider an example where the parameter $\alpha = 1$ is assumed to be known. This corresponds to a linear inverse calculation where we solve for ψ as a function of time given the observations. The results are presented in Fig. 12.3.

The representer solution is the maximum likelihood solution and can be used a reference. Note that, due to the use of white noise this curve will have discontinuous time derivatives at the measurement locations, a property of the representer and EnKS solutions when white model errors are used.

The EnKF estimate has discontinuities at the measurement locations due to the analysis updates. During the integration between the measurement locations the ensemble mean satisfies the dynamical part of the model equation, i.e. the time derivative of the solution is zero. The ensemble standard deviation is reduced at every measurement time, and increases according to the stochastic forcing term during the integration between the measurements.

The EnKS provides a continuous curve and is thus a more realistic estimate than the EnKF solution. It is clear that the EnKS solution is very similar to the representer solution, and it only differs due to the use of a finite ensemble size. Note that from the central limit theorem, we could run the EnKS experiments many times, and the resulting estimates would be normally distributed with a standard deviation given by $\sigma/\sqrt{N}$. A quick estimate is computed by setting $\sigma \approx 2$ and $N = 1000$, and we get a standard deviation of 0.06 which seems to be consistent with the difference between the EnKS and representer solution in this case and the cases to follow.

From the ensemble standard deviation for the EnKS, there is clearly an impact backward in time from the measurements and the overall error estimate is smaller than for the EnKF. The minimum errors are found at the measurement locations as expected. After the final measurement update the EnKF and EnKS solutions are identical, thus, for forecasting purposes it suffices to compute the EnKF solution.

12.5.2 Case A1

This experiment is similar to Case A0 but we have now introduced time correlated model noise. The first impression from the lower plot in Fig. 12.3 is that all curves are smoother and less noisy in this case. The EnKS and representer solutions are now also smooth at the measurement locations as is expected when time correlated model noise is used.

An important difference between this and the previous case is that now the EnKF solution sometimes shows a positive or negative trend during the integration between the measurements. This is caused by the assimilation updates of the model noise which introduces a "bias" in the stochastic forcing. This is seen in upper plot of Fig. 12.4 which plots the EnKF solution as the blue line, the EnKF estimate for the model noise as the red line, and the

12.5 Examples   187

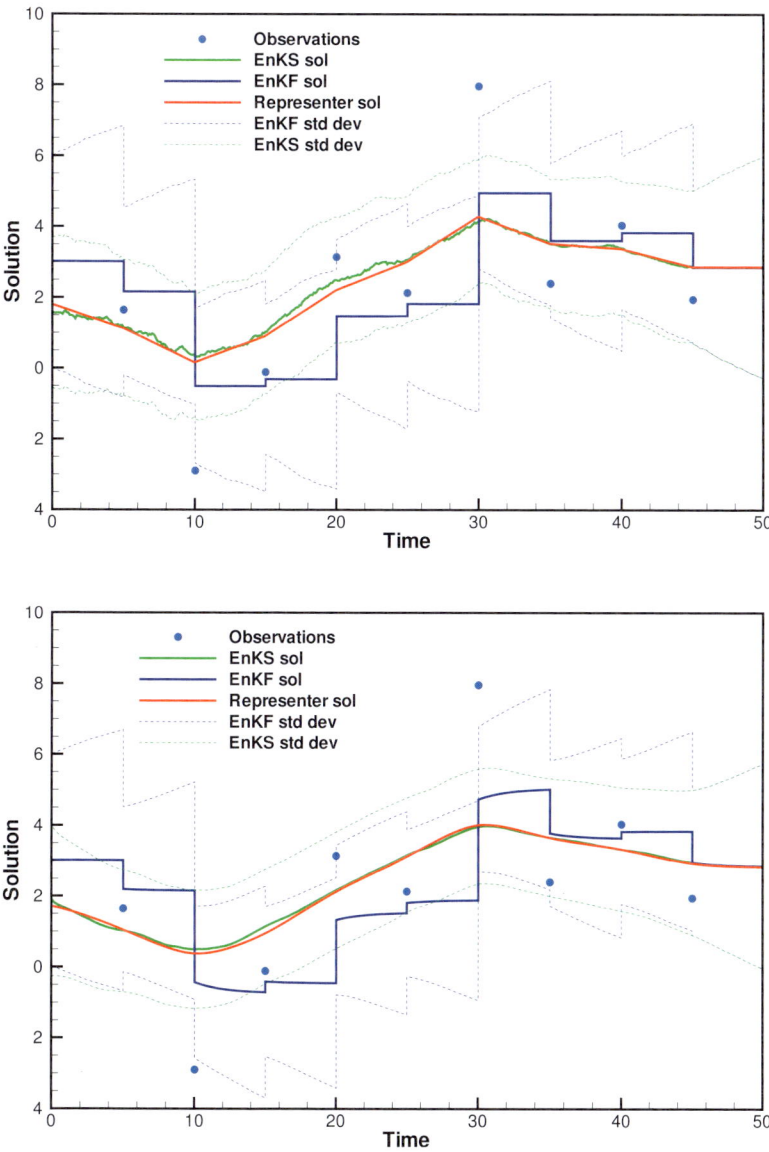

Fig. 12.3. Cases A0 and A1: Pure state estimation with unbiased model. The upper plot shows the results from Case A0 with white model noise while the lower plots shows the results from Case A1 where coloured model noise is used

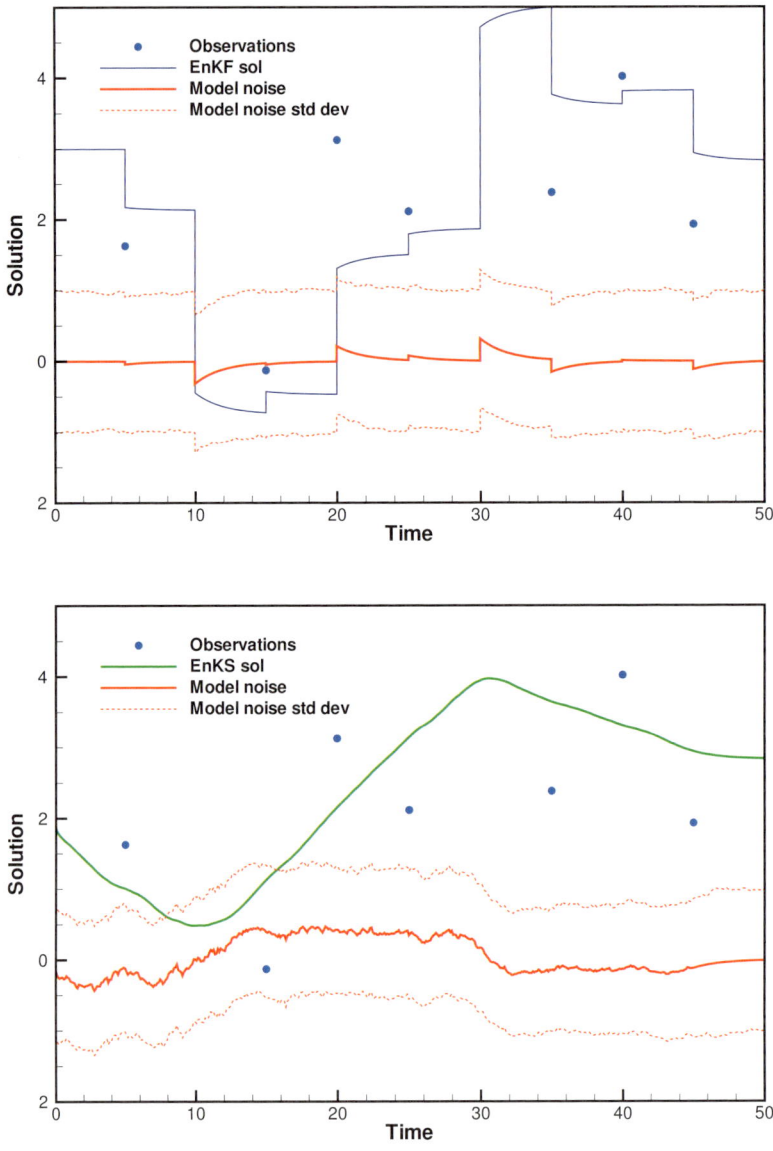

Fig. 12.4. Case A1: The system noise estimated by the EnKF (upper plot) and the EnKS (lower plot)

standard deviation of the model noise as the red dashed lines. It is clearly seen that the model noise is being updated at the assimilation steps, e.g. note the large updates at the second and sixth measurements. These updates introduce a bias in the system noise which helps relaxing the solution in the direction of the measurements. Thus, as we will see below, the model noise can help counteract a bias in model. Note that during the integration between the measurements the bias slowly relaxes back toward zero in agreement with the equation used for the simulation of the model noise.

The estimated EnKS system noise is presented as the red solid line in the lower plot of Fig. 12.4 and also here the time derivatives are continuous at the measurement locations. In fact, this estimated model noise is the forcing needed to reproduce the EnKS solution when a single model is integrated forward in time starting from the initial condition estimated by the EnKS; i.e. the solution of

$$\psi_k = \psi_{k-1} + \sqrt{\Delta t}\sigma c \hat{q}_k,$$
$$\psi_0 = \hat{\psi}_0, \qquad (12.55)$$

with $\hat{q}_k$ and $\hat{\psi}_0$ being the EnKS estimated model noise and initial condition respectively, will exactly reproduce the EnKS estimate. Obviously, the estimated model noise is the same as is computed and used in the forward Euler Lagrange equation in the representer method. This points to the similarity between the EnKS and the representer method, which for linear models will give identical results when the ensemble size becomes infinite.

12.5.3 Case B

We now consider a case where we have an erroneous value, $\alpha = 0$, and the model thus contains a bias, always predicting a line with slope equal to one, while the true solution should have zero slope. In this case we are not attempting to estimate the parameter, but rather trying to solve the inverse problem in the case when the model contains a bias. The results are presented in Fig. 12.5 for the cases with $\tau = 0$ and $\tau = 2$. Again it is seen that the EnKS and representer solutions are nearly identical and they provide a good estimate of the true solution over most of the time interval. There is an exception for the estimate near the beginning and end of the time interval where the bias in the model cannot be corrected for. It is clear that the EnKS provides a significantly better result than the EnKF for this particular case. This is partly related to the measurement frequency and the fact that the information from only past and present measurements is insufficient to properly constrain the evolution of the filter.

The reason that the EnKS and the representer methods provide good results for most of the time interval is that they are both finding good estimates of the model error, i.e. q_i from (12.54) for the EnKS, and $\lambda(t)$ for the representer method, which corrects for the bias. However, this correction is not

190   12 Model errors

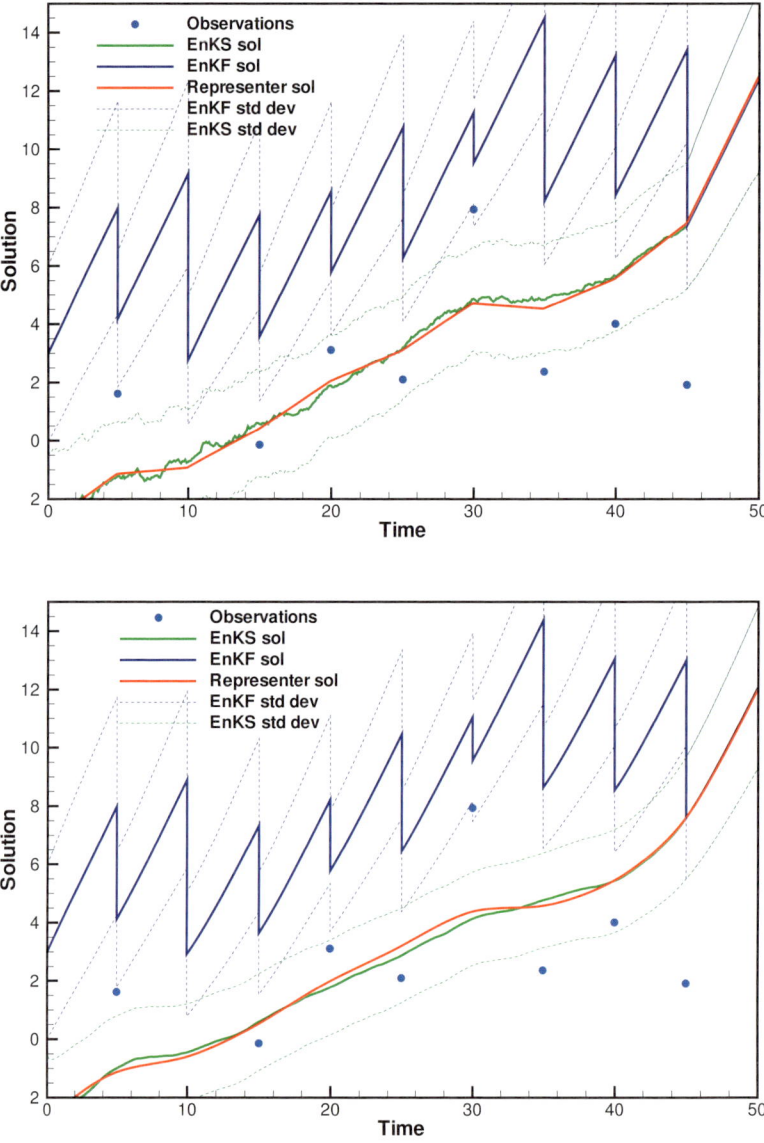

Fig. 12.5. Cases B0 and B1: Pure state estimation with biased model. The upper plot shows the results from Case B0 with white model noise while the lower plot shows the results from Case B1 where coloured model noise is used

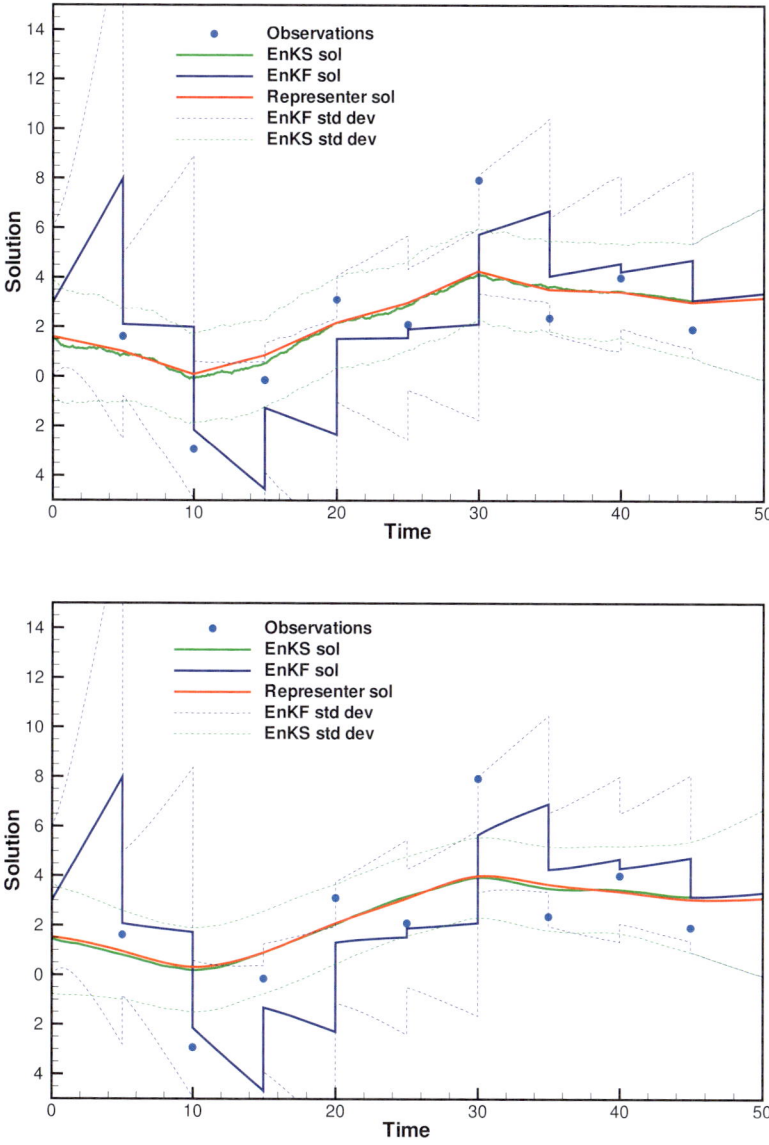

Fig. 12.6. Cases C0 and C1: Combined parameter and state estimation with biased model. The upper plot shows the results from Case C0 with white model noise while the lower plot shows the results from Case C1 where coloured model noise is used

192   12 Model errors

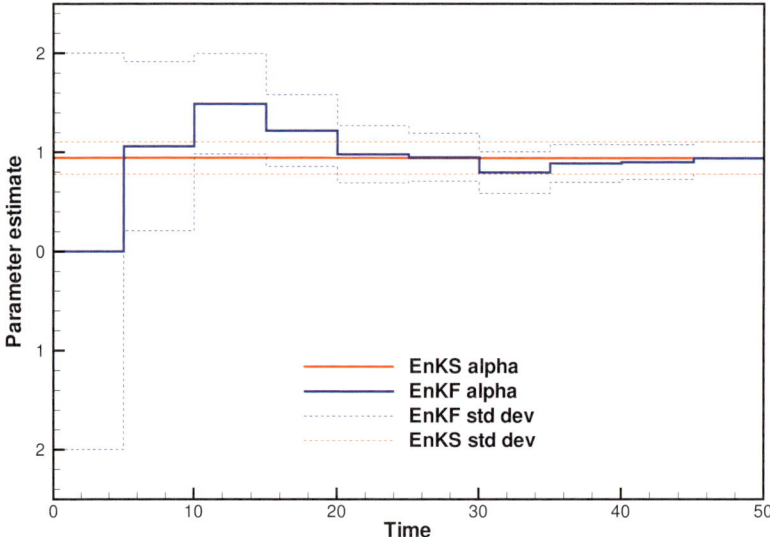

Fig. 12.7. Case C1. Convergence of parameter value over time using the EnKF and EnKS

maintained after the final measurement due to the limited time correlation specified.

12.5.4 Case C

In this case we also start out with an erroneous value, $\alpha = 0$, but assume that the parameter contains an error of variance equal to 4. The inverse solution is given in Fig. 12.6. It is clear that both the representer method and the EnKS provide realistic and very similar results. Further, the bias observed from the previous case is entirely eliminated since we have now also computed an estimate of α which in the representer method converged to 0.96 and in the EnKF and EnKS we obtained the value 0.94. Thus, as expected we obtained values in between the first guess of $\alpha = 0$ and the true value of $\alpha = 1$. We cannot expect to converge exactly to the true value since we have included a prior error statistics for the parameter. This prior is needed to ensure the existence of a solution independent of the number of measurements assimilated. We also observe that the EnKF solution initially shows a strong bias, but this is quickly reduced after a few updates with measurements.

In Fig. 12.7 we have plotted the estimated value for α as a function of time for the EnKF and EnKS. We have also included the one standard deviation of the errors in the plot. We started out with a value of α equal to zero and set the prior variance for the parameter equal to four. It is seen that the

EnKF provides an update of the parameter at each measurement time, and at the same time the estimated error variance is reduced. In this example the parameter estimate converges quickly, and the standard deviation of the error is reduced at each update with measurements. The final estimate for the error standard deviation of the parameter is 0.16 corresponding to an error variance of 0.026, so a significant improvement is obtained in the parameter estimate. Note also that the EnKS propagates information backward in time and thus provides a time independent estimate of α which is identical to the final estimate from the EnKF.

Using the representer method, the iterations on α in (12.34) converged quickly in around 10 iterations when an iteration step, $\gamma = 0.01$, was used, and we did not attempt to optimize or tune this value further.

12.5.5 Discussion

The conclusion from these experiments is that the EnKF, EnKS and representer method all provide the same solution of the inverse state and parameter estimation problem as long as the model is linear and the assumption of Gaussian priors is valid. It should be emphasized that this example has used a very simple linear model and we do expect that the associated inverse formulation is fairly well posed and easy to solve. Hence, for the representer method, the penalty function for each linear iterate is quadratic without local minima, and a unique solution is always obtained.

For the EnKF, we do not have to consider effects of non-Gaussian error statistics since the model is linear. Thus, we have considered a very simple problem where we would expect both the representer method and the EnKF/EnKS methods to work well.

It is also interesting to see that the case with no measurements are accounted for using both the representer and the ensemble methods. In the representer method the solution then becomes the first guess solution ψ_F. This corresponds to the mode, or modal trajectory, of the joint pdf defined by, e.g. (7.10), and the value of α becomes the prior value α_0.

The Ensemble methods provide a pure ensemble integration when no measurements are available. Clearly, we can store the ensemble at all times and compute the modal trajectory as well. However, we believe that the mode of the marginal pdf would be a better estimator. An argument for this is that a single model realization from a nonlinear model does not make any statistical sense. It is just one out of infinitively many possible realizations.

In the ensemble methods the mean is used as the best estimator. This is mostly a practical choice since the estimation of the mode will require the use of a much larger ensemble. Thus, the estimate from the EnKF and EnKS when no measurements are assimilated is just the evolution of ensemble mean. This corresponds to the mean of the marginal pdf which also happens to be equal to the mean of the joint pdf. Thus, in the ensemble methods the ensemble mean is the best estimate and it comes with an associated error covariance estimate.

13
Square Root Analysis schemes

It has been suggested that further improvement can be obtained in the computation of the analysis in the EnKF if the sampling errors introduced by the perturbation of measurements are eliminated. The works by *Anderson* (2001), *Whitaker and Hamill* (2002), *Bishop et al.* (2001), the review by *Tippett et al.* (2003), and the paper by *Evensen* (2004) have developed implementations of the analysis scheme where the perturbation of measurements is avoided. The new algorithms are named "square root" schemes and the algorithm from *Evensen* (2004) will be derived and explained below. The methods are intuitively very appealing but there are also some pitfalls as pointed out by *Lawson and Hansen* (2004) and further explained by *Leeuwenburgh et al.* (2006). A simple linear advection model will be used to demonstrate the impact of the different analysis schemes as well as the impact of using the improved sampling technique from the previous chapter when generating the initial ensemble and measurement perturbations.

13.1 Square root algorithm for the EnKF analysis

The square root schemes presented by *Anderson* (2001), *Whitaker and Hamill* (2002), *Bishop et al.* (2001) and in the review by *Tippett et al.* (2003), all introduced some kind of approximation to make them efficient, e.g. the use of a diagonal measurement error covariance matrix or knowledge of the inverse of the measurement error covariance matrix. Here the simpler and more direct variant of the square root analysis scheme, by *Evensen* (2004), is derived, which solves for the analysis without imposing any additional approximations.

The square root algorithm is used to update the ensemble perturbations and is derived starting from the traditional analysis equation for the covariance update in the Kalman Filter (9.6). The time index has in the remainder of this chapter been dropped for convenience. When using the ensemble representation for the error covariance matrix, $C_{\psi\psi}$, as defined in (9.14), (9.6) can be written

$$\boldsymbol{A}^{a\prime}\boldsymbol{A}^{a\prime\mathrm{T}} = \boldsymbol{A}' \left(\boldsymbol{I} - \boldsymbol{S}^{\mathrm{T}}\boldsymbol{C}^{-1}\boldsymbol{S}\right) \boldsymbol{A}'^{\mathrm{T}}, \qquad (13.1)$$

where we have used the definitions of $\boldsymbol{S}$ and $\boldsymbol{C}$ from (9.33) and (9.34), i.e. $\boldsymbol{S} = \mathcal{M}[\boldsymbol{A}']$ is the measurement of the ensemble perturbations and $\boldsymbol{C} = \boldsymbol{S}\boldsymbol{S}^{\mathrm{T}} + (N-1)\boldsymbol{C}_{\epsilon\epsilon}$, with $\boldsymbol{C}_{\epsilon\epsilon}$ being the measurement error covariance matrix. We have for simplicity dropped the 'f' superscript on $\boldsymbol{A}^{\mathrm{f}}$ and $\boldsymbol{A}^{\mathrm{f}\prime}$.

13.1.1 Updating the ensemble mean

The analyzed ensemble mean is computed from the standard Kalman filter analysis equation for the mean which may be obtained by multiplication of (9.39) from the right with $\boldsymbol{1}_N$, i.e. each column is the resulting equation for the mean and can be written

$$\overline{\boldsymbol{\psi}^{\mathrm{a}}(\boldsymbol{x})} = \overline{\boldsymbol{\psi}^{\mathrm{f}}(\boldsymbol{x})} + \boldsymbol{A}'\boldsymbol{S}^{\mathrm{T}}\boldsymbol{C}^{-1}\left(\boldsymbol{d} - \mathcal{M}\left[\overline{\boldsymbol{\psi}^{\mathrm{f}}(\boldsymbol{x})}\right]\right). \qquad (13.2)$$

The following derives an equation for the ensemble analysis by defining a factorization of (13.1) where there are no references to the measurements or measurement perturbations.

13.1.2 Updating the ensemble perturbations

We start by forming $\boldsymbol{C}$ as defined above. We will also for now assume that $\boldsymbol{C}$ is of full rank such that $\boldsymbol{C}^{-1}$ exists. In general this is satisfied if the ensemble is of full rank and the number of measurements is less than or equal to $N-1$. The low rank case with many measurements is discussed in the next chapter. Note also that the use of a full rank $\boldsymbol{C}_{\epsilon\epsilon}$ can result in a full rank $\boldsymbol{C}$ even when $m \geq N$.

We can then compute the eigenvalue decomposition, $\boldsymbol{Z}\boldsymbol{\Lambda}\boldsymbol{Z}^{\mathrm{T}} = \boldsymbol{C}$, and we get the inverse of $\boldsymbol{C}$ defined by (11.43). The eigenvalue decomposition may be the most demanding computation required in the analysis when m is large. However, a more efficient inversion algorithm is presented in the next chapter.

We now write (13.1) as follows:

$$\begin{aligned}\boldsymbol{A}^{a\prime}\boldsymbol{A}^{a\prime\mathrm{T}} &= \boldsymbol{A}'\left(\boldsymbol{I} - \boldsymbol{S}^{\mathrm{T}}\boldsymbol{Z}\boldsymbol{\Lambda}^{-1}\boldsymbol{Z}^{\mathrm{T}}\boldsymbol{S}\right)\boldsymbol{A}'^{\mathrm{T}} \\ &= \boldsymbol{A}'\left(\boldsymbol{I} - (\boldsymbol{\Lambda}^{-\frac{1}{2}}\boldsymbol{Z}^{\mathrm{T}}\boldsymbol{S})^{\mathrm{T}}(\boldsymbol{\Lambda}^{-\frac{1}{2}}\boldsymbol{Z}^{\mathrm{T}}\boldsymbol{S})\right)\boldsymbol{A}'^{\mathrm{T}} \qquad (13.3) \\ &= \boldsymbol{A}'\left(\boldsymbol{I} - \boldsymbol{X}_2^{\mathrm{T}}\boldsymbol{X}_2\right)\boldsymbol{A}'^{\mathrm{T}},\end{aligned}$$

where we have defined $\boldsymbol{X}_2 \in \Re^{m \times N}$ as

$$\boldsymbol{X}_2 = \boldsymbol{\Lambda}^{-\frac{1}{2}}\boldsymbol{Z}^{\mathrm{T}}\boldsymbol{S}, \qquad (13.4)$$

where $\mathrm{rank}(\boldsymbol{X}_2) = \min(m, N-1)$. Thus, $\boldsymbol{X}_2$ is a projection of $\boldsymbol{S}$ onto the eigenvectors of $\boldsymbol{C}$ scaled by the square root of the eigenvalues of $\boldsymbol{C}$.

13.1 Square root algorithm for the EnKF analysis

When computing the singular value decomposition of $\boldsymbol{X}_2$,

$$\boldsymbol{U}_2 \boldsymbol{\Sigma}_2 \boldsymbol{V}_2^\mathrm{T} = \boldsymbol{X}_2, \tag{13.5}$$

with $\boldsymbol{U}_2 \in \Re^{m \times m}$, $\boldsymbol{\Sigma}_2 \in \Re^{m \times N}$ and $\boldsymbol{V}_2 \in \Re^{N \times N}$, (13.3) can be written

$$\begin{aligned}
\boldsymbol{A}^{\mathrm{a}\prime} \boldsymbol{A}^{\mathrm{a}\prime \mathrm{T}} &= \boldsymbol{A}' \left(\boldsymbol{I} - [\boldsymbol{U}_2 \boldsymbol{\Sigma}_2 \boldsymbol{V}_2^\mathrm{T}]^\mathrm{T} [\boldsymbol{U}_2 \boldsymbol{\Sigma}_2 \boldsymbol{V}_2^\mathrm{T}] \right) \boldsymbol{A}'^\mathrm{T} \\
&= \boldsymbol{A}' \left(\boldsymbol{I} - \boldsymbol{V}_2 \boldsymbol{\Sigma}_2^\mathrm{T} \boldsymbol{\Sigma}_2 \boldsymbol{V}_2^\mathrm{T} \right) \boldsymbol{A}'^\mathrm{T} \\
&= \boldsymbol{A}' \boldsymbol{V}_2 \left(\boldsymbol{I} - \boldsymbol{\Sigma}_2^\mathrm{T} \boldsymbol{\Sigma}_2 \right) \boldsymbol{V}_2^\mathrm{T} \boldsymbol{A}'^\mathrm{T} \\
&= \left(\boldsymbol{A}' \boldsymbol{V}_2 \sqrt{\boldsymbol{I} - \boldsymbol{\Sigma}_2^\mathrm{T} \boldsymbol{\Sigma}_2} \right) \left(\boldsymbol{A}' \boldsymbol{V}_2 \sqrt{\boldsymbol{I} - \boldsymbol{\Sigma}_2^\mathrm{T} \boldsymbol{\Sigma}_2} \right)^\mathrm{T}.
\end{aligned} \tag{13.6}$$

Thus, a solution for the analysis ensemble perturbations is

$$\boldsymbol{A}^{\mathrm{a}\prime} = \boldsymbol{A}' \boldsymbol{V}_2 \sqrt{\boldsymbol{I} - \boldsymbol{\Sigma}_2^\mathrm{T} \boldsymbol{\Sigma}_2}. \tag{13.7}$$

This can be considered as a standard form of the square root analysis scheme, and it will produce an ensemble of perturbations with the correct covariance. The previous square root schemes reviewed by *Tippett et al.* (2003) are all on a form similar to (13.7).

13.1.3 Randomization of the analysis update

The update computed through (13.7) is purely deterministic, and the stochastic component introduced by the perturbation of measurements in the original EnKF analysis scheme is completely eliminated.

Analysis update using a single measurement

We will now look at the special case where a single measurement ($m = 1$) is used. In this case we will have that $\boldsymbol{Z} = 1$ and $\boldsymbol{\Lambda}$ becomes a scalar $\lambda = \boldsymbol{S} \boldsymbol{S}^\mathrm{T} + (N-1) \boldsymbol{C}_{\epsilon\epsilon}$. Thus, we get $\boldsymbol{X}_2 = \lambda^{-\frac{1}{2}} \boldsymbol{S}$.

The singular value decomposition (13.5) of $\boldsymbol{X}_2$ then equals $\lambda^{-\frac{1}{2}}$ times the singular value decomposition of $\boldsymbol{S}$,

$$\lambda^{-\frac{1}{2}} \boldsymbol{U} \boldsymbol{\Sigma} \boldsymbol{V}^\mathrm{T} = \lambda^{-\frac{1}{2}} \boldsymbol{S}. \tag{13.8}$$

Here we must have $\boldsymbol{U} = \boldsymbol{U}_2 = 1$ and $\boldsymbol{\Sigma} \in \Re^{1 \times N}$ has the value σ in the first location and zero in the remainder locations. Further, $\boldsymbol{V} \in \Re^{N \times N}$ will then have $\boldsymbol{S}/\sqrt{\boldsymbol{S}\boldsymbol{S}^\mathrm{T}}$ in the first column and vectors orthogonal to $\boldsymbol{S}$ in the other columns. Note that $\sigma = \sqrt{\boldsymbol{S}\boldsymbol{S}^\mathrm{T}}$. Thus, we can write the singular value decomposition (13.7) of $\boldsymbol{X}_2$ as

$$X_2 = \Sigma_2 V_2^T, \tag{13.9}$$

where we have

$$\Sigma_2 = (\lambda^{-\frac{1}{2}}\sigma, 0, \ldots, 0), \tag{13.10}$$

and $V_2 = V$.

The final analysis equation (13.7) then gives the following at the measurement location

$$S^a = SV_2\sqrt{I - \Sigma_2^T\Sigma_2}. \tag{13.11}$$

Here we note that the matrix $\sqrt{I - \Sigma_2^T\Sigma_2}$, is diagonal with ones on the diagonal except for the first element which is $\sqrt{1 - \sigma^2/\lambda}$. Further, since all columns of V_2, except the first, are orthogonal to S, all elements of S^a, except the first, will be zero.

The first element of S^a then becomes equal to $\sigma\sqrt{1 - \sigma^2/\lambda}$ while the rest are zero. Thus, the first element contains all of the variance of the analysis at the measurement location.

We also note that $\lambda = \sigma^2 + (N-1)C_{\epsilon\epsilon}$, thus the variance at the measurement location becomes

$$\frac{S^a S^{aT}}{N-1} = \frac{\sigma^2}{N-1}\left(1 - \frac{\sigma^2/(N-1)}{\sigma^2/(N-1) + C_{\epsilon\epsilon}}\right), \tag{13.12}$$

which is identical to (3.15).

For state spaces where $n > 1$ the rank of the ensemble is reduced to one at the measurement locations, while the rows of A' corresponding to other grid points will generally not be parallel to S and the rank will be maintained. Note, however, that imposed spatial correlations will lead to poor conditioning of the ensemble at grid points close to the measurement location.

Analysis update using a diagonal $C_{\epsilon\epsilon}$

With more than one measurement the situation changes slightly but the same problem will occur if $C_{\epsilon\epsilon}$ is diagonal. We now consider the case with $1 < m < N$. Then the eigenvectors Z, will be identical to the singular vectors U, of $S = U\Sigma V^T$, and we can write

$$X_2 = \Lambda^{-\frac{1}{2}}Z^T S = \Lambda^{-\frac{1}{2}}\Sigma V^T. \tag{13.13}$$

Thus, the singular value decomposition of X_2 becomes again (13.5) but with

$$\Sigma_2 = \Lambda^{-\frac{1}{2}}\Sigma, \tag{13.14}$$

containing m nonzero elements on the diagonal, $V_2 = V$ and $U_2 = I$.

Then each of the m columns in S^T will be contained in the space defined by the first m columns of V_2. Thus, in the update, the first m ensemble

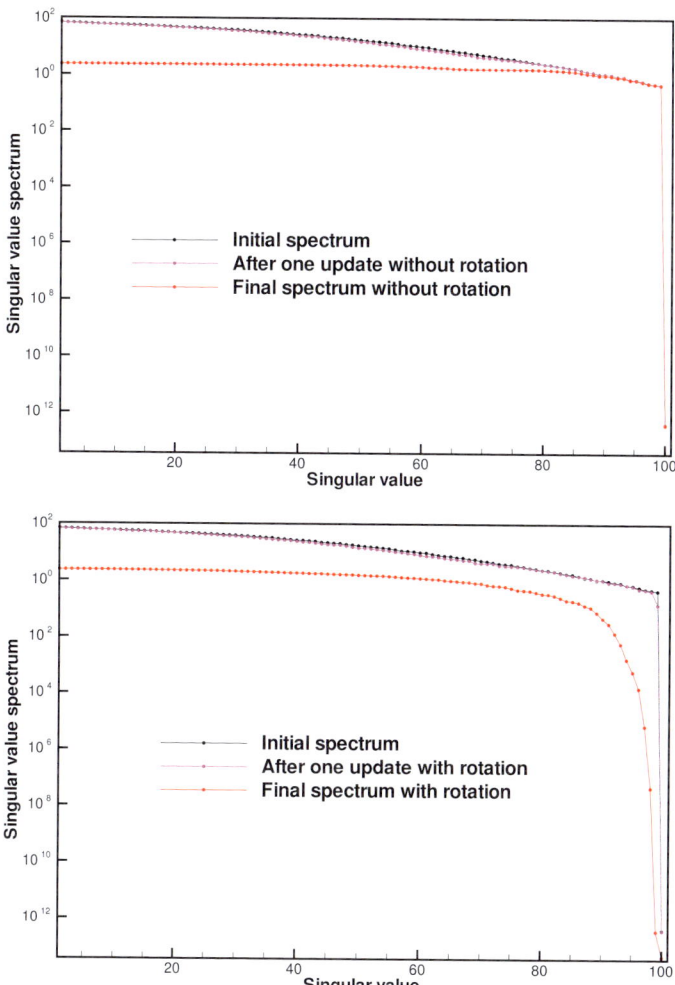

Fig. 13.1. Singular value spectra resulting from the square root scheme without *(upper)* and with *(lower)* a random rotation

perturbations will represent the analysis variance while the remainder will be zero.

Thus, it is clear that the square root analysis scheme (13.7) results in an updated ensemble where the ensemble variance is reduced in directions defined by the rotation $\boldsymbol{V}_2$. In cases when $\boldsymbol{S}^{\mathrm{T}}$ is fully represented by a selection of singular vectors, as is the case when a single measurement is used and if m measurements are used with a diagonal $\boldsymbol{C}_{\epsilon\epsilon}$, then the ensemble variance at the measurement locations is represented by the first m ensemble members.

Introduction of a random rotation

The problem discussed above can be avoided by the insertion of a random orthogonal matrix product $\boldsymbol{I} = \boldsymbol{\Theta}^\mathrm{T}\boldsymbol{\Theta}$ in (13.6), i.e.

$$\boldsymbol{A}^{\mathrm{a}\prime}\boldsymbol{A}^{\mathrm{a}\prime\mathrm{T}} = \boldsymbol{A}'\boldsymbol{Z}\sqrt{\boldsymbol{I} - \boldsymbol{\Lambda}}\boldsymbol{\Theta}^\mathrm{T}\boldsymbol{\Theta}\left(\boldsymbol{A}'\boldsymbol{Z}\sqrt{\boldsymbol{I}-\boldsymbol{\Lambda}}\right)^\mathrm{T}, \qquad (13.15)$$

which leads to the following update equation

$$\boldsymbol{A}^{\mathrm{a}\prime} = \boldsymbol{A}'\boldsymbol{V}_2\sqrt{\boldsymbol{I} - \boldsymbol{\Sigma}_2^\mathrm{T}\boldsymbol{\Sigma}_2}\boldsymbol{\Theta}^\mathrm{T}. \qquad (13.16)$$

This is equivalent to a random rotation which has the effect of randomly distributing the variance reduction among all the ensemble members, and according to (13.15) this also results in a valid solution.

The matrix $\boldsymbol{\Theta}^\mathrm{T}$ is easily constructed, e.g. by using the result from a QR decomposition of an $N \times N$ matrix of independent random normal distributed numbers.

It is unfortunate that the randomization leads to a reduction of the conditioning of the ensemble perturbations. This is seen from Fig. 13.1 where the plots show the ensemble perturbation spectrum at the initial time, after the first update and after 60 updates for the advection problem from Sect. 4.1.3. The upper plot is with no rotation, and the lower plot is for the case where the random rotation is included. It is seen that after one update including the rotation there is a reduction of the singular value $N - 1$. The cause for this is that the rank of the updated ensemble perturbations, before rotation, is $N - 1$ and the rotation includes N columns of $\boldsymbol{A}'$, thus the variance reduction is also distributed along the dependent direction in $\boldsymbol{A}'$. During a longer time integration with many updates this leads to a decreasing tail of the spectrum that is even more pronounced than for the original EnKF analysis scheme as was used in Sect. 11.5. Note that the spectra in the upper plot are identical to the results from *Evensen* (2004) which did not include the random rotation in the corresponding experiments.

13.1.4 Final update equation in the square root algorithms

In Chap. 9 it was shown that the EnKF analysis update can be written as

$$\boldsymbol{A}^\mathrm{a} = \boldsymbol{A}\boldsymbol{X}, \qquad (13.17)$$

where $\boldsymbol{X}$ is an $N \times N$ matrix of coefficients. The square root schemes can also be written in the same simple form. This is illustrated by writing the analysis as the updated ensemble mean plus the updated ensemble perturbations,

$$\boldsymbol{A}^\mathrm{a} = \overline{\boldsymbol{A}}^\mathrm{a} + \boldsymbol{A}^{\mathrm{a}\prime}. \qquad (13.18)$$

The updated mean can, using (13.2), be written as

Experiment	N	β_{ini}	Residual	Std. dev.
F	100	1	0.731	0.075
G	100	6	0.656	0.066

Table 13.1. Summary of experiments. The first column is the experiment name, in the second column N is the ensemble size used, β_{ini} is a number which defines the size, $\beta_{\text{ini}} N$, of the start ensemble when using the improved sampling scheme from section 11.4 for generating the initial ensemble. The two last columns contain the average RMS errors of the 50 simulations in each experiment and the standard deviation of these

$$\begin{aligned}\overline{\boldsymbol{A}^{\text{a}}} &= \overline{\boldsymbol{A}} + \boldsymbol{A}' \boldsymbol{S}^{\text{T}} \boldsymbol{C}^{-1} \left(\overline{\boldsymbol{D}} - \mathcal{M}\left[\overline{\boldsymbol{A}}\right]\right) \\ &= \boldsymbol{A} \boldsymbol{1}_N + \boldsymbol{A}(\boldsymbol{I} - \boldsymbol{1}_N) \boldsymbol{S}^{\text{T}} \boldsymbol{C}^{-1} \left(\boldsymbol{D} - \mathcal{M}[\boldsymbol{A}]\right) \boldsymbol{1}_N \\ &= \boldsymbol{A} \boldsymbol{1}_N + \boldsymbol{A} \boldsymbol{S}^{\text{T}} \boldsymbol{C}^{-1} \left(\boldsymbol{D} - \mathcal{M}[\boldsymbol{A}]\right) \boldsymbol{1}_N,\end{aligned} \qquad (13.19)$$

and the updated perturbations are, from (13.16),

$$\begin{aligned}\boldsymbol{A}^{\text{a}\prime} &= \boldsymbol{A}' \boldsymbol{V}_2 \sqrt{\boldsymbol{I} - \boldsymbol{\Sigma}_2^{\text{T}} \boldsymbol{\Sigma}_2} \boldsymbol{\Theta}^{\text{T}} \\ &= \boldsymbol{A}(\boldsymbol{I} - \boldsymbol{1}_N) \boldsymbol{V}_2 \sqrt{\boldsymbol{I} - \boldsymbol{\Sigma}_2^{\text{T}} \boldsymbol{\Sigma}_2} \boldsymbol{\Theta}^{\text{T}}.\end{aligned} \qquad (13.20)$$

Combining the previous equations we get (13.17) with $\boldsymbol{X}$ defined as

$$\boldsymbol{X} = \boldsymbol{1}_N + \boldsymbol{S}^{\text{T}} \boldsymbol{C}^{-1} \left(\boldsymbol{D} - \mathcal{M}[\boldsymbol{A}]\right) \boldsymbol{1}_N + (\boldsymbol{I} - \boldsymbol{1}_N) \boldsymbol{V}_2 \sqrt{\boldsymbol{I} - \boldsymbol{\Sigma}_2^{\text{T}} \boldsymbol{\Sigma}_2} \boldsymbol{\Theta}^{\text{T}}. \qquad (13.21)$$

Thus, we still search for the solution as a combination of ensemble members, and it also turns out that forming $\boldsymbol{X}$ and then computing the matrix multiplication in (13.17) is the most efficient algorithm for computing the analysis when many measurements are used. Note that we already have $\boldsymbol{C}^{-1}$ from (11.43).

13.2 Experiments

The impact of using the square root analysis scheme from the previous section will now be examined in some detail using the model and configuration from Sect. 11.5.

13.2.1 Overview of experiments

Several experiments have been carried out as listed in Table 13.1. For each of the experiments, 50 EnKF simulations were performed to allow for a statistical comparison. In each simulation, the only difference is the random seed used. Thus, every simulation will have a different and random true state, first guess, initial ensemble and set of measurements.

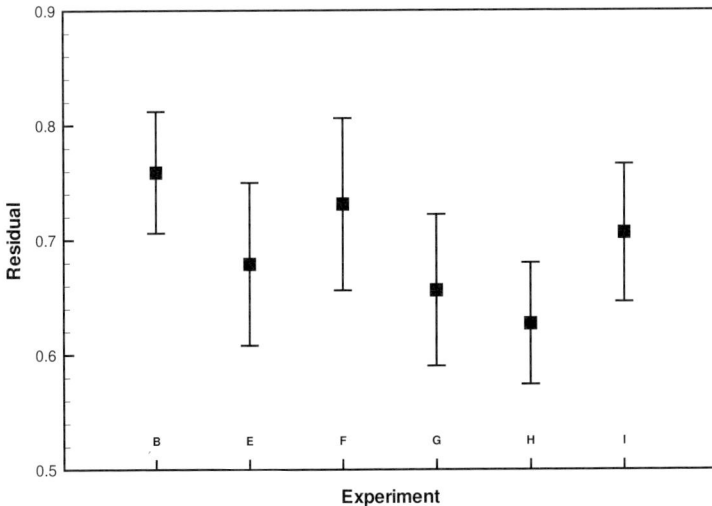

Fig. 13.2. Mean and standard deviation of the residuals from each of the experiments. Included are also *Exps. B,E,H* and *I* from Sect. 11.5

As in Sect. 11.5 a full rank matrix $C = SS^{\mathrm{T}} + (N-1)C_{\epsilon\epsilon}$ is specified and then factorized by computing the eigenvalue decomposition. The analysis is then computed from the square root implementation of the analysis scheme using (13.21) with (11.43) in (13.17), i.e.

$$A^{\mathrm{a}} = A\Big(\mathbf{1}_N + S^{\mathrm{T}} Z \Lambda^{-1} Z^{\mathrm{T}} (D - \mathcal{M}[A]) \mathbf{1}_N \\ + (I - \mathbf{1}_N) V_2 \sqrt{I - \Sigma_2^{\mathrm{T}} \Sigma_2} \Theta^{\mathrm{T}}\Big). \tag{13.22}$$

In all the experiments the residuals were computed as the Root Mean Square (RMS) errors of the difference between the estimate and the true solution taken over the complete space and time domain. For each of the experiments we have plotted the mean and standard deviation of the residuals in Fig. 13.2.

The Table 13.2 gives the probabilities that the average residuals from the experiments are equal, as computed from the Student's t-test. Probabilities lower than, say 0.1, indicate statistically that the distributions from two experiments are significantly different. When selecting a method or approach, one should use the one which has a distribution with the lowest average residual, and possibly also a low variance of the residuals.

It is also of interest to examine how well the predicted errors represent the actual residuals (RMS as a function of time). In the summary Figs. 13.3 we have plotted the average of the predicted errors from the 50 simulations

Exp	B	B150	B200	B250	C	D	E	F	G	H	I
A	0.86	0	0	0	0	0	0	0.05	0	0	0
B		0	0	0	0	0	0	0.03	0	0	0
B150			0.01	0	0.01	0.86	0.86	0	0.04	0	0.03
B200				0.04	0	0.01	0.04	0	0.73	0.04	0
B250					0	0	0	0	0.02	0.91	0
C						0.01	0.01	0.25	0	0	0.48
D							0.75	0	0.03	0	0.06
E								0	0.09	0	0.04
F									0	0	0.07
G										0.02	0
H											0

Table 13.2. Statistical probability that two experiments provide an equal mean for the residuals as computed using the Student's t-test. A probability close to one indicates that it is likely that the two experiments provide distributions of residuals with similar mean

as the thick full line. The thin full lines indicate the one standard deviation spread of the predicted errors from the 50 simulations. The average of the RMS errors from the 50 simulations is plotted as the thick dotted line, with associated one standard deviation spread shown by the dotted thin lines.

The further details of the different experiments are as follows:

Exp. F is an experiment where a standard Monte Carlo ensemble is used for generating the 100 member initial ensemble without improved sampling. It is thus similar to and can be compared with *Exp. B* from Sect. 11.5.

Exp. G is similar to *Exp. F* except that the initial ensemble is sampled from a start ensemble of 600 members as in *Exp. E* from Sect. 11.5. It examines the benefit of combined use of improved initial sampling and the square root algorithm.

13.2.2 Impact of the square root analysis algorithm

Let us use the *Exps. B, E, H* and *I*, from Sect. 11.5, as the original EnKF reference cases. The *Exp. B* did not use improved sampling, *Exp. E* used improved sampling for the initial ensemble, *Exp. I* used improved sampling for the measurement perturbations, and *Exp. H* used improved sampling both for the initial ensemble and the measurement perturbations.

Then we run two experiments, *Exp. F* and *Exp. G*, using the square root algorithm. *Exp. F* uses standard sampling of 100 members as in *Exp. B* while *Exp. G* uses a 600 member start ensemble as in *Exp. E*.

Referring again to the residuals plotted in Fig. 13.2, it is clear that the *Exp. F* provides a marginal improvement compared to *Exp. B* but *Exp. I* is still slightly better. Similarly, *Exp. G* provides a marginal improvement compared

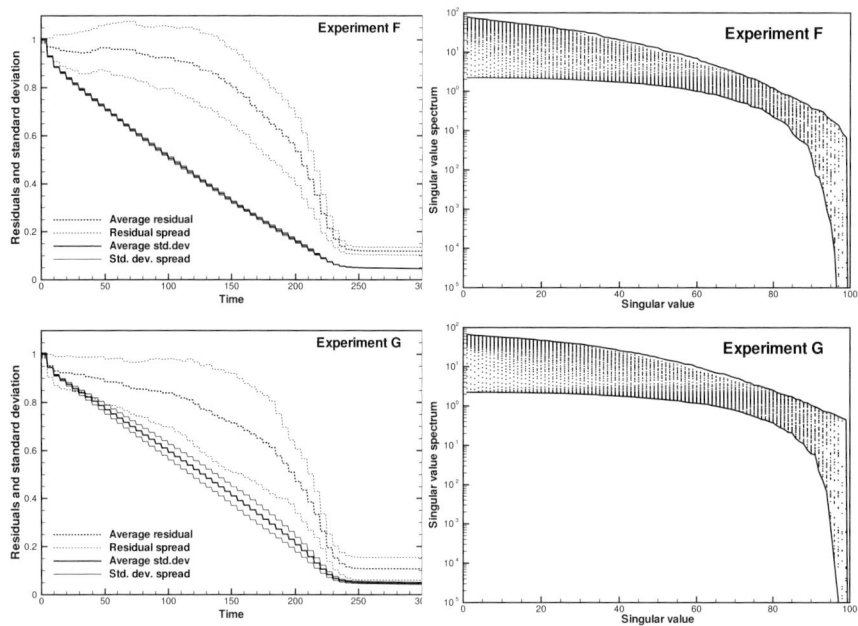

Fig. 13.3. Left column shows the time evolution for RMS residuals *(dashed lines)* and estimated standard deviations *(full lines)*. The thick lines show the means over the 50 simulations and the thin lines show the means plus/minus one standard deviation. The right column shows the time evolution of the ensemble singular value spectra for some of the experiments

to *Exp. E*, but *Exp. H* is still a little better. Thus, the standard EnKF scheme with improved sampling for the measurement perturbations still provides the results with the lowest residuals. This result is contrary to what was found in *Evensen* (2004) where the square root algorithm gave the best results. The reason for this is that the random rotation was not used in the *Exps. F* and *G* in *Evensen* (2004), and as we found above, the random rotation also leads to some rank loss. The random rotation may not be needed in all linear systems, but it is included here to ensure that we obtain a randomized ensemble which properly represents the error statistics.

The time evolutions of the residuals in *Exps. F* and *G*, plotted in Fig. 13.3, show fairly good consistency with the estimated standard deviations, and are located somewhere between what was found in *Exps. B* and *I*, and *Exps. E* and *H* respectively.

Finally, it is of interest to examine how the rank and conditioning of the ensemble evolves in time and is impacted by the computation of the analysis. In the right column of Fig. 13.3 we have plotted the singular values for the ensemble at each analysis time. The initial singular spectrum of the ensemble is plotted as the upper thick line. Then the dotted lines indicate the reduction of the ensemble variance introduced at each analysis update, until the end of the experiment where the singular spectrum is given by the lower thick line.

The initial spectra from *Exps. F* and *G* are similar to the ones from *Exps. B* and *E*, but now the reduction of variance is more confined to the dominant singular values. There is also an additional strong reduction for the least significant singular values indicating a rank loss in the ensemble. It was explained above that this is caused by the introduction of the random rotation. At the final time the singular spectrum is almost flat up to a certain singular value, which is in contrast to the singular spectra obtained when measurements were perturbed. This shows that the square root scheme weights the singular vectors more equally.

14
Rank issues

It has been stated in the previous chapters that the EnKF analysis scheme may have some problems in cases where the number of measurements is larger than the number of members in the ensemble or when the matrix C for some reason has poor conditioning. In this chapter we will discuss these difficulties and propose algorithms that still makes it possible to use the EnKF analysis schemes in these cases. Thus, we provide an extended discussion of the rank problem as was introduced in *Evensen* (2004).

14.1 Pseudo inverse of C

The matrix C which must be inverted in the analysis schemes was in (9.26) defined as
$$C = SS^{\mathrm{T}} + (N-1)C_{\epsilon\epsilon}. \tag{14.1}$$
As in the previous chapters we have defined $S = \mathcal{M}[A^{\mathrm{f}\prime}]$ as the measurements of the ensemble perturbations, and $C_{\epsilon\epsilon}$ is the measurement error covariance matrix.

The analysis scheme for the EnKF was given in (11.44) as
$$A^{\mathrm{a}} = A^{\mathrm{f}}\left(I + S^{\mathrm{T}}C^{-1}(D - \mathcal{M}[A^{\mathrm{f}}])\right), \tag{14.2}$$
with D being the ensemble of perturbed measurements.

In the square root scheme we compute the update of the mean which is derived from (14.2) by multiplication from the right by $\mathbf{1}_N$, where $\mathbf{1}_N$ is an N-dimensional quadratic matrix with all elements equal to $1/N$. Thus, we get the update for the mean (13.2), written as
$$\overline{A}^{\mathrm{a}} = A^{\mathrm{f}}\left(\mathbf{1}_N + S^{\mathrm{T}}C^{-1}(\overline{D} - \mathcal{M}[\overline{A}^{\mathrm{f}}])\right). \tag{14.3}$$

The perturbations are updated according to (13.16), i.e.

$$\boldsymbol{A}^{\mathrm{a}\prime} = \boldsymbol{A}^{\mathrm{f}\prime}\boldsymbol{V}_2\sqrt{\boldsymbol{I} - \boldsymbol{\Sigma}_2^{\mathrm{T}}\boldsymbol{\Sigma}_2}\boldsymbol{\Theta}^{\mathrm{T}}, \tag{14.4}$$

which is derived from a factorization of (13.1), i.e.

$$\boldsymbol{A}^{\mathrm{a}\prime}\boldsymbol{A}^{\mathrm{a}/\mathrm{T}} = \boldsymbol{A}^{\mathrm{f}\prime}\left(\boldsymbol{I} - \boldsymbol{S}^{\mathrm{T}}\boldsymbol{C}^{-1}\boldsymbol{S}\right)\boldsymbol{A}^{\mathrm{f}/\mathrm{T}}. \tag{14.5}$$

Equations (14.3) and (14.4) can be combined into one single equation, similar to (13.22), as

$$\begin{aligned}\boldsymbol{A}^{\mathrm{a}} = \boldsymbol{A}^{\mathrm{f}}&\bigg(\mathbf{1}_N + \boldsymbol{S}^{\mathrm{T}}\boldsymbol{C}^{-1}(\boldsymbol{D} - \mathcal{M}[\boldsymbol{A}^{\mathrm{f}}])\mathbf{1}_N \\ &+ (\boldsymbol{I} - \mathbf{1}_N)\boldsymbol{V}_2\sqrt{\boldsymbol{I} - \boldsymbol{\Sigma}_2^{\mathrm{T}}\boldsymbol{\Sigma}_2}\boldsymbol{\Theta}^{\mathrm{T}}\bigg).\end{aligned} \tag{14.6}$$

For the definition of the various matrices we refer to Chap. 13 where the square root scheme was derived.

It is seen that in both the EnKF and the square root algorithm we need to compute the inverse of $\boldsymbol{C}$. In the previous discussion this was done using an eigenvalue factorization. When the dimension of $\boldsymbol{C}$ is large, or if nearly dependent measurements are assimilated, it is possible that $\boldsymbol{C}$ becomes numerically singular. When $\boldsymbol{C}$ is singular it is possible to compute the pseudo inverse $\boldsymbol{C}^+$ of $\boldsymbol{C}$. It is convenient to formulate the analysis schemes in terms of the pseudo inverse, since we have $\boldsymbol{C}^+ \equiv \boldsymbol{C}^{-1}$, when $\boldsymbol{C}$ is of full rank. The algorithm will then be valid in the general case.

14.1.1 Pseudo inverse

The pseudo inverse of the quadratic matrix $\boldsymbol{C}$ with eigenvalue factorization

$$\boldsymbol{C} = \boldsymbol{Z}\boldsymbol{\Lambda}\boldsymbol{Z}^{\mathrm{T}}, \tag{14.7}$$

is defined as

$$\boldsymbol{C}^+ = \boldsymbol{Z}\boldsymbol{\Lambda}^+\boldsymbol{Z}^{\mathrm{T}}. \tag{14.8}$$

The matrix $\boldsymbol{\Lambda}^+$ is diagonal and with $p = \mathrm{rank}(\boldsymbol{C})$ it is defined as

$$\mathrm{diag}(\boldsymbol{\Lambda}^+) = (\lambda_1^{-1}, \ldots, \lambda_p^{-1}, 0, \ldots, 0), \tag{14.9}$$

with the eigenvalues $\lambda_i \geq \lambda_{i+1}$.

The pseudo inverse has the following properties

$$\boldsymbol{C}\boldsymbol{C}^+\boldsymbol{C} = \boldsymbol{C}, \qquad \boldsymbol{C}^+\boldsymbol{C}\boldsymbol{C}^+ = \boldsymbol{C}^+, \tag{14.10}$$

$$(\boldsymbol{C}^+\boldsymbol{C})^{\mathrm{T}} = \boldsymbol{C}^+\boldsymbol{C}, \qquad (\boldsymbol{C}\boldsymbol{C}^+)^{\mathrm{T}} = \boldsymbol{C}\boldsymbol{C}^+. \tag{14.11}$$

Furthermore,

$$\boldsymbol{x} = \boldsymbol{C}^+\boldsymbol{b}, \tag{14.12}$$

is the least squares solution of the problem

$$\boldsymbol{C}\boldsymbol{x} = \boldsymbol{b}, \tag{14.13}$$

when $\boldsymbol{C}$ is singular.

14.1.2 Interpretation

It is useful to attempt an interpretation of the algorithm when using the pseudo inverse for C. We start by storing the p nonzero elements of $\text{diag}(\Lambda^+)$ on the diagonal of $\Lambda_p^{-1} \in \Re^{p \times p}$, i.e.

$$\text{diag}(\Lambda_p^{-1}) = (\lambda_1^{-1}, \ldots, \lambda_p^{-1}). \tag{14.14}$$

We then define the matrix containing the first p eigenvectors in Z as $Z_p = (z_1, \ldots, z_p) \in \Re^{m \times p}$. It is clear that the product $Z_p \Lambda_p^{-1} Z_p^\mathrm{T}$ is the Moore-Penrose or pseudo inverse of the original matrix C.

We now define the projected measurement operator $\widetilde{\mathcal{M}} \in \Re^{p \times n}$ as

$$\widetilde{\mathcal{M}} = Z_p^\mathrm{T} \mathcal{M}, \tag{14.15}$$

the ensemble of p projected measurements

$$\widetilde{D} = Z_p^\mathrm{T} D, \tag{14.16}$$

and the p projected measurements of the ensemble perturbations $\widetilde{S} \in \Re^{p \times N}$, as

$$\widetilde{S} = Z_p^\mathrm{T} \mathcal{M}[A'] = \widetilde{\mathcal{M}}[A'] = Z_p^\mathrm{T} S. \tag{14.17}$$

This corresponds to the use of a measurement antenna which is oriented along the p dominant principal directions of C (see *Bennett*, 1992, Chap. 6). The analysis equation in the original EnKF analysis scheme then becomes

$$A^\mathrm{a} = A^\mathrm{f} \left(I + \widetilde{S}^\mathrm{T} \Lambda_p^{-1} (\widetilde{D} - \widetilde{\mathcal{M}}[A^\mathrm{f}]) \right). \tag{14.18}$$

Thus, the analysis is just the assimilation of the p rotated and projected measurements in the space where $\widetilde{C} = \Lambda_p$ is diagonal.

14.1.3 Analysis schemes using the pseudo inverse of C

The modification required for the EnKF and square root analysis schemes to use the pseudo inverse of C is minor. The same equations and derivation are used, it is only necessary to perform a truncation of the spectrum at the desired variance level, i.e. one need to decide how many eigenvalues to include and set the remainder to zero. Then Λ^+ is defined and used instead of Λ^{-1}.

14.1.4 Example

The advection example from Sects. 11.5 and 13.2 is now used to illustrate the importance of being able to handle a rank deficient C. We first construct a case where we have five measurements located at neighbouring grid points. The measurement error covariance matrix is also nondiagonal, and it

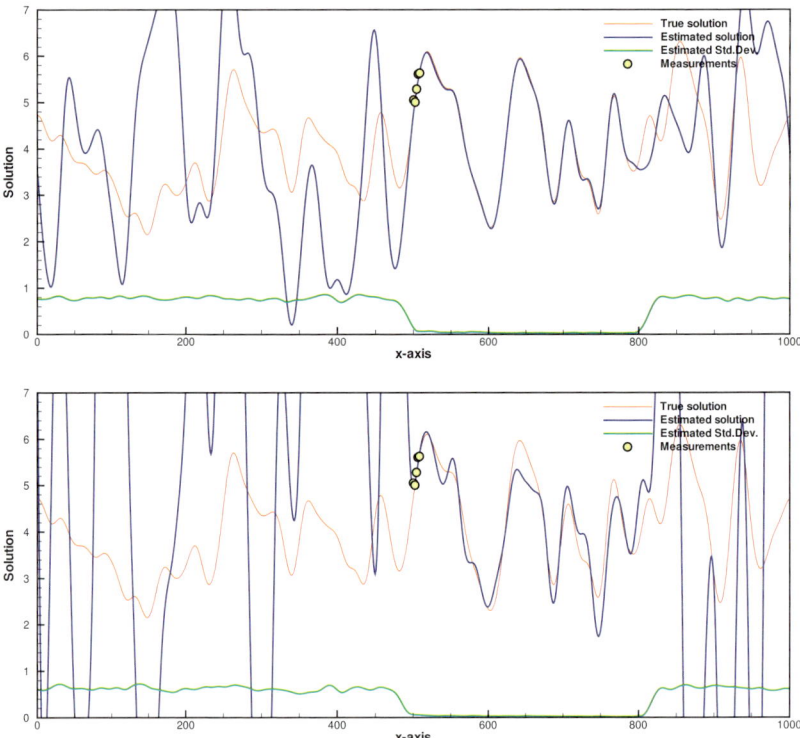

Fig. 14.1. Solution at the final time using the traditional EnKF analysis scheme with pseudo inversion of C. The upper plot is the solution with truncation at 90% of the variance of the eigenvalue spectrum, while the lower plot is with truncation at 99.9% of the variance

is assumed that the measurement errors are correlated with a Gaussian covariance function of de-correlation length equal to 20. This leads to a matrix C with a ratio of the largest over smallest eigenvalue of order 10^5. Thus, the conditioning of C is rather poor and the use of a pseudo inversion may be advantageous.

We now run two experiments similar to *Exp. E* from Sect. 11.5, and *Exp. G* from Sect. 13.2, and plot the solution at the final time $t = 300$, for different truncations of the eigenvalue spectrum. The results are plotted in respectively Figs. 14.1 and 14.2 for the traditional EnKF and the square root analysis algorithms. It is seen that the inversion, using a truncation of the eigenvalue spectrum where 90% of the variance, corresponding to a single eigenvalue, is retained, leads to stable solutions. On the other hand, when the truncation is accounting for 99.9% of the variance, which retains four eigenvalues, both the traditional EnKF and square root scheme result in unstable inversions.

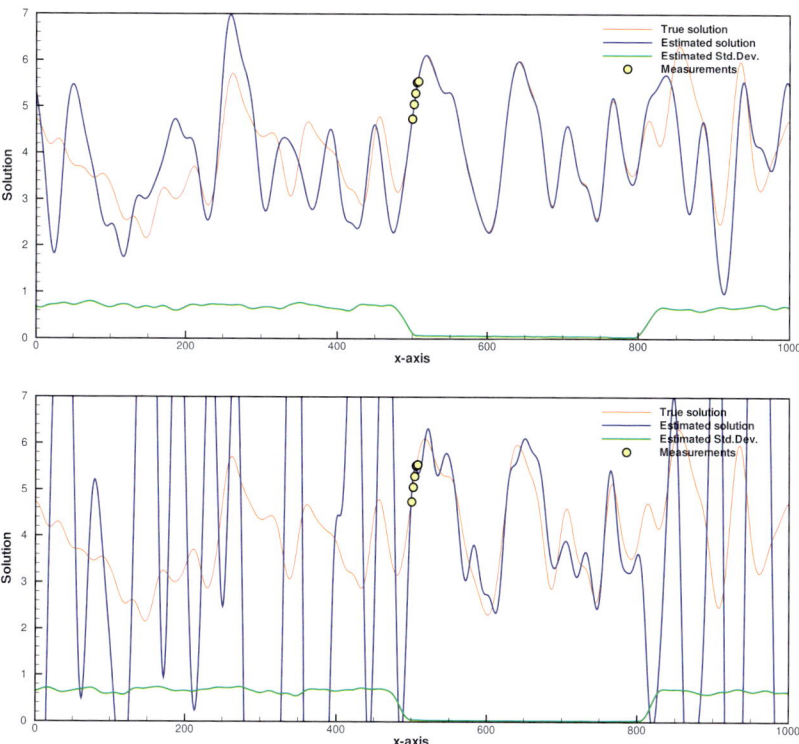

Fig. 14.2. Solution at the final time using the square root analysis scheme with pseudo inversion of C. The upper plot is the solution with truncation at 90% of the variance of the eigenvalue spectrum, while the lower plot is with truncation at 99.9% of the variance

We now increase the number of measurements to 200, and use the same Gaussian error covariance matrix for the measurement errors. The results at $t = 25$, after 5 updates with measurements, are plotted in Fig. 14.3 for the traditional EnKF analysis and the square root analysis. In this case around 40 significant eigenvalues were included when a truncation at 99% of the variance was specified. It is clear that both schemes produce a stable inversion which is consistent with the measurements. For this case we also plotted the eigenvalue spectrum of C at each of the updates in Fig. 14.4. It is seen that there are around 40–50 significant eigenvalues for all the updates, and there is a reduction of the variance for all of the significant eigenvalues, corresponding to the reduction of ensemble variance at the measurement locations.

Thus, it is clear that both the EnKF and the square root scheme can handle cases with dependent measurements and a larger number of measurements than ensemble members. Note that we may expect problems if the number of significant eigenvalues becomes larger than the number of ensemble members.

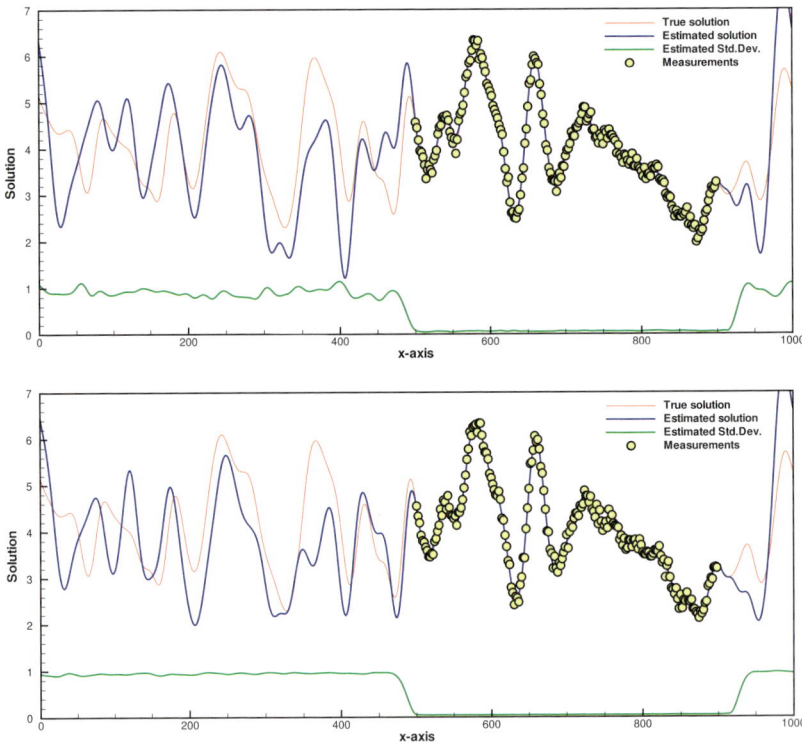

Fig. 14.3. Solution at time $t = 25$ using the EnKF scheme in the upper plot and square root scheme in the lower plot, and with a truncation accounting for 99% of the variance of the eigenvalue spectrum

14.2 Efficient subspace pseudo inversion

In cases with many measurements the computational cost becomes large since Nm^2 operations are required to form the matrix C and the eigenvalue decomposition requires $\mathcal{O}(m^3)$ operations. An alternative inversion algorithm which reduces the factorization of the $m \times m$ matrix to a factorization of an $N \times N$ matrix is now presented. The algorithm computes the inverse in the N-dimensional ensemble space rather than the m-dimensional measurement space.

14.2.1 Derivation of the subspace pseudo inverse

We start by assuming that S has rank $p \leq \min(m, N-1)$. The equality can be satisfied when the ensemble consists of linearly independent members and the measurement operator has full rank, i.e. the measurements are independent. The SVD of S is

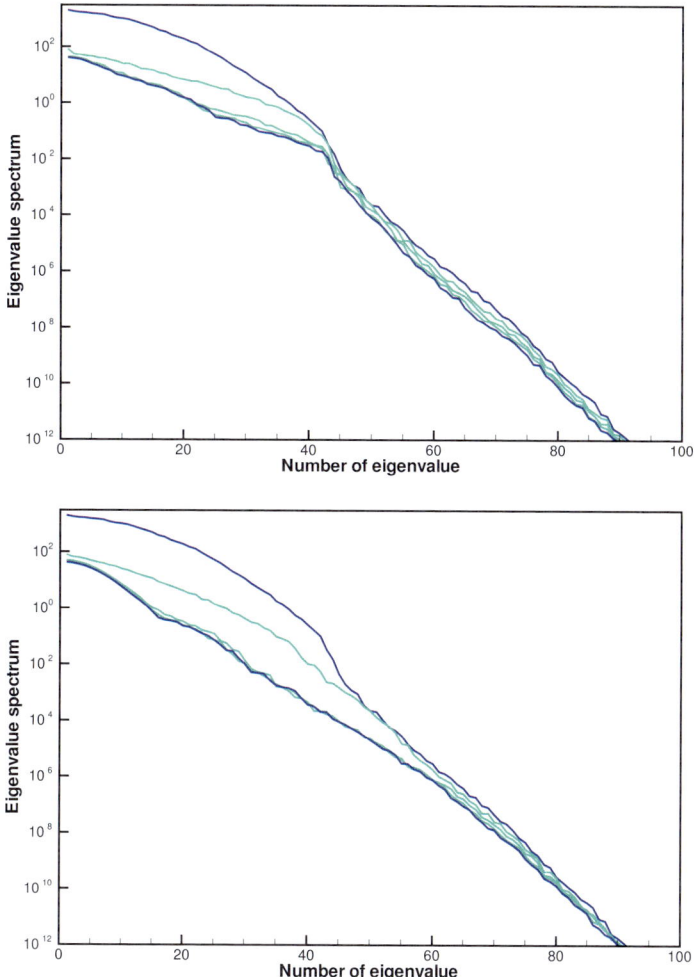

Fig. 14.4. Eigenvalue spectrum of C in the cases shown in Fig. 14.3. Results from the EnKF and square root schemes are shown in the left and right plot respectively

$$U_0 \Sigma_0 V_0^T = S, \quad (14.19)$$

with $U_0 \in \Re^{m \times m}$, $\Sigma_0 \in \Re^{m \times N}$ and $V_0 \in \Re^{N \times N}$. The subspace $\mathcal{S}$ is now defined by the first p singular vectors of S as contained in U_0.

The pseudo inverse of S is defined as

$$S^+ = V_0 \Sigma_0^+ U_0^T, \quad (14.20)$$

where $\Sigma_0^+ \in \Re^{N \times m}$ is a diagonal matrix with elements defined as $\mathrm{diag}(\Sigma_0^+) = (\sigma_1^{-1}, \sigma_2^{-1}, \ldots, \sigma_p^{-1}, \ldots, 0)$. Thus, by computing the pseudo inversion in (14.20)

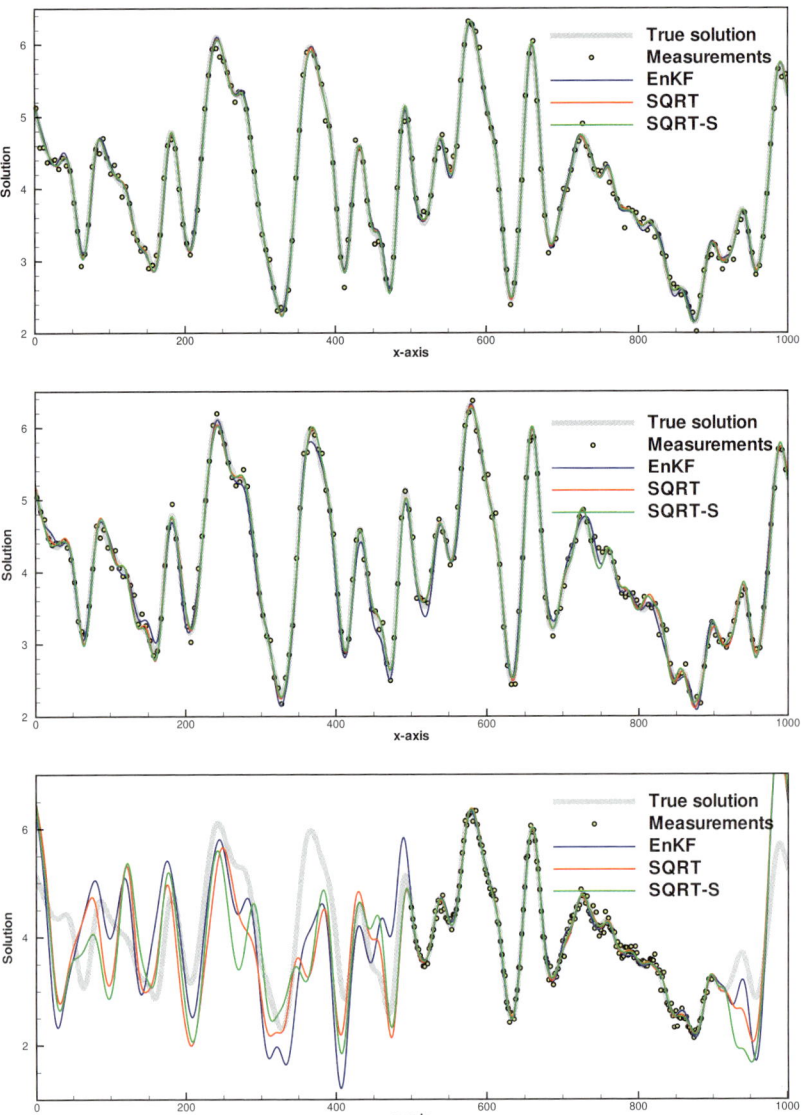

Fig. 14.5. Solution after 5 updates using the traditional EnKF, the square root scheme, and the new square root scheme with subspace projection of $\boldsymbol{C}_{\epsilon\epsilon}$ and pseudo inversion of $\boldsymbol{C}$. The upper plot shows the solution with uniform distribution of 200 measurements and a diagonal $\boldsymbol{C}_{\epsilon\epsilon}$. The middle plot is similar to the upper one but with a nondiagonal $\boldsymbol{C}_{\epsilon\epsilon}$. The lower plot has clustered the measurements and also uses a nondiagonal $\boldsymbol{C}_{\epsilon\epsilon}$

14.2 Efficient subspace pseudo inversion

it is also possible to use the algorithm with the number of measurements being less than $N-1$ and also with dependent measurements or dependent ensemble members.

The matrix product $\boldsymbol{\Sigma}_0 \boldsymbol{\Sigma}_0^+ = \tilde{\boldsymbol{I}}_p \in \Re^{m \times m}$ where $\tilde{\boldsymbol{I}}_p$ has the first p diagonal elements equal to one and the remainder of the elements in the matrix are zero.

We now use this in the expression for $\boldsymbol{C}$, as defined in (14.1), to get

$$\boldsymbol{C} = \left(\boldsymbol{U}_0 \boldsymbol{\Sigma}_0 \boldsymbol{\Sigma}_0^{\mathrm{T}} \boldsymbol{U}_0^{\mathrm{T}} + (N-1)\boldsymbol{C}_{\epsilon\epsilon}\right) \tag{14.21}$$

$$= \boldsymbol{U}_0 \left(\boldsymbol{\Sigma}_0 \boldsymbol{\Sigma}_0^{\mathrm{T}} + (N-1)\boldsymbol{U}_0^{\mathrm{T}} \boldsymbol{C}_{\epsilon\epsilon} \boldsymbol{U}_0\right) \boldsymbol{U}_0^{\mathrm{T}} \tag{14.22}$$

$$\approx \boldsymbol{U}_0 \boldsymbol{\Sigma}_0 \left(\boldsymbol{I} + (N-1)\boldsymbol{\Sigma}_0^+ \boldsymbol{U}_0^{\mathrm{T}} \boldsymbol{C}_{\epsilon\epsilon} \boldsymbol{U}_0 \boldsymbol{\Sigma}_0^{+\mathrm{T}}\right) \boldsymbol{\Sigma}_0^{\mathrm{T}} \boldsymbol{U}_0^{\mathrm{T}} \tag{14.23}$$

$$= \boldsymbol{S}\boldsymbol{S}^{\mathrm{T}} + (N-1)(\boldsymbol{S}\boldsymbol{S}^+)\boldsymbol{C}_{\epsilon\epsilon}(\boldsymbol{S}\boldsymbol{S}^+)^{\mathrm{T}}. \tag{14.24}$$

In (14.22) the matrix $\boldsymbol{U}_0^{\mathrm{T}} \boldsymbol{C}_{\epsilon\epsilon} \boldsymbol{U}_0$ is the projection of the measurement error covariance matrix $\boldsymbol{C}_{\epsilon\epsilon}$ onto the space spanned by the m singular vectors of $\boldsymbol{S}$, contained in the columns of $\boldsymbol{U}_0$.

Then in (14.23) we introduce an approximation by effectively multiplying $\boldsymbol{U}_0^{\mathrm{T}} \boldsymbol{C}_{\epsilon\epsilon} \boldsymbol{U}_0$ from left and right by the matrix $\boldsymbol{\Sigma}_0 \boldsymbol{\Sigma}_0^+ = \tilde{\boldsymbol{I}}_p \in \Re^{m \times m}$. Thus, we extract the part of $\boldsymbol{C}_{\epsilon\epsilon}$ contained in the subspace consisting of the p dominant directions in $\boldsymbol{U}_0$, i.e. the subspace $\mathcal{S}$.

The matrix $\boldsymbol{S}\boldsymbol{S}^+ = \boldsymbol{U}_0 \tilde{\boldsymbol{I}}_p \boldsymbol{U}_0^{\mathrm{T}}$ in (14.24) is a Hermitian and normal matrix. It is also an orthogonal projection onto $\mathcal{S}$. Thus, we essentially adopt a low-rank representation for $\boldsymbol{C}_{\epsilon\epsilon}$ which is contained in the same subspace as the ensemble perturbations in $\boldsymbol{S}$.

We use the expression for $\boldsymbol{C}$ as given in (14.23), i.e.

$$\boldsymbol{C} \approx \boldsymbol{U}_0 \boldsymbol{\Sigma}_0 (\boldsymbol{I} + \boldsymbol{X}_0) \boldsymbol{\Sigma}_0^{\mathrm{T}} \boldsymbol{U}_0^{\mathrm{T}}, \tag{14.25}$$

where we have defined

$$\boldsymbol{X}_0 = (N-1)\boldsymbol{\Sigma}_0^+ \boldsymbol{U}_0^{\mathrm{T}} \boldsymbol{C}_{\epsilon\epsilon} \boldsymbol{U}_0 \boldsymbol{\Sigma}_0^{+\mathrm{T}}, \tag{14.26}$$

which is an $N \times N$ matrix with rank equal to p and it requires $m^2 N + mN^2 + mN$ floating point operations to form it. We then proceed with an eigenvalue decomposition

$$\boldsymbol{Z}_1 \boldsymbol{\Lambda}_1 \boldsymbol{Z}_1^{\mathrm{T}} = \boldsymbol{X}_0, \tag{14.27}$$

where all matrices are $N \times N$, and insert this in (14.25) to get

$$\begin{aligned} \boldsymbol{C} &\approx \boldsymbol{U}_0 \boldsymbol{\Sigma}_0 (\boldsymbol{I} + \boldsymbol{Z}_1 \boldsymbol{\Lambda}_1 \boldsymbol{Z}_1^{\mathrm{T}}) \boldsymbol{\Sigma}_0^{\mathrm{T}} \boldsymbol{U}_0^{\mathrm{T}} \\ &= \boldsymbol{U}_0 \boldsymbol{\Sigma}_0 \boldsymbol{Z}_1 (\boldsymbol{I} + \boldsymbol{\Lambda}_1) \boldsymbol{Z}_1^{\mathrm{T}} \boldsymbol{\Sigma}_0^{\mathrm{T}} \boldsymbol{U}_0^{\mathrm{T}}. \end{aligned} \tag{14.28}$$

Now the pseudo inverse of $\boldsymbol{C}$ becomes

$$\begin{aligned} \boldsymbol{C}^+ &\approx (\boldsymbol{U}_0 \boldsymbol{\Sigma}_0^{+\mathrm{T}} \boldsymbol{Z}_1)(\boldsymbol{I} + \boldsymbol{\Lambda}_1)^{-1}(\boldsymbol{U}_0 \boldsymbol{\Sigma}_0^{+\mathrm{T}} \boldsymbol{Z}_1)^{\mathrm{T}} \\ &= \boldsymbol{X}_1 (\boldsymbol{I} + \boldsymbol{\Lambda}_1)^{-1} \boldsymbol{X}_1^{\mathrm{T}}, \end{aligned} \tag{14.29}$$

where we have defined $X_1 \in \Re^{m \times N}$ of rank $N-1$ as

$$X_1 = U_0 \Sigma_0^{+\mathrm{T}} Z_1. \tag{14.30}$$

14.2.2 Analysis schemes based on the subspace pseudo inverse

By replacing C^{-1} in (14.2) with the pseudo inverse C^+, from (14.29), we can easily compute the EnKF analysis using the subspace pseudo inversion by carrying out the matrix multiplications in

$$A^{\mathrm{a}} = A^{\mathrm{f}} \left(I + S^{\mathrm{T}} X_1 (I + \Lambda_1)^{-1} X_1^{\mathrm{T}} (D - \mathcal{M}[A^{\mathrm{f}}]) \right). \tag{14.31}$$

Similarly the square root algorithm uses (14.3) with C^{-1} replaced by C^+, from (14.29),

$$\overline{A}^{\mathrm{a}} = A^{\mathrm{f}} \left(1_N + S^{\mathrm{T}} X_1 (I + \Lambda_1)^{-1} X_1^{\mathrm{T}} (\overline{D} - \mathcal{M}[\overline{A}^{\mathrm{f}}]) \right), \tag{14.32}$$

to compute the updated ensemble mean.

Using the expression (14.5) together with the pseudo inverse from (14.29) we can derive the update equation for the analysis perturbations in the square root scheme

$$\begin{aligned} A^{\mathrm{a}\prime} A^{\mathrm{a}\prime\mathrm{T}} &= A^{\mathrm{f}\prime} \left(I - S^{\mathrm{T}} C^+ S \right) A^{\mathrm{f}\prime\mathrm{T}} \\ &= A^{\mathrm{f}\prime} \left(I - S^{\mathrm{T}} X_1 (I + \Lambda_1)^{-1} X_1^{\mathrm{T}} S \right) A^{\mathrm{f}\prime\mathrm{T}} \\ &= A^{\mathrm{f}\prime} \left(I - \left((I + \Lambda_1)^{-\frac{1}{2}} X_1^{\mathrm{T}} S \right)^{\mathrm{T}} \left((I + \Lambda_1)^{-\frac{1}{2}} X_1^{\mathrm{T}} S \right) \right) A^{\mathrm{f}\prime\mathrm{T}} \\ &= A^{\mathrm{f}\prime} \left(I - X_2^{\mathrm{T}} X_2 \right) A^{\mathrm{f}\prime\mathrm{T}}, \end{aligned} \tag{14.33}$$

where we have defined X_2 as

$$X_2 = (I + \Lambda_1)^{-\frac{1}{2}} X_1^{\mathrm{T}} S = (I + \Lambda_1)^{-\frac{1}{2}} Z_1^{\mathrm{T}} \tilde{I}_p V_0^{\mathrm{T}}, \tag{14.34}$$

which also has rank equal to p. We then end up with the final update equation (14.4) by following the derivation defined in (13.5–13.6).

Equations (14.32) and (14.4) can be combined into one single equation, similar to (14.6), as

$$\begin{aligned} A^{\mathrm{a}} = A^{\mathrm{f}} \Big(& 1_N + S^{\mathrm{T}} X_1 (I + \Lambda_1)^{-1} X_1^{\mathrm{T}} (D - \mathcal{M}[A^{\mathrm{f}}]) 1_N \\ & + (I - 1_N) V_2 \sqrt{I - \Sigma_2^{\mathrm{T}} \Sigma_2} \Theta^{\mathrm{T}} \Big). \end{aligned} \tag{14.35}$$

It is clear that, for $m > p$, this subspace algorithm will be an approximation except for some special cases. First, if $C_{\epsilon\epsilon}$ is diagonal, then the matrix

SS^T and C will have the same eigenvectors as defined by U_0, thus there is no approximation involved. On the other hand if $C_{\epsilon\epsilon}$ is nondiagonal the eigenvectors will differ and the projection onto the $\mathcal{S}$-space eliminates the part of C which is orthogonal to the $\mathcal{S}$-space. Fortunately, in many applications this is a modest approximation.

Interestingly, the update of the perturbations in the square root algorithm does not suffer from this approximation since C^{-1} is already projected onto the $\mathcal{S}$-space through the matrix product $S^T C^{-1} S$ in (14.5).

14.2.3 An interpretation of the subspace pseudo inversion

A simple interpretation of this subspace pseudo inversion for the case when $m \gg N$ was given by *Skjervheim et al.* (2006). We start by computing the singular value factorization of S as in (14.19), and realize that Σ_0 is diagonal and only the first $p \leq N-1$ singular values are larger than zero, i.e. the rank of S equals p. We then write the EnKF analysis scheme (14.2), with $D' = (D - \mathcal{M}[A^f])$, as

$$A^a = A^f \left(I + S^T \left(SS^T + (N-1)C_{\epsilon\epsilon} \right)^{-1} D' \right) \tag{14.36}$$

$$= A^f \left(I + S^T \left(U_0 \Sigma_0 \Sigma_0^T U_0^T + (N-1)C_{\epsilon\epsilon} \right)^{-1} D' \right) \tag{14.37}$$

$$= A^f \left(I + S^T \left(U_0 (\Sigma_0 \Sigma_0^T + (N-1) U_0^T C_{\epsilon\epsilon} U_0) U_0^T \right)^{-1} D' \right) \tag{14.38}$$

$$= A^f \left(I + S^T U_0 \left(\Sigma_0 \Sigma_0^T + (N-1) U_0^T C_{\epsilon\epsilon} U_0 \right)^{-1} U_0^T D' \right) \tag{14.39}$$

$$= A^f \left(I + \widehat{S}^T \left(\Sigma_0 \Sigma_0^T + (N-1) U_0^T C_{\epsilon\epsilon} U_0 \right)^{-1} \widehat{D}' \right). \tag{14.40}$$

Here we have defined the rotated operators

$$\widehat{D}' = U_0^T D', \tag{14.41}$$

$$\widehat{\mathcal{M}} = U_0^T \mathcal{M}, \tag{14.42}$$

$$\widehat{S} = U_0^T S = \widehat{\mathcal{M}} A^{f'}. \tag{14.43}$$

It is clear that the original assimilation of m measurements in (14.36) is equivalent to the assimilation of m rotated measurements in (14.40), where the rotation is defined such that the matrix product $\widehat{S}\widehat{S}^T$ becomes diagonal.

We now take this one step further and define the projection operator $U_{0p} = SS^+$ which consists of the first p columns of U. We can then define the projections

$$\widehat{\boldsymbol{D}}'_p = \boldsymbol{U}_{0p}^{\mathrm{T}} \boldsymbol{D}', \qquad (14.44)$$

$$\widehat{\boldsymbol{\mathcal{M}}}_p = \boldsymbol{U}_{0p}^{\mathrm{T}} \boldsymbol{\mathcal{M}}, \qquad (14.45)$$

$$\widehat{\boldsymbol{S}}_p = \boldsymbol{U}_{0p}^{\mathrm{T}} \boldsymbol{S} = \widehat{\boldsymbol{\mathcal{M}}}_p \boldsymbol{A}^{\mathrm{f'}}, \qquad (14.46)$$

all of dimension $\Re^{p \times N}$, and in addition we define $\boldsymbol{\Sigma}_{0p} \in \Re^{p \times p}$ to hold the p significant singular values on the diagonal. We can then write an approximate EnKF analysis equation as

$$\boldsymbol{A}^{\mathrm{a}} = \boldsymbol{A}^{\mathrm{f}} \left(\boldsymbol{I} + \widehat{\boldsymbol{S}}_p^{\mathrm{T}} \left(\boldsymbol{\Sigma}_{0p} \boldsymbol{\Sigma}_{0p}^{\mathrm{T}} + (N-1) \boldsymbol{U}_{0p}^{\mathrm{T}} \boldsymbol{C}_{\epsilon\epsilon} \boldsymbol{U}_{0p} \right)^{-1} \widehat{\boldsymbol{D}}'_p \right). \qquad (14.47)$$

It is left as an exercise to show that this equation is identical to (14.31). Thus, we can interpret the subspace EnKF analysis scheme as the assimilation of a set of measurements after they have been projected onto the subspace $\mathcal{S}$ as defined by the first p singular vectors of $\boldsymbol{S}$. This projection allows us to assimilate very large data sets to a low cost in a stable algorithm. However, one can imagine cases where the subspace $\mathcal{S}$ is to small to properly represent the data. This can be resolved by either using a larger ensemble size or one may use a local analysis update as will be discussed in the Appendix.

14.3 Subspace inversion using a low-rank $C_{\epsilon\epsilon}$

With large data sets one will have to generate and store the measurement error covariance matrix, $\boldsymbol{C}_{\epsilon\epsilon} \in \Re^{m \times m}$, and multiply it with the singular vectors in $\boldsymbol{U}_0$ at the cost of Nm^2 floating point operations. In the EnKF we have simulated measurement perturbations which reflect the error statistics of the measurement errors. It is clear that given the measurement perturbations we can use these to represent a low-rank approximation of the measurement error covariance matrix.

14.3.1 Derivation of the pseudo inverse

We now replace $\boldsymbol{C}_{\epsilon\epsilon}$ with a low-rank version $\boldsymbol{C}_{\epsilon\epsilon}^{\mathrm{e}} = \boldsymbol{E}\boldsymbol{E}^{\mathrm{T}}/(N-1)$, in (14.24) to get

$$\begin{aligned} \boldsymbol{C} &= \boldsymbol{S}\boldsymbol{S}^{\mathrm{T}} + \boldsymbol{E}\boldsymbol{E}^{\mathrm{T}} \\ &\approx \boldsymbol{S}\boldsymbol{S}^{\mathrm{T}} + (\boldsymbol{S}\boldsymbol{S}^{+})\boldsymbol{E}\boldsymbol{E}^{\mathrm{T}}(\boldsymbol{S}\boldsymbol{S}^{+}) \\ &= \boldsymbol{S}\boldsymbol{S}^{\mathrm{T}} + \widehat{\boldsymbol{E}}\widehat{\boldsymbol{E}}^{\mathrm{T}}, \end{aligned} \qquad (14.48)$$

where $\widehat{\boldsymbol{E}} = (\boldsymbol{S}\boldsymbol{S}^{+})\boldsymbol{E}$ is the projection of $\boldsymbol{E}$ onto the first p singular vectors in $\boldsymbol{U}_0$, with p still being the rank of $\boldsymbol{S}$. When we project $\boldsymbol{E}$ onto $\mathcal{S}$ we reject all possible contributions in $\mathcal{S}^{\perp}$, and we can only account for the measurement variance contained in $\mathcal{S}$.

Replacing $C_{\epsilon\epsilon}$ with $EE^T/(N-1)$ in (14.23) we get

$$C \approx U_0 \Sigma_0 (I + \Sigma_0^+ U_0^T E E^T U_0 \Sigma_0^{+T}) \Sigma_0^T U_0^T \qquad (14.49)$$
$$= U_0 \Sigma_0 (I + X_0 X_0^T) \Sigma_0^T U_0^T, \qquad (14.50)$$

where we have defined

$$X_0 = \Sigma_0^+ U_0^T E, \qquad (14.51)$$

which is an $N \times N$ matrix with rank equal to $N-1$ and it requires $mN^2 + N^2$ floating point operations to form it. The approximate equality sign introduced in (14.49) just denotes that all components in E contained in $\mathcal{S}^\perp$ have now been removed.

We then proceed with a singular value decomposition

$$U_1 \Sigma_1 V_1^T = X_0, \qquad (14.52)$$

where all matrices are $N \times N$, and insert this in (14.50) to get

$$C \approx U_0 \Sigma_0 (I + U_1 \Sigma_1^2 U_1^T) \Sigma_0^T U_0^T$$
$$= U_0 \Sigma_0 U_1 (I + \Sigma_1^2) U_1^T \Sigma_0^T U_0^T. \qquad (14.53)$$

Now the pseudo inverse of C becomes

$$C^+ \approx (U_0 \Sigma_0^{+T} U_1)(I + \Sigma_1^2)^{-1} (U_0 \Sigma_0^{+T} U_1)^T$$
$$= X_1 (I + \Sigma_1^2)^{-1} X_1^T, \qquad (14.54)$$

where we have defined $X_1 \in \Re^{m \times N}$ of rank $N-1$ as

$$X_1 = U_0 \Sigma_0^{+T} U_1. \qquad (14.55)$$

14.3.2 Analysis schemes using a low-rank $C_{\epsilon\epsilon}$

By replacing C^{-1} in (14.2) with the pseudo inverse C^+, from (14.54), we can easily compute the EnKF analysis using the subspace pseudo inversion by carrying out the matrix multiplications in

$$A^a = A^f \left(I + S^T X_1 (I + \Sigma_1^2)^{-1} X_1^T (D - \mathcal{M}[A^f]) \right). \qquad (14.56)$$

Similarly the square root algorithm uses (14.3) with C^{-1} replaced by C^+, from (14.54),

$$\overline{A}^a = A^f \left(1_N + S^T X_1 (I + \Sigma_1^2)^{-1} X_1^T (\overline{D} - \mathcal{M}[\overline{A}^f]) \right), \qquad (14.57)$$

to compute the updated ensemble mean.

Using the expression (14.54) for the inverse in (14.5) we get the following derivation of the perturbation updates in the square root analysis scheme,

$$\begin{aligned}
A^{a\prime} A^{a\prime T} &= A' \left(I - S^T C^+ S \right) A'^T \\
&= A' \left(I - S^T X_1 (I + \Sigma_1^2)^{-1} X_1^T S \right) A'^T \\
&= A' \Bigg(I - \left((I + \Sigma_1^2)^{-\tfrac{1}{2}} X_1^T S \right)^T \\
&\qquad \left((I + \Sigma_1^2)^{-\tfrac{1}{2}} X_1^T S \right) \Bigg) A'^T \\
&= A' \left(I - X_2^T X_2 \right) A'^T,
\end{aligned} \qquad (14.58)$$

where we have defined X_2 as

$$X_2 = (I + \Sigma_1^2)^{-\tfrac{1}{2}} X_1^T S = (I + \Sigma_1^2)^{-\tfrac{1}{2}} U_1^T \tilde{I}_p V_0^T. \qquad (14.59)$$

We then end up with the same final update equation (14.4) by following the derivation defined in (13.5–13.6).

Thus, we have replaced the explicit factorization of $C \in \Re^{m \times m}$, with an SVD of $S \in \Re^{m \times N}$, and this is a significant saving when $m \gg N$. Further, by using a low-rank version for $C_{\epsilon\epsilon}$ we replace the matrix multiplication $\Sigma_0^+ U_0^T C_{\epsilon\epsilon}$ in (14.23) with the less expensive $\Sigma_0^+ U_0^T E$. Thus, there are none matrix operations which requires $\mathcal{O}(m^2)$ floating point operations in the new algorithm.

Equations (14.57) and (14.4) can be combined into one single equation, similar to (14.35), as

$$A^a = A^f \Bigg(\mathbf{1}_N + S^T X_1 (I + \Sigma_1^2)^{-1} X_1^T \left(D - \mathcal{M}[A^f] \right) \mathbf{1}_N \\
+ (I - \mathbf{1}_N) V_2 \sqrt{I - \Sigma_2^T \Sigma_2} \Theta^T \Bigg). \qquad (14.60)$$

Note that if we set $\Lambda_1 = \Sigma_1^2$ in (14.56) and (14.60) these equations becomes identical to respectively (14.31) and (14.35). Similarly, by replacing the expressions $X_1 (I + \Sigma_1^2)^{-1} X_1^T$ and $X_1 (I + \Lambda_1)^{-1} X_1^T$ in these equations, with $Z \Lambda^+ Z^T$ or $Z \Lambda^{-1} Z^T$ they become identical to the analysis equations (14.2) and (14.6).

14.4 Implementation of the analysis schemes

For the practical implementation we first note that we can choose from three different algorithms when computing the pseudo inverse of C. We can use a standard pseudo inversion based on an eigenvalue decomposition of C, or we can use the subspace pseudo inversion with either a full measurement error covariance matrix $C_{\epsilon\epsilon}$, or with a low-rank representation of the measurement error covariance matrix $C_{\epsilon\epsilon}^e = E E^T / (N - 1)$. From the standard eigenvalue factorization we obtain $\boldsymbol{Z}$ and $\boldsymbol{\Lambda}$. For the two subspace algorithms we obtain $\boldsymbol{X}_1$ and either $(\boldsymbol{I} + \boldsymbol{\Sigma}_1^2)$ or $(\boldsymbol{I} + \boldsymbol{\Lambda}_1^2)$.

Thereafter, we can choose between the computation of a traditional EnKF analysis or a square root analysis. Each of these requires the evaluation of the matrix multiplied with $\boldsymbol{A}$ in one of (14.2), (14.31) or (14.56) for the EnKF and one of (14.6), (14.35) or (14.60) for the square root algorithm. The final multiplication with $\boldsymbol{A}$ to compute the updated ensemble is the same for all of the algorithms.

Thus, it is clear that it is possible to combine all of these algorithms into one efficient routine where the user can choose between different pseudo inversions and analysis schemes. In this routine one should also include specific code for handling the case with a single observation where a scalar inverse can be used. Note also that in the EnKF with few observations, it is more efficient to reorder the matrix multiplications and rewrite (14.2) as

$$\boldsymbol{A}^{\mathrm{a}} = \boldsymbol{A}^{\mathrm{f}} + \left(\boldsymbol{A}^{\mathrm{f}} \boldsymbol{S}^{\mathrm{T}}\right) \left(\boldsymbol{C}^{-1}(\boldsymbol{D} - \mathcal{M}[\boldsymbol{A}^{\mathrm{f}}])\right). \tag{14.61}$$

The standard analysis scheme needs to compute a matrix multiplication for the final update which requires nN^2 floating point operations. When $n > m$ this becomes the most expensive computation in the analysis scheme. Note also that, in the standard scheme, mN^2 operations are required when $\boldsymbol{S}^{\mathrm{T}}$ is multiplied with the $m \times N$ matrix $\boldsymbol{C}^{-1}(\boldsymbol{D} - \mathcal{M}[\boldsymbol{A}^{\mathrm{f}}])$.

However, with few observations it is more efficient to first compute the product $\boldsymbol{A}^{\mathrm{f}} \boldsymbol{S}^{\mathrm{T}}$, which requires nmN floating point operations. The additional multiplication with the matrix $\boldsymbol{C}^{-1}(\boldsymbol{D} - \mathcal{M}[\boldsymbol{A}^{\mathrm{f}}])$ requires another nmN operations. Thus, when $2nmN < (n+m)N^2$ this procedure is more efficient. For the assimilation of a single observation this reduces the computation by a factor $N/2$.

14.5 Rank issues related to the use of a low-rank $C_{\epsilon\epsilon}$

It has recently been shown by *Kepert* (2004) that the use of an ensemble representation $C_{\epsilon\epsilon}^{\mathrm{e}}$ for $C_{\epsilon\epsilon}$, in some cases leads to a loss of rank in the ensemble when $m > N$. The rank problem may occur both using the EnKF analysis scheme with perturbation of measurements and using the square root algorithm. However, it is not obvious that the case with $m > N$ and the use of a low-rank representation $C_{\epsilon\epsilon}^{\mathrm{e}}$ of $C_{\epsilon\epsilon}$, should pose a problem. After all, the final coefficient matrix which is multiplied with the ensemble forecast to produce the analysis, is an $N \times N$ matrix.

The following will revisit the analysis by *Kepert* (2004) and extend it to a more general situation. Further, it will be shown that the rank problem can be avoided when the measurement perturbations, used to represent the low-rank measurement error covariance matrix, are sampled under specific constraints.

The EnKF analysis equation (14.2) can be rewritten as

$$\begin{aligned} \boldsymbol{A} &= \overline{\boldsymbol{A}} + \boldsymbol{A}'\boldsymbol{S}^{\mathrm{T}}\left(\boldsymbol{S}\boldsymbol{S}^{\mathrm{T}} + \boldsymbol{E}\boldsymbol{E}^{\mathrm{T}}\right)^{+}\left(\overline{\boldsymbol{D}} - \mathcal{M}[\overline{\boldsymbol{A}}^{\mathrm{f}}]\right) \\ &+ \boldsymbol{A}' + \boldsymbol{A}'\boldsymbol{S}^{\mathrm{T}}\left(\boldsymbol{S}\boldsymbol{S}^{\mathrm{T}} + \boldsymbol{E}\boldsymbol{E}^{\mathrm{T}}\right)^{+}(\boldsymbol{E} - \boldsymbol{S}), \end{aligned} \quad (14.62)$$

where the first line is the update of the mean and the second line is the update of the ensemble perturbations. Thus, for the standard EnKF is suffices to show that rank($\boldsymbol{W}$) = $N-1$ to conserve the full rank of the state ensemble, with $\boldsymbol{W}$ defined as

$$\boldsymbol{W} = \boldsymbol{I} - \boldsymbol{S}^{\mathrm{T}}\left(\boldsymbol{S}\boldsymbol{S}^{\mathrm{T}} + \boldsymbol{E}\boldsymbol{E}^{\mathrm{T}}\right)^{+}(\boldsymbol{S} - \boldsymbol{E}). \quad (14.63)$$

Similarly, for the square root algorithm $\boldsymbol{W}$ is redefined from (14.5) as

$$\boldsymbol{W} = \boldsymbol{I} - \boldsymbol{S}^{\mathrm{T}}\left(\boldsymbol{S}\boldsymbol{S}^{\mathrm{T}} + \boldsymbol{E}\boldsymbol{E}^{\mathrm{T}}\right)^{+}\boldsymbol{S}. \quad (14.64)$$

We consider the case when $m > N-1$ which was shown to cause problems in *Kepert* (2004). Define $\boldsymbol{S} \in \Re^{m \times N}$ with rank($\boldsymbol{S}$) = $N-1$, where the columns of $\boldsymbol{S}$ span a subspace $\mathcal{S}$ of dimension $N-1$. Further, we define $\boldsymbol{E} \in \Re^{m \times q}$ with rank($\boldsymbol{E}$) = $\min(m, q-1)$, where $\boldsymbol{E}$ contains an arbitrary number q, of measurement perturbations.

As in *Kepert* (2004) one can define the matrix $\boldsymbol{Y} \in \Re^{m \times (N+q)}$ as

$$\boldsymbol{Y} = (\boldsymbol{S}, \boldsymbol{E}), \quad (14.65)$$

and the matrix $\boldsymbol{C}$ becomes

$$\boldsymbol{C} = \boldsymbol{Y}\boldsymbol{Y}^{\mathrm{T}}, \quad (14.66)$$

with rank

$$p = \text{rank}(\boldsymbol{Y}) = \text{rank}(\boldsymbol{C}). \quad (14.67)$$

Dependent on the definition of $\boldsymbol{E}$ we have $\min(m, N-1) \leq p \leq \min(m, N+q-2)$. One extreme is the case where $q \leq N$ and $\boldsymbol{E}$ is fully contained in $\mathcal{S}$, in which case we have $p = N-1$.

The case considered in *Kepert* (2004) is another extreme which has $q = N$, and $p = \min(m, 2N-2)$. This corresponds to a situation which is likely to occur when $\boldsymbol{E}$ is sampled randomly and includes components along $N-1$ directions in $\mathcal{S}^{\perp}$.

We define the SVD of $\boldsymbol{Y}$ as

$$\boldsymbol{U}\boldsymbol{\Sigma}\boldsymbol{V}^{\mathrm{T}} = \boldsymbol{Y}, \quad (14.68)$$

with $\boldsymbol{U} \in \Re^{m \times m}$, $\boldsymbol{\Sigma} \in \Re^{m \times (N+q)}$ and $\boldsymbol{V} \in \Re^{(N+q) \times (N+q)}$.

The pseudo inverse of $\boldsymbol{Y}$ is defined as

$$\boldsymbol{Y}^{+} = \boldsymbol{V}\boldsymbol{\Sigma}^{+}\boldsymbol{U}^{\mathrm{T}}, \quad (14.69)$$

where $\boldsymbol{\Sigma}^{+} \in \Re^{(N+q) \times m}$ is a diagonal matrix with the diagonal defined as diag($\boldsymbol{\Sigma}^{+}$) = $(\sigma_1^{-1}, \sigma_2^{-1}, \ldots, \sigma_p^{-1}, 0, \ldots, 0)$.

Both the equations for $\boldsymbol{W}$ in (14.63) and (14.64) can be rewritten in a form similar to what was used by *Kepert* (2004). Introducing the expressions (14.68) and (14.69) in (14.64), and defining $\boldsymbol{I}_N$ to be the N-dimensional identity matrix, we get

$$\begin{aligned}
\boldsymbol{W} &= \boldsymbol{I}_N - (\boldsymbol{I}_N, \boldsymbol{0}) \boldsymbol{Y}^{\mathrm{T}} (\boldsymbol{Y}\boldsymbol{Y}^{\mathrm{T}})^{+} \boldsymbol{Y} (\boldsymbol{I}_N, \boldsymbol{0})^{\mathrm{T}} \\
&= \boldsymbol{I}_N - (\boldsymbol{I}_N, \boldsymbol{0}) \boldsymbol{V} \boldsymbol{\Sigma}^{\mathrm{T}} \boldsymbol{\Sigma}^{+\mathrm{T}} \boldsymbol{\Sigma}^{+} \boldsymbol{\Sigma} \boldsymbol{V}^{\mathrm{T}} (\boldsymbol{I}_N, \boldsymbol{0})^{\mathrm{T}} \\
&= (\boldsymbol{I}_N, \boldsymbol{0}) \boldsymbol{V} \left\{ \boldsymbol{I}_{N+q} - \begin{pmatrix} \boldsymbol{I}_p & \boldsymbol{0} \\ \boldsymbol{0} & \boldsymbol{0} \end{pmatrix}_{N+q} \right\} \boldsymbol{V}^{\mathrm{T}} (\boldsymbol{I}_N, \boldsymbol{0})^{\mathrm{T}} \qquad (14.70) \\
&= (\boldsymbol{I}_N, \boldsymbol{0}) \boldsymbol{V} \begin{pmatrix} \boldsymbol{0} & \boldsymbol{0} \\ \boldsymbol{0} & \boldsymbol{I}_{N+q-p} \end{pmatrix}_{N+q} \boldsymbol{V}^{\mathrm{T}} (\boldsymbol{I}_N, \boldsymbol{0})^{\mathrm{T}}.
\end{aligned}$$

The similar expression for $\boldsymbol{W}$ in (14.63) is obtained by replacing the matrix, $(\boldsymbol{I}_N, \boldsymbol{0}) \in \Re^{N \times (N+q)}$, with $(\boldsymbol{I}_N, -\boldsymbol{I}_N, \boldsymbol{0}) \in \Re^{N \times (N+q)}$.

We need the $N+q$ matrix in (14.70) to have a rank of at least $N-1$ to maintain the rank of the updated ensemble perturbations. Thus, we require that $N + q - p \geq N - 1$ and get the general condition

$$p \leq q + 1. \qquad (14.71)$$

With $q = N$ this condition requires $p \leq N + 1$. This is only possible when all singular vectors of $\boldsymbol{E}$, except two, are contained in $\mathcal{S}$. Thus, it is clear that a low-rank representation of $\boldsymbol{C}_{\epsilon\epsilon}$ using N measurement perturbations $\boldsymbol{E}$, can be used as long as the selected perturbations do not increase the rank of $\boldsymbol{Y}$ to more than $N + 1$.

It is also clear that if the constrained low-rank representation $\boldsymbol{E} \in \Re^{m \times N}$, is unable to properly represent the real measurement error covariance, it is possible to increase the number of perturbations to an arbitrary number $q > N$ as long as the rank p satisfies the condition (14.71).

In *Kepert* (2004) it was assumed that the rank $p = 2N - 2$. That is, $\boldsymbol{E}$ has components in $N - 1$ directions of $\mathcal{S}^{\perp}$. Then, clearly, the condition (14.71) is violated and this results in a loss of rank. It was showed that this can be resolved using a full rank measurement error covariance matrix (corresponding to the limiting case when $q \geq m + 1$). Then, $p = \mathrm{rank}(\boldsymbol{Y}) = \mathrm{rank}(\boldsymbol{C}_{\epsilon\epsilon}^{\mathrm{e}}) = m$ and the condition (14.71) is always satisfied.

As an example, assume now that we have removed r columns from the matrix $\boldsymbol{E} \in \Re^{m \times (q=m+1)}$. We then get the reduced $\boldsymbol{E} \in \Re^{m \times (q=m+1-r)}$ of rank equal to $m - r$. In this situation we can consider two cases. First, if the removed perturbations are also fully contained in $\mathcal{S}$, then this does not lead to a reduction of p which still equals m. In this case we can write the condition (14.71), for $r \leq N - 1$, as

$$p = m \leq m + 2 - r, \qquad (14.72)$$

which is violated for $r > 2$. Secondly, assume that the removed perturbations are fully contained in $\mathcal{S}^{\perp}$. Then the rank p will be reduced with r and we write the condition (14.71) as

Exp. 1	EnKF	EIGC	$C_{\epsilon\epsilon}$	Exp. 6	SQRT	EIGC	$C_{\epsilon\epsilon}$
Exp. 2	EnKF	EIGC	$C_{\epsilon\epsilon}^{\text{e}}$	Exp. 7	SQRT	EIGC	$C_{\epsilon\epsilon}^{\text{e}}$
Exp. 3	EnKF	SUBC	$C_{\epsilon\epsilon}$	Exp. 8	SQRT	SUBC	$C_{\epsilon\epsilon}$
Exp. 4	EnKF	SUBC	$C_{\epsilon\epsilon}^{\text{e}}$	Exp. 9	SQRT	SUBC	$C_{\epsilon\epsilon}^{\text{e}}$
Exp. 5	EnKF	SUBE	$\boldsymbol{E}$	Exp. 10	SQRT	SUBE	$\boldsymbol{E}$

Table 14.1. List of experiments. See explanation in text

$$p = m - r \leq m + 2 - r. \tag{14.73}$$

We can continue to remove columns of $\boldsymbol{E}$ contained in $\mathcal{S}^{\perp}$, without violating the condition (14.71), until there are only $N-1$ columns left in $\boldsymbol{E}$, all contained in $\mathcal{S}$.

From this discussion, it is clear that we need the measurement error perturbations to explain variance within $\mathcal{S}$. Note that the subspace pseudo inversion schemes automatically projects the measurement error covariance matrix or the measurement perturbations onto $\mathcal{S}$.

14.6 Experiments with $m \gg N$

The following experiments were performed to evaluate the properties of the analysis schemes in the case where $m \gg N$. An experimental setup, similar to the advection example from Sect. 4.1.3 is used. However, now 500 measurements are used, i.e. there is a measurement at every second grid point. The measurements have correlated errors of de-correlation length equal to 20 m. The error variance of the measurements is set to 0.09, corresponding to a standard deviation of 0.30 and the number of assimilation steps is 5.

Ten experiments are run which differ in the choice of analysis scheme (EnKF or SQRT) and inversion algorithm for $\boldsymbol{C}$. In addition, both an exact and a low-rank representation of $\boldsymbol{C}_{\epsilon\epsilon}$ are used. The experiments are summarized in Table 14.1 where EnKF and SQRT denote the analysis scheme used. EIGC denote the inversion algorithm based on the eigenvalue factorization from Sect. 14.1, SUBC denote the subspace inversion discussed in Sect. 14.2, and SUBE means the subspace inversion using the measurement perturbations rather than the full measurement error covariance matrix, as presented in Sect. 14.3. In the different experiments we have specified either a full rank measurement error covariance matrix $\boldsymbol{C}_{\epsilon\epsilon}$, or a low-rank version defined as $\boldsymbol{C}_{\epsilon\epsilon}^{\text{e}} = \boldsymbol{E}\boldsymbol{E}^{\text{T}}/(N-1)$.

Note that it is straight-forward to sample normal correlated perturbations for each element of $\boldsymbol{E}$ with the correct statistics. This is done using the same sampling scheme as was used to generate the initial ensemble and then measuring each member to create the columns in $\boldsymbol{E}$. In all the experiments we used improved sampling of order six for the initial ensemble and order four for the measurement perturbations.

14.6 Experiments with $m \gg N$ 225

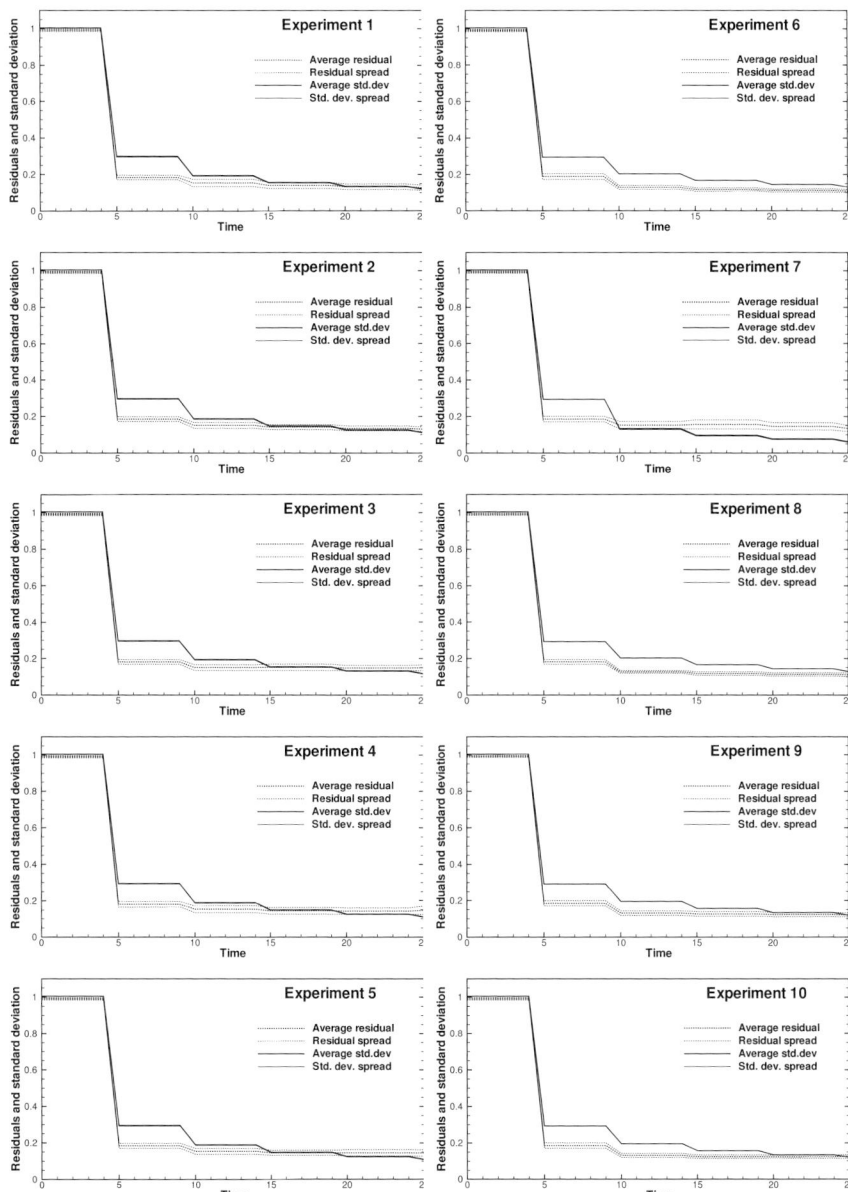

Fig. 14.6. Time evolution for RMS residuals *(dotted lines)* and estimated standard deviations *(full lines)* for all 50 simulations in the respective experiments

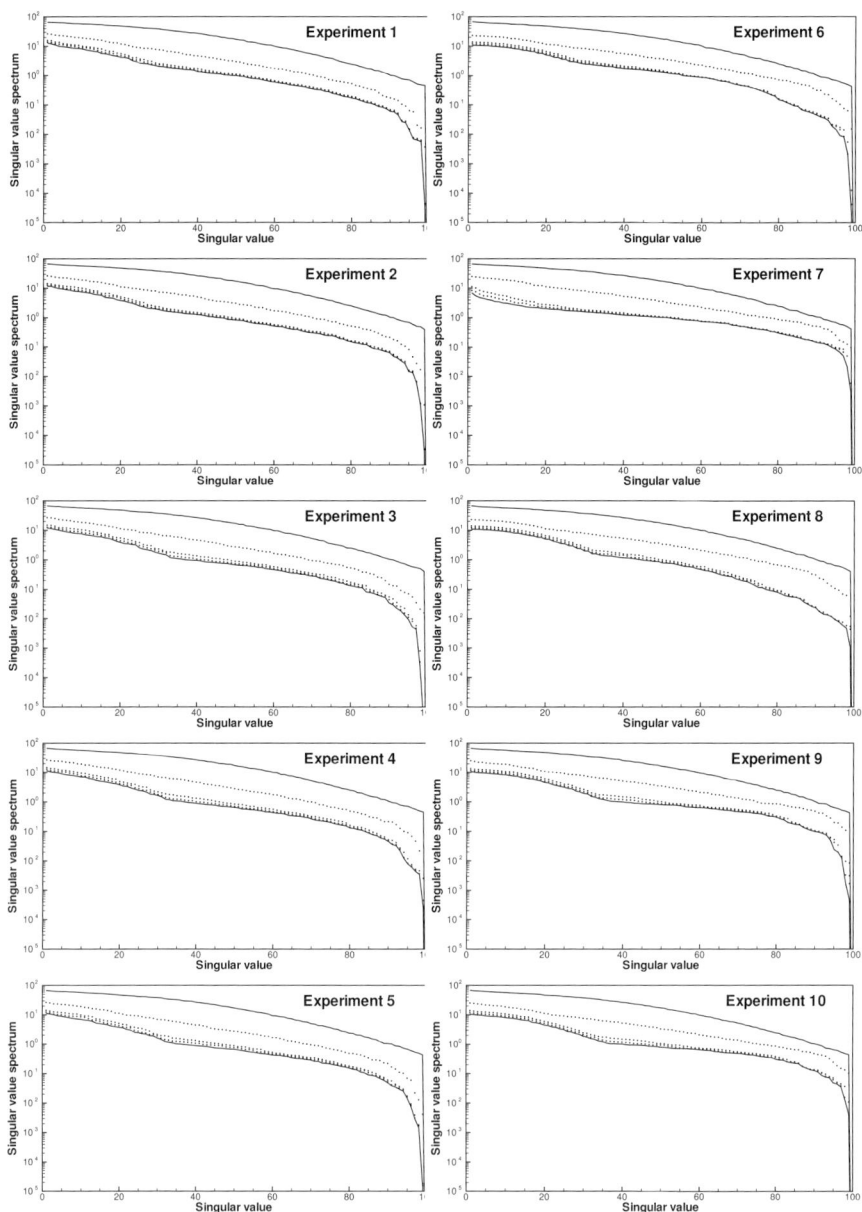

Fig. 14.7. Time evolution of the ensemble singular value spectra for some of the experiments

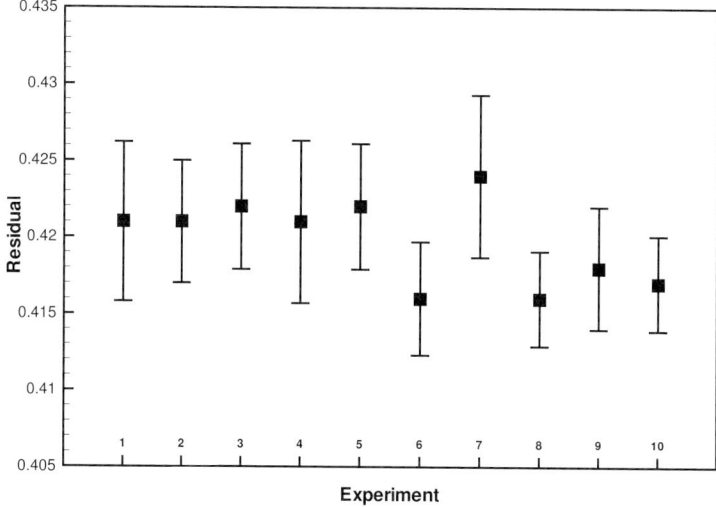

Fig. 14.8. Average residual and standard deviation for the 10 cases

Note that $\boldsymbol{E}$ is sampled with rank equal to $N-1$. When projected onto $\boldsymbol{U}_{0p}$, i.e. the sub-space $\mathcal{S}$ spanned by the first p singular vectors in $\boldsymbol{U}_0$, we are not guaranteed that the rank of $\boldsymbol{U}_{0p}^{\mathrm{T}}\boldsymbol{C}_{\epsilon\epsilon}^{\mathrm{e}}\boldsymbol{U}_{0p}$ or $\boldsymbol{U}_{0p}\boldsymbol{E}$ is equal to $N-1$. If $\boldsymbol{E}$ has columns which are orthogonal to $\boldsymbol{U}_{0p}$, these do not contribute when projected onto $\boldsymbol{U}_{0p}$. This corresponds to the assimilation of perfect measurements and will lead to a corresponding loss of rank in the updated ensemble. We did not experience this to be a problem in the present experiments.

The use of a low-rank representation for $\boldsymbol{C}_{\epsilon\epsilon}$ is valid, and the results will be the same as obtained using a full rank $\boldsymbol{C}_{\epsilon\epsilon}$ if $\boldsymbol{U}_{0p}^{\mathrm{T}}\boldsymbol{C}_{\epsilon\epsilon}^{\mathrm{e}}\boldsymbol{U}_{0p} = \boldsymbol{U}_{0p}^{\mathrm{T}}\boldsymbol{C}_{\epsilon\epsilon}\boldsymbol{U}_{0p}$. This equality is nearly satisfied here since the random sampling of $\boldsymbol{E}$ used the same correlation functions as was used to generate the initial ensemble. Probably, in this case, the use of a diagonal error covariance matrix would be more difficult to represent properly by a low-rank random ensemble of smooth members.

It is also clear that the projection of $\boldsymbol{C}_{\epsilon\epsilon}$ onto the $\mathcal{S}$-space may lead to a lower measurement variance than specified in the full rank $\boldsymbol{C}_{\epsilon\epsilon}$, thus there may be a need to rescale $\boldsymbol{C}_{\epsilon\epsilon}^{\mathrm{e}}$ to avoid over-fitting the data, in which case the EnKF will predict too low estimated standard deviations.

As before we have run 50 assimilation simulations for each experiment to be able to give a statistical comparison of results between the different experiments. The time evolution of the residuals and singular spectra are presented in Figs. 14.6 and 14.7. It is clear that these are rather similar between all the different experiments, and all appear to provide consistent solutions.

In Fig. 14.8 we have plotted the mean and standard deviation of the residuals as predicted by the 50 assimilation simulations in each experiment. The

Exp	2	3	4	5	6	7	8	9	10
1	0.96	0.27	0.71	0.35	0	0.02	0	0	0
2		0.23	0.72	0.31	0	0.01	0	0	0
3			0.50	0.85	0	0.10	0	0	0
4				0.61	0	0.04	0	0	0
5					0	0.07	0	0	0
6						0	0.57	0.10	0.11
7							0	0	0
8								0.02	0.02
9									0.78

Table 14.2. Statistical probability that two experiments provide an equal mean for the residuals as computed using the Student's t-test. A probability close to one indicates that it is likely that the two experiments provide distributions of residuals with similar mean

five EnKF experiments are all very similar, and by examining the statistical test presented in Table 14.2, it is clear that these are all likely to be sampled from the same distribution. Thus, in the EnKF, the different inversion schemes and the use of a full versus low-rank representation for $C_{\epsilon\epsilon}$ do not significantly influence the results. However, we do not expect this to be a general result and a different configuration of the experiment may lead to a different conclusion.

The experiments using the SQRT scheme seem to do a slightly better job in this experiment. An exception is *Exp. 7* which used the EIGC inversion together with a low-rank $C_{\epsilon\epsilon}^{\mathrm{e}}$. It is not clear why this experiment performed so poorly compared to the other experiments. All experiments were rerun starting from different random seeds and this confirmed the results. From Table 14.2 it is clear that *Exps. 6,8,9* and *10* are all internally consistent, they are all different from *Exp. 7*, and they are all different from the EnKF experiments *Exp. 1–5*.

In the *Exps. 5* and *10* we used the algorithm as defined in Sect. 14.3 where we avoid the formation of the full measurement error covariance matrix. In this case we obtain results which are almost identical to the results from respectively *Exps. 4* and *9* where the low-rank measurement error covariance matrix $C_{\epsilon\epsilon}^{\mathrm{e}}$ is used. In fact, the only difference between these cases is related to the use of different random seeds.

Further examination of these schemes in more realistic settings is clearly required before they are adapted in operational systems. From the previous theoretical analysis, the new low-rank square root scheme introduces an approximation by projecting the measurements onto the $\mathcal{S}$ sub-space, and it was seen that this both stabilises the computation of the analysis and also makes it computationally much more efficient. However, when a low-rank $C_{\epsilon\epsilon}^{\mathrm{e}}$ is used, a scheme is required for the proper sampling of measurement perturbations in $\mathcal{S}$.

14.7 Summary

A comprehensive analysis has been given on the use of the EnKF and square root analysis schemes when used with large data sets. It was seen that in this case, the inversion of C may become poorly conditioned, and a pseudo inversion may be required. The analysis schemes were reformulated using a standard pseudo inversion based on an eigenvalue factorization of C followed by a truncation of the eigenvalue spectrum to only account for the significant eigenvalues. This algorithm seems to work well in many cases. However, when the number of measurements becomes large it is very inefficient, since a matrix of dimension $m \times m$ needs to be factorized at a cost proportional to $\mathcal{O}(m^3)$.

An alternative pseudo inversion was derived where the measurements are projected onto a sub-space $\mathcal{S}$, spanned by the measured ensemble perturbations. It was seen that this may introduce an approximation in some cases. In particular, if the measurement error covariance matrix is diagonal then the eigenvectors of SS^T and C are identical and there is no approximation introduced. On the other hand, if $C_{\epsilon\epsilon}$ is nondiagonal the eigenvectors will differ and the projection onto the $\mathcal{S}$-space eliminates the part of C which is orthogonal to the $\mathcal{S}$-space. Fortunately, this is mostly noise in many applications.

The sub-space pseudo inversion can be computed at a cost of $\mathcal{O}(Nm^2)$ which is a significant saving when $m \gg N$. However, it was also seen that a further speedup is possible if a low-rank representation is used for the measurement error covariance matrix. In particular, if we write the measurement error covariance matrix as $(N-1)C_{\epsilon\epsilon}^{\text{e}} = EE^T$, and represent it by the measurement perturbations E, it is possible to compute the analysis without forming $C_{\epsilon\epsilon}^{\text{e}}$. This further reduces the cost of the inversion to be proportional to $\mathcal{O}(N^2 m)$, and this algorithm allows us to compute the analysis update using very large data sets. An important point in this case is that the measurement perturbations must be sampled to span $\mathcal{S}$ to avoid a loss of rank in the updated ensemble.

15
An ocean prediction system

The ocean modelling community has been in the forefront when it comes to developing advanced data assimilation systems and taking these into use in real applications. This chapter will briefly present one such system, named TOPAZ, forming the North Atlantic and Arctic component of the European "MERSEA" integrated system, and being one of the contributors to the international Global Ocean Data Assimilation Experiment (GODAE). The system is based on the latest scientific developments in terms of ocean modelling with the Hybrid Coordinate Ocean Model (HYCOM) and data assimilation with the EnKF.

15.1 Introduction

The need for high quality predictions of marine parameters has been well identified. During recent years, offshore oil-exploration activities have expanded off the continental shelfs to deeper waters. Drilling and production of oil and gas at depths of 2000 meters or more are ongoing at several locations, and the Arctic Shelf contains considerable gas resources in ice-covered areas. This has introduced a need for real time forecasts of oceanic currents and sea-ice which in some cases may have severe impact on the safety related to drilling, production and critical operations. In addition, sustainable exploitation of marine resources through commercial fisheries and fish farming are becoming increasingly important. Fisheries management systems will benefit from accurate prediction of marine parameters such as nutrient and plankton concentrations, and this will lead to more accurate monitoring and prediction of fish stocks. Thus, there are needs for operational monitoring and prediction of both physical and biological marine parameters.

An ocean data assimilation system allows for the integration of remote-sensing and in situ observations of ocean, ice, biological, and chemical variables, with coupled marine ecosystem (*Natvik and Evensen*, 2003a,b) and ice-ocean general circulation models (*Brusdal et al.*, 2003, *Lisæter et al.*, 2003).

This integration can best be done using advanced data assimilation techniques. In the ocean community there has been a strong focus on the development and implementation of consistent data assimilation techniques that can be used with primitive equation models and also models of the marine ecosystem. Further, the real time processing and flow of observational data have now been developed to a degree where both satellite and in situ data are available in near real time. Several ocean forecasting systems are exploiting this real time flow of observed information in data assimilation systems and provide operational ocean forecasts.

The TOPAZ system consists of the HYCOM ocean model (*Bleck*, 2002) which has been coupled to two different sea-ice models, one is a simple model for ice-thickness and ice-concentration while the other is multi-category sea-ice model which represents ice-thickness distributions. Further, four ecosystem models of increasing complexity have been integrated in the system.

The TOPAZ system has been developed to meet the needs from future users of marine parameters. The system development has been supported by two previous European Commission funded projects, i.e. the DIADEM and TOPAZ projects, and current work is aimed at integration into the European MERSEA system within the MERSEA Integrated Project. TOPAZ results are displayed on the web-page http://topaz.nersc.no as well as validation statistics against in situ data provided by the Coriolis center.

15.2 System configuration and EnKF implementation

The model domain used for the TOPAZ prediction system is shown in Fig. 15.1. The grid is created using a conformal mapping of the poles to two new locations using the algorithm outlined in *Bentsen et al.* (1999). The horizontal model resolution varies from 11 km in the Arctic to 18 km near the Equator.

The TOPAZ system has a huge state vector consisting of 79.6 million variables just for the physical ocean parameters. The inclusion of the marine ecosystem multiplies the number of unknowns by a factor 2 to 3, depending on the ecosystem model formulation used. The system uses 100 members in the ensemble, thus the computational cost of running the system is 100 times the cost of running a single model. Fortunately, the members evolve completely independently of each other and the new parallel clusters with multiple CPUs are very well tailored to this kind of application. Clearly, to a similar computational cost it is possible to run a single model with quadruple resolution. On the other hand, we then lose the opportunity to update this single model consistently with the observations, and simplified and less consistent assimilation schemes need to be used. We would also lose the possibility to generate error estimates for the predictions.

The number of observations assimilated is huge. It consists of satellite observed sea level anomalies merged from four satellites (ERS2, Jason1, EN-

15.2 System configuration and EnKF implementation   233

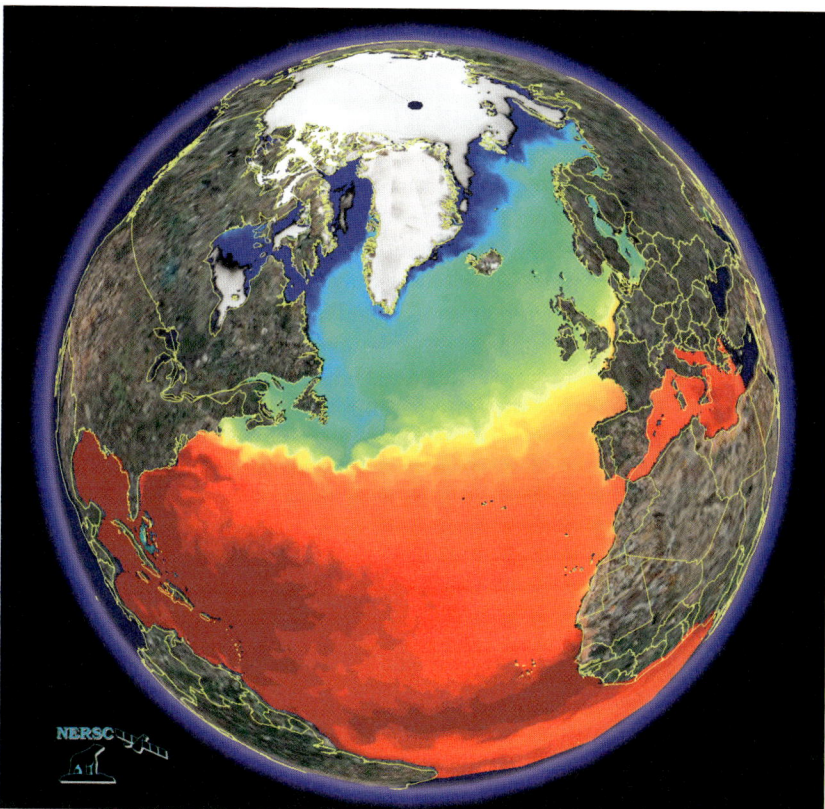

Fig. 15.1. Surface temperature and sea ice concentrations in the North Atlantic and Arctic Ocean with the TOPAZ system as viewed in Google Earth.

VISAT and GEOSAT follow on), available from Collecte Localisation Satellites (CLS) on a grid containing 100 000 observations in the North Atlantic at each assimilation cycle. In addition, TOPAZ assimilates 40 000 gridded ice concentration data from SSM/I and 8000 sea surface temperature observations (Reynolds SST), still with relatively low resolution (120 km at the Equator). When higher resolution products (25 km) will be available from the Medspiration project the number of SST data assimilated will increase to around 200 000 observations depending on cloud coverage.

Clearly, it is a challenge to represent the solution search space for such a large state vector and when assimilating this many measurements using only a limited ensemble size. It is possible to use the sophisticated analysis schemes discussed in the previous chapters, but for this particular system a slight modification is required. In *Haugen and Evensen* (2002), *Brusdal et al.* (2003), *Evensen* (2003), *Ott et al.* (2004) an algorithm named "local analysis" was used. This is a rather simple approach where the analysis update is computed grid point by grid point, and using only observations located within a certain distance from the grid point, see Appendix A.1 for a detailed discussion. In an ocean model it is convenient to consider this as an update of grid column by grid column since the depth is much less than the horizontal scale of the model.

The local analysis is spatially discontinuous and the updated ensemble members may not represent solutions of the original model equations, but the deviation should not be to large as long as the range of influence is large enough. In addition the updated ensemble members are not represented in the space spanned by the predicted ensemble. In fact, the use of an update matrix which varies smoothly throughout the grid effectively reduces the dimension of the problem. That is, in an ocean model where we update the solution grid column by grid column, we are solving many small problems instead of one large. In the TOPAZ system the number of unknowns in each grid column is of the same order as the number of ensemble members (113 for 22 hybrid vertical layers), as well as the number of local observations (50 at most).

The quality of the EnKF analysis is clearly connected to the ensemble size used. We expect that a larger ensemble is needed for the global analysis than the local analysis to achieve the same quality of the result. That is, in the global analysis a large ensemble is needed to properly explore the state space and to provide a consistent result for the global analysis. We expect this to be application dependent. Note also that the use of a local analysis scheme is likely to introduce non-dynamical modes, although the amplitudes of these will be small if a large enough influence radius is used when selecting measurements. In dynamical models with large state spaces, the local analysis allows for the computation of a realistic analysis result while still using a relatively small ensemble of model states. This also relates to the discussions on localization and filtering of long range correlations by *Mitchell et al.* (2002).

The TOPAZ system is run every week and produces two weeks forecasts. The propagation and analysis steps are orchestrated by a collection of scripts in the following way: every Tuesday the observations are collected and the analysis is run sequentially for each observed variable[1], then a single member forecast is run until the two-weeks forecast, initialized by the ensemble average, the whole ensemble is then propagated by the model with perturbed forcing fields (winds and thermodynamic forcing). The communication between the analysis and propagation steps is done by files so that both executables are distinct and mostly independent. This allows separate upgrades of the model and analysis codes. The propagation step requires 1200 CPU hours per week but is "embarrassingly parallel" and the hundred independent jobs are easily patched into the supercomputer idle time. TOPAZ runs on the super-

---

[1] This is meant to avoid scaling issues when assimilating different types of observations and it is in theory correct in the Gaussian case as all statistics (mean and variance-covariance) are updated by each observation set. This is not the case with OI-type of methods because the background covariance remains unchanged.

computing facilities of Parallab at the University of Bergen that are shared with many other users but the privileges required by the operational system are relatively small and do not represent a nuisance to other users.

The single member forecast dumps boundary conditions for nested models. Running an ensemble forecast is also possible starting from the latest analysis ensemble.

15.3 Nested regional models

To meet the end users needs of high resolution accurate information, regional models with very high resolution are embedded into the TOPAZ system in the target areas where mesoscale processes must be properly resolved. The nested models depend on the basin-scale model but the global system is not dependent on the regional models, thus each nested system can be tuned on purpose to satisfy one application without disturbing the globality of the system.

With the inclusion of a nesting capability and the assimilation of both in situ data and data from a variety of satellite sensors, the TOPAZ system constitutes a state of the art and flexible operational ocean prediction system. The model system has been designed to be easily extensible to other geographical areas including the global domain and it allows for nesting of an arbitrary number of regional high resolution models with arbitrary orientation and horizontal resolution.

Regional high-resolution models covering the Gulf of Mexico, the North Sea and the Barents Sea are currently receiving boundary conditions from TOPAZ and are run in real time. The Gulf of Mexico model uses data assimilation based on the ensemble OI method presented in Appendix A.5. It is used to predict the location of the Loop Current and the formation and propagation of rings in the Gulf of Mexico, and thus provides valuable information related to deep water drilling and oil production facilities in the Gulf of Mexico.

The only observations assimilated in the regional model are the sea surface heights from satellite altimeters, that are available with three days delay. The data assimilation is therefore performed one week back in time and assimilates gridded maps that are representative of a weekly average. The model is then integrated over the past week and two weeks forecast in the future. When necessary, e.g. when the situation is particularly dynamic, the nested system can be updated twice a week, independently from the updates of the outer model.

Figure 15.2 shows the observed limits of the Loop Current and two rings in the Northern and Northwestern Gulf of Mexico. The loop current and its detached rings may have large velocities, and when these exceed 1.5 m/s, the security of the staff and equipment is threatened and many operations have to be postponed, causing major financial losses.

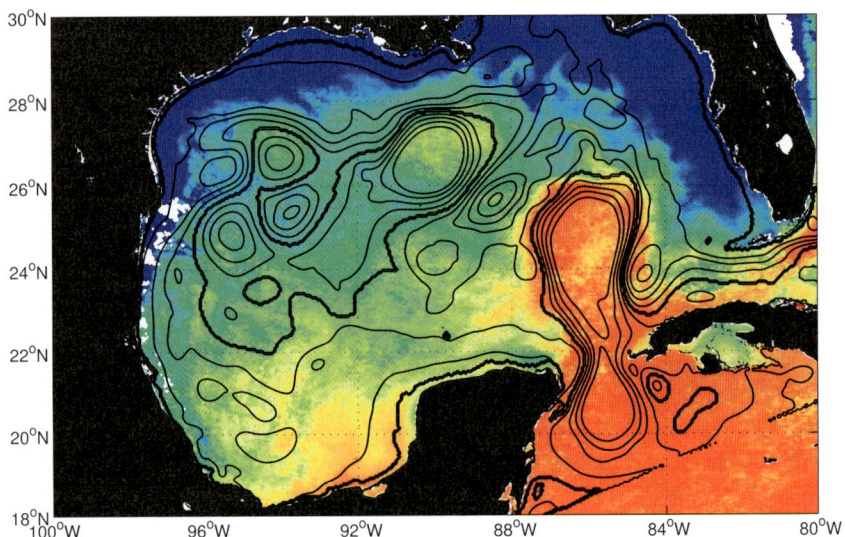

Fig. 15.2. Predicted sea-surface heights (*isolines*) overlaid a map of satellite observed sea surface temperatures (not yet assimilated) for the Gulf of Mexico on 29th March 2006, showing accurate positioning of the Loop Current and a detached ring. Red colours indicate high temperatures and the blue colours denote cold water.

The model nowcast (i.e. estimate at the current date) represents well the Loop Current and the two detached rings and agree well with the measured current directions, but some inaccuracies remain in the locations and extents of these features. We expect that the remaining errors are not far from being irreducible with respect to the chaotic behaviour of the small scales features, their representation by the model and in the observations. The next major improvement of the user product would therefore be a probabilistic forecast based on an ensemble. It would indicate the areas where the forecast can be given with some confidence and those where the situation is too chaotic to be predicted.

15.4 Summary

The real time operation of the system has proved to be feasible and relies on the availability of remote sensing products in near real time, and atmospheric forcing fields from the meteorological forecasting centers. The forecasts of eddies in the Gulf of Mexico have been presented to potential users in the offshore oil industry by Ocean Numerics Ltd., revealing their strong interest in the way the problem is tackled and providing useful feedback for the future product developments. Oil companies have also invested into the Barents Sea high-resolution model which is nested into the TOPAZ system in the perspective of offshore exploration and production in the ice-covered Shtokman field. The latter system provides information on ice-ocean conditions and will be the basis for an ice and iceberg forecasting system.

There is now a strong consensus in the offshore industry, within funding agencies and among ocean researchers, on the need for development of operational ocean prediction systems. It is expected that several such systems will be established in the near future, covering the global ocean and providing valueable information about the state of the ocean both to commercial users and the public.

16
Estimation in an oil reservoir simulator

The EnKF has recently been taken into use with simulation models for oil and gas reservoirs, with the purpose of estimating poorly known parameters and to improve the predictive capability of the models. There are economical benefits of obtaining a model which best possible represents the reservoir. Optimally, it could be used for predicting the future production and to assist in the planning of new production and injection wells. A better model also provides insight and understanding regarding the properties of the reservoir.

Parameter estimation in reservoir simulation models is often named "history matching" by reservoir engineers, and the purpose is to find model parameters that result in simulations which better match the production history. History matching has traditionally been considered as a manual process where the engineer wisely tunes parameters and the impact is examined through model simulations.

Recently, there has been a growing interest in more mathematical and statistical methods for history matching. These involve both brute force direct minimization techniques and gradient methods based on the use of adjoints. Common for these is that they have all considered a pure parameter estimation problem, and not the combined parameter and state estimation problem as was advocated in the previous chapters.

An alternative approach based on the EnKF was proposed by *Nævdal et al.* (2003), where the reservoir model state and parameters were updated sequentially in time, using the information contained in pressure and rate measurements from production wells. There are now several groups continuing this work and below an application of the EnKF for history mathing in an oil reservoir model is discussed.

16.1 Introduction

An oil reservoir often consists of layers of sand and shale, each characterized by their respective porosity and permeability. The sands and shales are sed-

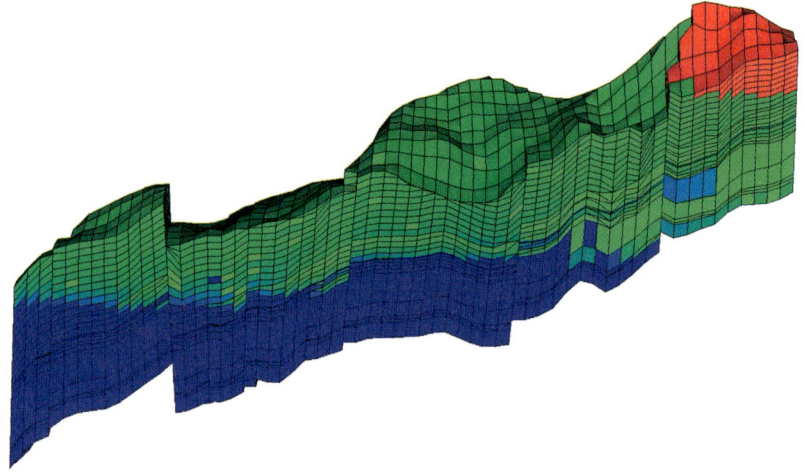

Fig. 16.1. Cross-section through the reservoir simulation model. Red colour represents gas, green denotes oil and blue is water

iments deposited on the seabed during different geological regimes, and are characterized by the porosity, $\phi(\boldsymbol{x})$, describing the fraction of a sand body which can accommodate fluids and the permeability, $\boldsymbol{k}_h(\boldsymbol{x})$, which describes how well fluids can flow in the reservoir. Normally the porosity of the reservoir sands is about 10–30 %, dependent on the grain size which varies for different depositional environments. The permeability is measured in a unit named Darcy, where 1 Darcy (D) is of order 10^{-12} m^2. Typical reservoirs have permeabilities in the range 0.1–10 D.

For reservoir sands to contain hydro-carbons, the permeable sands must be overlaid by an impermeable shale, or cap-rock, which prevents the oil and gas from escaping the reservoir. During geological time the sand layers fold and tilt, and faults may develop. The faults may become impermeable as well. Thus, the reservoir boundaries consist of the cap-rock and impermeable faults which enclose the oil and gas.

The density of gas is much less than the density of oil and water, and the mobility of gas is also much higher than for oil and water. Oil is also lighter than water and in a hydrostatic equilibrium we find gas overlaying oil and water below the oil. Fig. 16.1 shows a cross section of an oil reservoir in the North Sea. This reservoir is limited by an upper impermeable layer of shale and the horizontal extension is determined by two sealing faults. The depths of the gas-oil contact (*GOC*) and water-oil contact (*WOC*) are clearly identified. Note also the four faults located within the reservoir.

A reservoir simulation model describes the flow of oil, gas and water in the reservoir. The state vector in a reservoir model consists of the reservoir pressure, P, and saturations of water, gas and oil; S_w, S_g and S_o. The knowledge of two saturations allows for the computation of the third one. In addition one often includes variables describing the amount of gas which is in a fluid state at reservoir conditions and which becomes gas at the surface, R_s, and also gas in the reservoir which condensates and becomes fluid at surface conditions, R_v. When a well is drilled into the reservoir and operated at a pressure lower than the reservoir pressure, this sets up gradients in the reservoir pressure and the reservoir fluids start flowing towards the well.

The reservoir model is coupled to a model describing the flow of fluids in the wells. There are both production wells where oil, gas and water are produced from the reservoir, and injection wells which are used to pump water, gass and sometimes other chemicals into the reservoir to maintain the reservoir pressure and to force the oil and gas towards the production wells. The wells are often controlled by valves at the surface which regulate the rate of flow in the well and thus the pressure in the well.

Recent studies with reservoir simulation models suggest that the EnKF can be used for improved reservoir management. This was first proposed by *Nævdal et al.* (2002, 2003) who used the EnKF in a simplified reservoir model to estimate the permeability of the reservoir. They showed that there could be a great benefit of using the EnKF to improve the model through parameter estimation, and that this could lead to improved predictions. These initial works have been followed by several more recent publications (see the listing in the Appendix). These have mostly considered simplified reservoirs and various test cases. The estimated parameters comprise porosity and permeability and the data assimilated have been well pressures and rates. An exception is *Skjervheim et al.* (2005) where seismic 4D-data were assimilated as well. In the next sections we describe an implementation of the EnKF with a reservoir simulator for a North Sea field example.

16.2 Experiment

It is clear that there are large uncertainties when it comes to defining the exact properties of the reservoir. Geologists and geophysicists start by estimating the location of the top of the reservoir. Then, using seismic data together with log-data from test wells, combined with a good geological understanding of the depositional processes, they develop a conceptual model for the layering of different sand types and shales in the reservoir. A structural geologist will analyse the presence of faults in the reservoir and develop a structural model. This will also be based on the relatively few test wells and the seismic data. Using data from the test wells one attempts to identify the locations of the fluid contacts, as well as the properties of the oil, gas and water in the reservoir. One can then build a set of initial models or realizations of the reservoir using various statistical simulation methods.

16.2.1 Parameterization

The first step in the history matching procedure is to identify the parameters which determines the uncertainty of the model and need to be estimated. We have now assumed that the structural model is fairly accurate, i.e. the locations of faults and layers in the model are reasonable. This may not be the case but it is currently not clear how the EnKF can be used to estimate structural parameters, since the update equation in the EnKF combines ensemble members, and these all need to be defined on the same numerical grid.

Fluid contacts

In the current application we have identified large initial uncertainties in the oil-water and gas-oil contacts, WOC and GOC. The reservoir consists of several compartments which are separated by more or less insulating faults. Unless we have vertical wells penetrating the contacts it is difficult to obtain good estimates of them. The depths of the contacts varies between different isolated regions and we only have information from wells drilled through a few of these. The initial uncertainty of the WOC had in some regions standard deviations of up to 30 m. Thus, a major set of parameters to be estimated is the WOC and GOC in the different regions of the model, since this determines the volume of oil in the reservoir as well as the optimal vertical location of horizontal production wells.

Fault transmissibilties

With a large number of faults and only few pressure measurements there is a large uncertainty in the assumed fault transmissibilities. Thus, we also include the set of transmissibilities, *multflt*, of the faults as parameters to be estimated.

Vertical layer transmissibilties

The vertical flow in the reservoir is normally determined by the vertical permeability. In the current experiment we set the vertical permeability equal to 10 % of the horizontal permeability which is included as a parameter to be estimated. Instead of estimating the vertical permeability directly we include a parameter, *multz*, which describes how well fluids will flow between model layers. This is a constant for each layer, which is multiplied with the vertical permeability to get the effective vertical communication between two layers. Some of the model layers are also assumed to be more or less impermeable for vertical flow and the estimates of *multz* should allow us to determine the layers with low vertical communication.

Porosity and permeability fields

We have also included the full three dimensional porosity and permeability fields, $\phi(\boldsymbol{x})$ and $\boldsymbol{k}_h(\boldsymbol{x})$, as variables to be estimated. The porosity is important to be able to estimate the volume of oil a part of the reservoir can contain, e.g. by increasing the porosity in a region we allow for more oil to be accommodated there. The permeability determines how well fluids are flowing through the reservoir and need to be adjusted to match the observed production rate as well as the timing of the water breakthrough.

16.2.2 State vector

For the combined parameter and state estimation problem we define the state vector to contain dynamic variables of the reservoir model, such as the pressure and saturations, and static variables as defined above. With the parameters included in this example the EnKF update of each ensemble member can be written in a simple form as

$$\underbrace{\left\{\begin{array}{c} P \\ S_\text{w} \\ S_\text{g} \\ R_\text{s} \\ \boldsymbol{k}_h \\ \phi \\ multz \\ multflt \\ WOC \\ GOC \end{array}\right\}_j}_{\text{Update}} = \underbrace{\left\{\begin{array}{c} P \\ S_\text{w} \\ S_\text{g} \\ R_\text{s} \\ \boldsymbol{k}_h \\ \phi \\ multz \\ multflt \\ WOC \\ GOC \end{array}\right\}_j}_{\text{Forecast}} + \sum_i \alpha_{ji} \underbrace{\left\{\begin{array}{c} C(P, d_i) \\ C(S_\text{w}, d_i) \\ C(S_\text{g}, d_i) \\ C(R_\text{s}, d_i) \\ C(\boldsymbol{k}_h, d_i) \\ C(\phi, d_i) \\ C(multz, d_i) \\ C(multflt, d_i) \\ C(WOC, d_i) \\ C(GOC, d_i) \end{array}\right\}_j}_{\text{Covariances}}, \quad (16.1)$$

where j is a counter for the ensemble members and i is a counter for the measurements. The coefficients, α_{ji}, define the impact each measurement has on the update of the ensemble members.

It is seen that the different dynamic and static variables are updated by adding weighted covariances between the modelled measurements and the variables, one for each measurement. Note that both the state variables and the various parameters are updated simultaneously.

The reason why it is possible to update the parameters given only rate information from the wells, is that the rates are dependent on the properties of the reservoir as given by the parameter set defined above. Thus, there will exist correlations between reservoir properties and the observed production rates.

Considering that the porosity and permeability are defined as 3D fields with one unknown on each grid node, there is a large number of parameters to be estimated in the current system. However, the number of degrees of freedom of the parameter space is much less than the actual number of parameters. The reason is that the porosity and permeability are smooth fields and do not consist of independent numbers in each grid node. The smoothness is prescribed from prior statistics through horizontal and vertical correlations which characterizes each depositional environment in the model. This effectively reduces the actual dimension of the problem and makes it tractable using a finite ensemble size in the EnKF.

In a particular application, where we are trying to estimate, e.g. the permeability, this implies that we can only expect to find corrections to the permeability estimates which can be represented in the space spanned by the initial permeability ensemble. This is, however, only a practical restriction since its impact can be reduced by either increasing the ensemble size or by chosing the initial ensemble wisely.

Another issue considers the scales which can be estimated for permeability. This is also clearly dependent on the initial choice of ensemble members. The "smoothness" of the members should be chosen to represent the true scales of the permeability field while keeping in mind that the limited number of wells and measurements certainly constrains the scales which can be resolved or estimated.

The model has about 82 000 active grid nodes, and the state vector then consists of 328 000 dynamic variables, 5 WOC and GOC contacts, 42 fault transmissibilities, 24 vertical multipliers, and 82 000 parameters for each of the porosity and permeability. An initial ensemble of 100 model states were generated.

Priors for the first guesses of the parameters are constructed based on the interpretation and information available from several data sources in the project. In particular the ensemble of contacts are simulated as independent numbers drawn from a Gaussian distribution with the mean equal to a best guess estimate and standard deviations of 20 m. Note that the contacts are only used initially to initialize the model, and then define the vertical saturation profile for each region. By including the contacts in the state vector, they will be updated in every assimilation step, although they are not used explicitly in the model but rather indirectly through the updates of the saturations. At the end of the assimilation experiment we have obtained improved estimates of the contacts, which can then be used in new model simulations or oil volume calculations.

The first guesses of the fault transmissibilities are set to either 1.0, 0.1 or 0.001 and with standard deviations of 20 %. This took into account knowledge about some of the faults that are known to be almost closed.

The vertical multipliers had first guesses equal to 1.0 except for three of the layers that were assumed to have low vertical permeability from the well-log data. Standard deviations were set to 10–20 %.

The porosity and permeability fields are simulated using the algorithm from Sect. 11.2, based on average values, uncertainties, and Gaussian variograms with horizontal and vertical de-correlation lengths, as specified from the geological interpretation of the reservoir.

16.3 Results

Initially we ran a pure ensemble integration of the prior ensemble. The spread of the results then provides an indication if the parameter space and the perturbations used, lead to a realistic representation of the uncertainty in the model predictions. In Fig. 16.2 we have plotted, as the red curves, the total accumulated oil production from the first 20 ensemble members together with the actual production. The upper plot shows the total accumulated field production while the middle and lower plots show the prediction of the accumulated production from the two individual production wells, P1 and P2. It is clear that the uncertainties in the initial parameter space leads to a large uncertainty in the model predictions. Without access to the production history it would not be possible to discriminate between the different realizations since all of them represent a statistically valid representation of the reservoir. From the individual wells it is also clear that there is a problem in the simulation of P1 where we have very little spread and much to large oil production. The simulation of P2 leads to a huge uncertainty, but it also captures the magnitude of the observed production.

In the EnKF experiment we have assimilated the production rates of oil (OPR), the gas-oil-ratio (GOR) and the water cut (WCT), from the two production wells. In the assimilation run we obtained rates of oil, water and gas which were in good agreement with the observations, as is expected since these are also the data assimilated. Another verification test was therefore performed. The ensemble of estimated parameters, i.e. porosity and permeability, fault and vertical multipliers, and the initial contacts, were all used in a new pure ensemble integration starting from time zero. The results from this simulation are plotted as the blue curves in Fig. 16.2. It is clear that the initially predicted uncertainties have been significantly reduced, and this must be attributed to the use of improved values of the static model parameters. Thus, we have successfully managed to compute improved estimates of a total of more than 164 000 poorly known model parameters.

The estimates of the porosity and permeability for one of the model layers are plotted in Fig. 16.3. The ensemble mean for the estimated porosity and permeability are given, respectively, in the upper and lower plots in the left column. It is clear that the estimated fields have developed clear and significant structures when compared with the first guess ensemble mean which was constant throughout the model layer. The standard deviations are reduced by approximately 25% during the assimilation updates.

Another test was also carried out where results were compared with a third production well, P3, which was excluded from the assimilation experiment. It was shown that the ensemble of improved model parameters resulted in

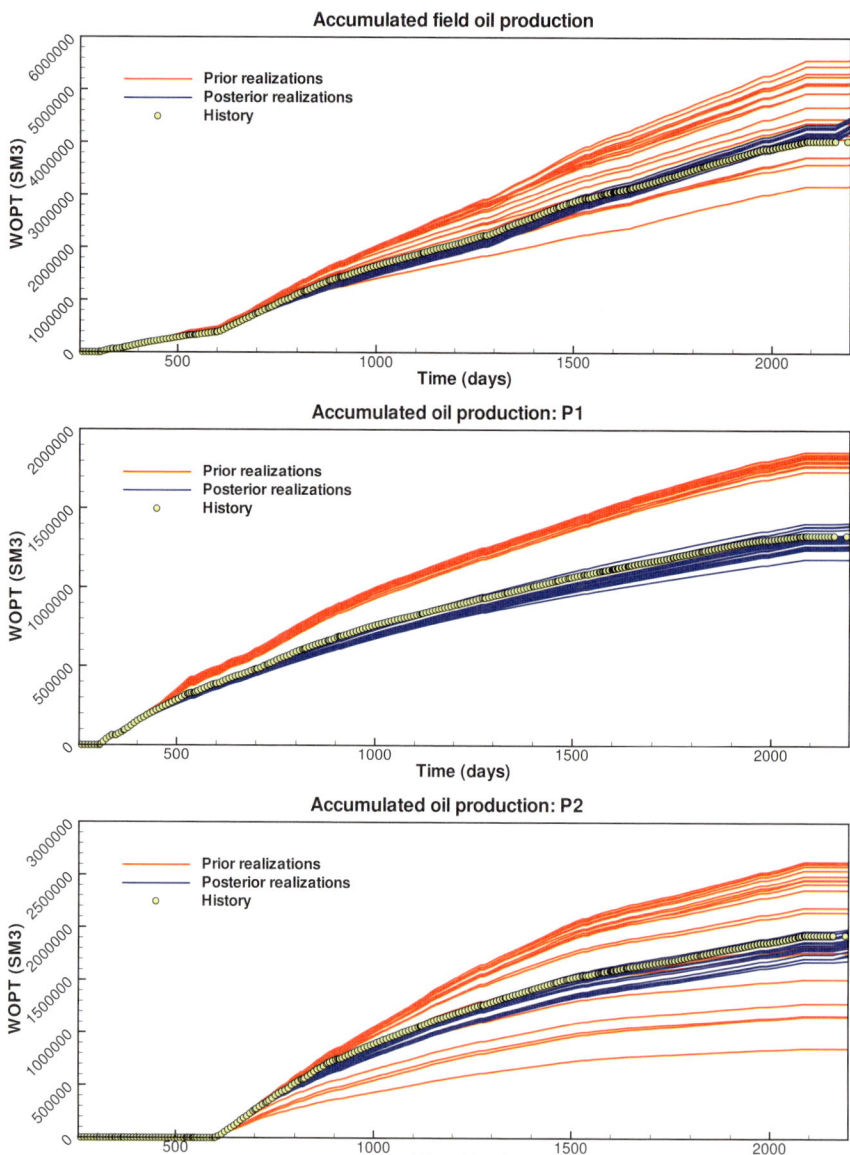

Fig. 16.2. Ensemble prediction based on initial ensemble of realizations. The total accumulated field oil production is shown in the upper plot. The middle and lower plots show the total accumulated oil production for the two wells

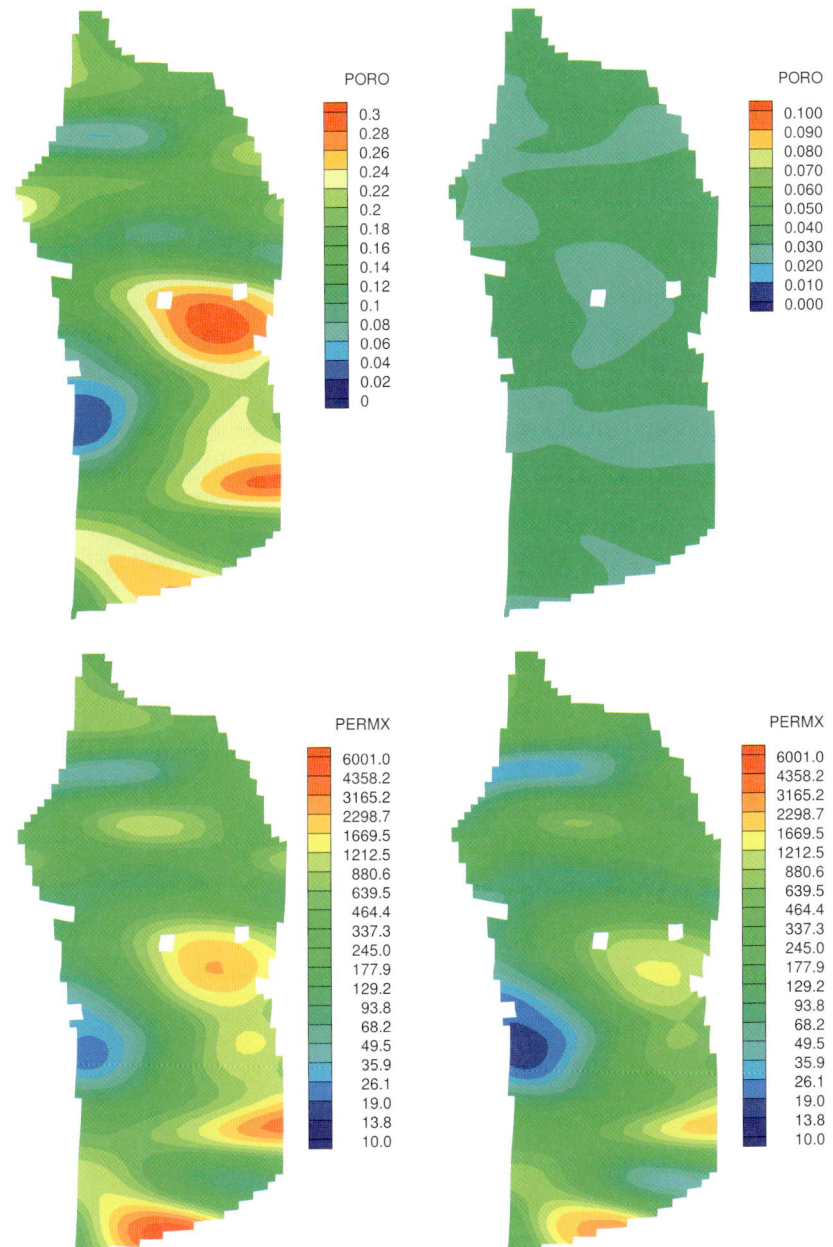

Fig. 16.3. Estimated porosity and permeability *(left column)* with standard deviations *(right column)* in one of the model layers

a significant improvement also for this well. This is an indication that the estimated model parameters are realistic and the improved realizations may then be used for the simulation and design of future wells.

16.4 Summary

The EnKF provides an ideal framework for real-time updating and prediction in reservoir simulation models. Every time new observations are available and are assimilated there is an improvement of the model parameters, and the associated model saturations and pressure. Thus, the analyzed ensemble provides optimal realizations which are conditioned on all previous data, and which can be used in a prediction of the future production. A single realization could be integrated forward in time starting from the ensemble mean or median, to obtain a quick forecast. Alternatively, the whole ensemble could be used in a forward integration to provide a future prediction with uncertainty estimates.

The EnKF has provided a tool for parameter estimation in cases with large number of poorly known parameters. It does not appear to suffer from the curse of dimensionality and multiple local minima, which have been observed in many other methods. This must be attributed to the sequential processing of observations, but also the fact that the EnKF also allows for model errors in addition to errors in the estimated parameters. Furthermore, the solution is searched for in the space spanned by the ensemble members rather than the high dimensional parameter space. Clearly, this approach should be examined in applications with other dynamical models as well.

A
Other EnKF issues

Below we have discussed some more specific issues related to the implementation and use of the EnKF. In particular we address the use of a local analysis scheme which updates the solution grid point by grid point using only nearby measurements and show how this allows for the solution to be searched for in a much larger space than originally represented by the N-member ensemble. The assimilation of nonlinear measurements is discussed and particular limitations of the method in this case is pointed out. The EnKF also allows for the assimilation of non-synoptic measurements meaning that, e.g. measurements which are arriving continuously in time, can be assimilated in batches at regular discrete time intervals. Finally, it is also possible to assimilate so-called time difference data, i.e. measurements which depend on the model state at two or more different times. A good example is the difference in the seismic response between surveys taken at different times, a commonly used data set in oil reservoir applications. Finally, we attempt to provide a chronological listing of the publications involving ensemble methods.

A.1 Local analysis

To avoid the problems associated with a large number of measurements, m, many operational assimilation schemes have made the assumption that only measurements located within a certain distance from a grid point will impact the analysis in this grid point. This allows for an algorithm where the analysis is computed grid point by grid point and only a subset of observations, located near the current grid point, is then used in the analysis. This algorithm has previously been discussed by *Haugen and Evensen* (2002), *Brusdal et al.* (2003), *Evensen* (2003), *Ott et al.* (2004).

In Chap. 14 it was shown that we could consistently compute the analysis update, with large number of measurements, using the subspace inversion schemes. It can also be shown that this is computationally more efficient than using a local scheme. However, there are other arguments for computing local analyses grid point by grid point. The analysis in the EnKF is computed in a space spanned by the ensemble members. This is a subspace which, in many cases, can be rather small compared to the total dimension of the model state. Computing the analysis grid point by grid point implies that, for each grid point, a small model state is solved for in a relatively large ensemble space. The analysis will then result from a different combination of ensemble members for each grid point, and this allows the analysis scheme to reach solutions not originally represented by the ensemble.

This algorithm is approximate and it does not solve the original problem posed. On the other hand, in the standard EnKF analysis we also introduce an approximation by using a limited ensemble size. With an infinite ensemble size there would be no need to use a local analysis scheme, since the whole solution space would be represented by the ensemble. The local analysis scheme will in many applications significantly reduce the impact of a limited ensemble size and allow for the use of the EnKF with high dimensional model systems like the ocean model application discussed in Chap. 15.

The degree of approximation introduced by the local analysis, is dependent on the range of influence defined for the observations. In the limit when this range becomes large enough to include all the data, for all the grid points, the solution becomes identical to the standard global analysis. Thus, the definition of the range parameter is a tuning problem. It should be large enough to include the information from measurements which contributes significantly, but small enough to eliminate the spurious impact of remote measurements.

The actual algorithm goes as follows. We first construct the input matrices to the global EnKF, i.e. the measured ensemble perturbations $\boldsymbol{S}$, the innovations $\boldsymbol{D}'$, the measurement perturbations $\boldsymbol{E}$, or the measurement error covariance matrix $\boldsymbol{C}_{\epsilon\epsilon}$. We then loop through the model grid, and for each grid point, e.g. (i,j) for a two dimensional model, we extract the rows from these matrices corresponding to measurements that will be used in the current update, and then compute the matrix $\boldsymbol{X}_{(i,j)}$ which defines the update for grid point (i,j).

The analysis at grid point (i,j), i.e. $\boldsymbol{A}^{\mathrm{a}}_{(i,j)}$, becomes

$$\boldsymbol{A}^{\mathrm{a}}_{(i,j)} = \boldsymbol{A}_{(i,j)} \boldsymbol{X}_{(i,j)} \qquad (\mathrm{A.1})$$
$$= \boldsymbol{A}_{(i,j)} \boldsymbol{X} + \boldsymbol{A}_{(i,j)}(\boldsymbol{X}_{(i,j)} - \boldsymbol{X}), \qquad (\mathrm{A.2})$$

where $\boldsymbol{X}$ is the global solution while $\boldsymbol{X}_{(i,j)}$ becomes the solution for a local analysis corresponding to grid point (i,j) where only the nearest measurements are used in the analysis. Thus, it is possible to compute the global analysis first, and then add the corrections from the local analysis if these are significant.

The quality of the EnKF analysis is clearly connected to the ensemble size used. We expect that a larger ensemble is needed for the global analysis than the local analysis to achieve the same quality of the result; i.e. in the global analysis a large ensemble is needed to properly explore the state space and to provide a consistent result for the global analysis which is as good as the local analysis. We expect this to be application dependent. Note also that the use of a local analysis scheme is likely to introduce non-dynamical modes, although the amplitudes of these will be small if a large enough influence radius is used when selecting measurements. This also relates to the discussions on localization and filtering of long range correlations by *Mitchell et al.* (2002).

Finally, note that it is not clear how the local analysis scheme is best implemented in the EnKS. In this case one would normally define the local analysis to use only measurements in a certain space-time domain, taking into account the propagation of information in the model together with the time scales of the model. Further, it is not practical to store $\boldsymbol{X}_{(i,j)}$ for all the grid points at each analysis step to compute a final EnKS update at the end of the simulation.

A.2 Nonlinear measurements in the EnKF

The original Kalman filter can only use measurements which are linearly related to the model state. The measurement operator is defined as a matrix, and this matrix needs to be multiplied with the error covariance matrix of the model state. If the observations are nonlinear functions of the model state this matrix formulation becomes invalid and the traditional solution is to linearize and iterate.

In the EnKF we take another approach, where we exploits that we never evaluate the full error covariance matrix, but rather work with the measurement of ensemble perturbations (*Evensen*, 2003). We start by augmenting the model state with a diagnostic variable which is the model prediction of the measurement. We first define the ensemble of m model predicted measurements as

$$\widehat{\boldsymbol{A}} = (\boldsymbol{m}(\boldsymbol{\psi}_1), \ldots, \boldsymbol{m}(\boldsymbol{\psi}_N)) \in \Re^{m \times N}, \qquad (A.3)$$

and the ensemble matrix

$$\left\{ \begin{matrix} \boldsymbol{A} \\ \widehat{\boldsymbol{A}} \end{matrix} \right\} = \left\{ \begin{matrix} \boldsymbol{A} \\ \boldsymbol{m}(\boldsymbol{A}) \end{matrix} \right\} \in \Re^{(n+m) \times N}, \qquad (A.4)$$

where m is the number of ensemble equivalents added to the original model state.

The EnKF analysis scheme can then be written as

$$\left\{ \begin{matrix} \boldsymbol{A}^{\mathrm{a}} \\ \widehat{\boldsymbol{A}}^{\mathrm{a}} \end{matrix} \right\} = \left\{ \begin{matrix} \boldsymbol{A} \\ \widehat{\boldsymbol{A}} \end{matrix} \right\} + \left\{ \begin{matrix} \boldsymbol{A}' \\ \widehat{\boldsymbol{A}}' \end{matrix} \right\} \widehat{\boldsymbol{A}}'^{\mathrm{T}} \left(\widehat{\boldsymbol{A}}' \widehat{\boldsymbol{A}}'^{\mathrm{T}} + (N-1) \boldsymbol{C}_{\epsilon\epsilon} \right)^{-1} (\boldsymbol{D} - \widehat{\boldsymbol{A}}). \qquad (A.5)$$

Normally we would just compute the EnKF analysis as

$$\boldsymbol{A}^{\mathrm{a}} = \boldsymbol{A} + \boldsymbol{A}' \widehat{\boldsymbol{A}}'^{\mathrm{T}} \left(\widehat{\boldsymbol{A}}' \widehat{\boldsymbol{A}}'^{\mathrm{T}} + (N-1) \boldsymbol{C}_{\epsilon\epsilon} \right)^{-1} (\boldsymbol{D} - \widehat{\boldsymbol{A}}), \qquad (A.6)$$

and we note that $\widehat{\boldsymbol{A}}^{\mathrm{a}}$ is never computed. Moreover, the analysis equation uses the covariance between $\boldsymbol{m}(\boldsymbol{\psi})$ and $\boldsymbol{\psi}$, through $\boldsymbol{A}'\widehat{\boldsymbol{A}}'^{\mathrm{T}}$.

The analysis is then a combination of model predicted error covariances between the observation equivalents, $\boldsymbol{m}(\boldsymbol{\psi})$, and all other model variables. Thus, we have a fully multivariate analysis scheme.

Note that the measurement of the analysis ensemble members $\boldsymbol{m}(\boldsymbol{A}^{\mathrm{a}})$, is not equal to the analyzed modelled measurement $\widehat{\boldsymbol{A}}^{\mathrm{a}} = \boldsymbol{m}(\boldsymbol{A})^{\mathrm{a}}$, the reason being the nonlinearity introduced by the nonlinear measurement functional. The residual between $\widehat{\boldsymbol{A}}^{\mathrm{a}}$ and $\boldsymbol{m}(\boldsymbol{A})^{\mathrm{a}}$ will thus serve as a measure of the approximation introduced using this algorithm.

In general we have seen that as long as the measurement functional is a monotonic function of the model state, and not too nonlinear, this procedure appears to work well. A non-monotonic function becomes problematic since it is then not clear whether an increase in the measurement value should lead to an increase or decrease in the update. Furthermore, a too nonlinear measurement functional may lead to a strongly non-Gaussian probability density function for the measured ensemble forecast, and the EnKF analysis scheme will fail.

This procedure has successfully been used for assimilation of sea level anomalies in an ocean modelling system in Chap. 15, where the sea level anomalies are weakly nonlinear functions of the model state. Further, in the reservoir application in Chap. 16, we assimilated rate measurements from production wells, which are nonlinearly related to the flow in the reservoir, and the flow in the reservoir is again nonlinearly related to the properties of the reservoir that we wish estimate. There is of course no guarantee that these measurements are monotonic functions, but it is anticipated that in a simplified picture, an increase in permeability will lead to an increase in the flow of oil in the reservoir, which leads to an increased production rate.

In Chap. 9 we defined the analysis equation as the minimum of the cost function (9.4). As long as the measurement operator is linear, the minimum of (9.4) is also the minimum variance estimate. This is no longer true for a nonlinear measurement operator, where the ensemble analysis equation introduces a linearization when computing the update. On the other hand, we now define N cost functions as

$$\mathcal{J}[\boldsymbol{\psi}_j^{\mathrm{a}}] = \left(\boldsymbol{\psi}_j^{\mathrm{a}} - \boldsymbol{\psi}_j^{\mathrm{f}}\right)^{\mathrm{T}} \bullet \left(\boldsymbol{C}_{\psi\psi}^{\mathrm{f}}\right)^{-1} \bullet \left(\boldsymbol{\psi}_j^{\mathrm{a}} - \boldsymbol{\psi}_j^{\mathrm{f}}\right) \\ + \left(\boldsymbol{d}_j - \boldsymbol{m}(\boldsymbol{\psi}_j^{\mathrm{a}})\right)^{\mathrm{T}} \boldsymbol{C}_{\epsilon\epsilon}^{-1} \left(\boldsymbol{d}_j - \boldsymbol{m}(\boldsymbol{\psi}_j^{\mathrm{a}})\right),$$
(A.7)

i.e. there is one cost function for each ensemble member. The bullets denote integration over the space where the state vector is defined. We can then derive the N analysis equations

$$\boldsymbol{\psi}_j^{\mathrm{a}} = \boldsymbol{\psi}_j^{\mathrm{f}} + \boldsymbol{A}' \boldsymbol{m}^{\mathrm{T}}(\boldsymbol{A}') \Big(\boldsymbol{m}(\boldsymbol{A}') \boldsymbol{m}^{\mathrm{T}}(\boldsymbol{A}') + (N-1)\boldsymbol{C}_{\epsilon\epsilon}\Big)^{-1} \Big(\boldsymbol{d}_j - \boldsymbol{m}(\boldsymbol{\psi}_j^{\mathrm{f}})\Big), \quad \text{(A.8)}$$

which constitutes the equations solved in the EnKF, and where the ensemble average provides the EnKF solution.

It is clear that the solution obtained from (A.8) does not exactly correspond to the minimum of (A.7), due to the linearization used when deriving (A.8) from (A.7). Thus, an alternative analysis scheme could be defined where (A.7) is solved directly, e.g. using a gradient method. Most effectively this may be done by solving (A.8) to get a first guess and then carrying out a few iterations using the gradient of (A.7) in a descent algorithm. The resulting algorithm will then to some degree resemble the randomized maximum likelihood algorithm (see e.g. *Gao and Reynolds*, 2005).

In the linear case these methods will both provide the same result, but (A.8) is a closed form solution and is the computationally most efficient. With modest nonlinearities in the measurement operator, the results should not be very different. With highly nonlinear measurement operators we also expect that the direct minimization of (A.7) may converge to local minima and the global solution may be hard to find. More research is needed to evaluate the use of these schemes when nonlinear measurement operators are used.

A.3 Assimilation of non-synoptic measurements

In some cases measurements occur with high frequency in time. An example is along track satellite data. It is not practical to perform an analysis every time there is a measurement. Further, the normal approach of assimilating, at one time instant, all data collected within a time interval, is not optimal. Based on the theory from *Evensen and van Leeuwen* (2000), it is possible to assimilate the non-synoptic measurements at one time instant by exploiting the time correlations in the ensemble. Thus, a measurement collected at a previous time allows for the computation of $\mathcal{M}[\boldsymbol{A}]$ at that time and thereby also the innovations. By treating these as augmented model variables, like in the case with nonlinear measurements, (A.6) can again be used but with $\widehat{\boldsymbol{A}}'$ now being interpreted as the measurements of the ensemble perturbations.

This procedure was presented in *Evensen* (2003) and has been further discussed by *Hunt et al.* (2004), where it is denoted four-dimensional ensemble Kalman filtering. The approach provides a simple and efficient approach for handling non-synoptic measurements which occur in many applications. The actual implementation only requires the measurements of the ensemble and the ensemble perturbations, which are evaluated and accumulated during the forward ensemble integration. Then, at the analysis time, this information is used to update the ensemble.

A.4 Time difference data

Time difference data, i.e. data which are related to the model state at two or more time instances, are difficult to assimilate using sequential assimilation methods. However, there is a way to do this properly when using the EnKF, and the algorithm is based on the formulation of the EnKS.

A detailed discussion of the algorithm is given by *Skjervheim et al.* (2006). It was shown that, given a data set which depends on the model state at two distinct times, t_k and t_j with $t_k < t_i < t_j$, and where t_i denotes update times using synoptic data in between t_k and t_j, one should proceed as follows:

During the forward integration, the updated ensemble at time t_k is augmented to the model state. Thus, at time t_k we start with the augmented ensemble

$$\left\{ \begin{array}{c} \boldsymbol{A}_k^{\mathrm{a}} \\ \boldsymbol{A}_k^{\mathrm{a}} \end{array} \right\}. \tag{A.9}$$

Then at time $t_i = t_{k+1}$ we will have integrated the dynamic part of the ensemble forward in time, while the augmented part is kept constant to get

$$\left\{ \begin{array}{c} \boldsymbol{A}_i^{\mathrm{f}} \\ \boldsymbol{A}_{kk}^{\mathrm{a}} \end{array} \right\}. \tag{A.10}$$

Here $\boldsymbol{A}_i^{\mathrm{f}}$ is just the forecast ensemble at time t_i while $\boldsymbol{A}_{kk}^{\mathrm{a}}$ is the analyzed ensemble at time t_k, updated with measurements up to time t_k.

An ensemble update at time t_i is computed using the EnKF analysis equation as

$$\left\{ \begin{array}{c} \boldsymbol{A}_i^{\mathrm{a}} \\ \boldsymbol{A}_{ki}^{\mathrm{a}} \end{array} \right\} = \left\{ \begin{array}{c} \boldsymbol{A}_i^{\mathrm{f}} \boldsymbol{X}_i \\ \boldsymbol{A}_{kk}^{\mathrm{a}} \boldsymbol{X}_i \end{array} \right\}, \tag{A.11}$$

where $\boldsymbol{A}_{ki}^{\mathrm{a}}$ is just the smoother solution at time t_k where the measurements at time t_i have been assimilated.

This procedure continues until time t_j where the time difference data is to be assimilated, and we have the augmented ensemble as

$$\left\{ \begin{array}{c} \boldsymbol{A}_j^{\mathrm{f}} \\ \boldsymbol{A}_{kj}^{\mathrm{a}} \end{array} \right\}. \tag{A.12}$$

We then use the time difference measurement operator which relates the measurements to both $\boldsymbol{A}_j^{\mathrm{f}}$ and $\boldsymbol{A}_{kj}^{\mathrm{a}}$ and compute a standard EnKF analysis. This procedure has proven to work well for the assimilation of seismic time difference data in *Skjervheim et al.* (2006), and should also be applicable to other types of data which are related to the model state at different time instants.

A.5 Ensemble Optimal Interpolation (EnOI)

Traditional optimal interpolation (OI) schemes have estimated or prescribed covariances using an ensemble of model states which has been sampled during a long time integration. Normally the estimated covariances are fitted to simple functional forms which are used uniformly throughout the model grid.

Based on the ensemble formulation used in the EnKF it is natural to derive an OI scheme where the analysis is computed in the space spanned by a stationary ensemble of model states sampled, e.g., during a long time integration. This approach is denoted ensemble optimal interpolation (EnOI), and was presented in *Evensen* (2003)

The EnOI analysis is computed by solving an equation similar to the update of the mean in the EnKF, see e.g. (13.2) but written as

$$\boldsymbol{\psi}_{\text{EnOI}}^{\text{a}}(\boldsymbol{x}) = \boldsymbol{\psi}_{\text{EnOI}}^{\text{f}}(\boldsymbol{x}) \\ + \alpha \boldsymbol{A}' \boldsymbol{S}^{\text{T}} (\alpha \boldsymbol{S} \boldsymbol{S}^{\text{T}} + (N-1)\boldsymbol{C}_{\epsilon\epsilon})^{-1} \left(\boldsymbol{d} - \mathcal{M}\left[\boldsymbol{\psi}_{\text{EnOI}}^{\text{f}}(\boldsymbol{x})\right] \right). \tag{A.13}$$

The analysis is now computed for only one single model state, and a parameter $\alpha \in (0,1]$ is introduced to allow for different weights on the ensemble versus measurements. Naturally, an ensemble consisting of model states sampled over a long time period will have a climatological variance which is too large to represent the actual error in the model forecast, and α is used to reduce the variance to a realistic level.

The EnOI method allows for the computation of a multivariate analysis in dynamical balance, just like the EnKF. However, a larger ensemble may be useful to ensure that it spans a large enough space to properly represent the correct analysis.

The EnOI can be an attractive approach to save computer time. Once the stationary ensemble is created, only a single model integration is required in addition to the analysis step where the final update cost is reduced to $\mathcal{O}(nN)$ floating point operations because only one model state is updated. The method is numerically extremely efficient but it will always provide a suboptimal solution compared to the EnKF. In addition it does not provide consistent error estimates for the solution.

A.6 Chronology of ensemble assimilation developments

This section attempts to provide a complete overview of the developments and applications related to the EnKF. In addition it also points to other recently proposed ensemble based methods and some smoother applications.

A.6.1 Applications of the EnKF

Applications involving the EnKF are numerous and started with the initial work by *Evensen* (1994a) and an additional example in *Evensen* (1994b) which showed that the EnKF resolved the closure problems reported from applications of the extended Kalman filter (EKF).

An application with assimilation of altimeter data for the Agulhas region was discussed in *Evensen and van Leeuwen* (1996) and later in a comparison with the ensemble smoother (ES) by *van Leeuwen and Evensen* (1996).

An example with the Lorenz equations was presented by *Evensen* (1997) where it was shown that the EnKF could track the phase transitions and find a consistent solution with realistic error estimates even for such a chaotic and nonlinear model.

Burgers et al. (1998) reviewed and clarified some points related to the perturbation of measurements in the analysis scheme, and also gave a nice interpretation supporting the use of the ensemble mean as the best estimate.

Houtekamer and Mitchell (1998) introduced a variant of the EnKF where two ensembles of model states are integrated forward in time, and statistics from one ensemble is used to update the other. The use of two ensembles was motivated by suggesting that this reduces possible inbreeding in the analysis. This has, however, lead to some dispute discussed in the comment by *van Leeuwen* (1999b) and the reply by *Houtekamer and Mitchell* (1999).

Miller et al. (1999) included the EnKF in a comparison with nonlinear filters and the extended Kalman filter, and concluded that it performed well, but could be beaten by a nonlinear and more expensive filter in difficult cases where the ensemble mean is not a good estimator.

Madsen and Cañizares (1999) compared the EnKF and the reduced rank square root implementation of the extended Kalman filter with a 2–D storm surge model. This is a weakly nonlinear problem and good agreement was found between the EnKF and the extended Kalman filter implementation.

Echevin et al. (2000) studied the EnKF with a coastal version of the Princeton Ocean Model and focussed in particular on the horizontal and vertical structure of multivariate covariance functions from sea surface height. It was concluded that the EnKF could capture anisotropic covariance functions resulting from the impact of coastlines and coastal dynamics, and had a particular advantage over simpler methodologies in such areas.

Evensen and van Leeuwen (2000) rederived the EnKF as a suboptimal solver for the general Bayesian problem of finding the posterior distribution given densities for the model prediction and the observations. From this formulation the general filter could be derived and the EnKF could be shown to be a suboptimal solver of the general filter where the prior densities are assumed to be Gaussian distributed.

Hamill and Snyder (2000) constructed a hybrid assimilation scheme by combining 3DVAR and the EnKF. The estimate is computed using the 3DVAR algorithm but the background covariance is a weighted average of the time evolving EnKF error covariance and the constant 3DVAR error covariance. A conclusion was that with increasing ensemble size the best results were found with larger weight on the EnKF error covariance.

Hamill et al. (2000) report from working groups in a workshop on ensemble methods.

Keppenne (2000) implemented the EnKF with a two layer shallow water model and examined the method in twin experiments assimilating synthetic altimetry data. He focused on the numerical implementation on parallel computers with distributed memory and found the approach efficient for such systems. He also examined the impact of ensemble size and concluded that realistic solutions could be found using a modest ensemble size.

Mitchell and Houtekamer (2000) introduced an adaptive formulation of the EnKF where the model error parameterization was updated by incorporating information from the innovations during the integration.

Park and Kaneko (2000) presented an experiment where the EnKF was used to assimilate acoustic tomography data into a barotropic ocean model.

Grønnevik and Evensen (2001) examined the EnKF for use in fish stock assessment, and also compared it with the ensemble smoother (ES) and the more recent ensemble Kalman smoother (EnKS).

Heemink et al. (2001) have been examining different approaches which combine ideas from Reduced Rank Square Root (RRSQRT) filtering and the EnKF to derive computationally more efficient methods.

Houtekamer and Mitchell (2001) continued the examination of the two-ensemble approach and introduced a technique for computing the global EnKF analysis in the case with many observations, and also a method for filtering of eventual long range spurious correlations caused by a limited ensemble size.

Pham (2001) reexamined the EnKF in an application with the Lorenz attractor and compared results with those obtained from different versions of the singular evolutive extended Kalman (SEEK) filter and a particle filter. Ensembles with very few members were used and this favoured methods like the SEEK where the "ensemble" of EOFs is selected to best possible represent the model attractor.

Verlaan and Heemink (2001) applied the RRSQRT and EnKF filters in test examples with the purpose of classifying and defining a measure of the degree of nonlinearity of the model dynamics. Such an estimate may have an impact on the choice of assimilation method.

Hansen and Smith (2001) proposed a method for producing analysis ensembles based on integrated use of the 4DVAR method and the EnKF. A probabilistic approach was used and lead to high numerical cost, but an improved estimate could be found compared to 4DVAR and the EnKF used separately.

Hamill et al. (2001) examined the impact of ensemble size on noise in distant covariances. They evaluated the impact of using an "inflation factor" as introduced by *Anderson and Anderson* (1999), and also the use of a Schur product of the covariance with a correlation function to localize the background covariances as previously discussed by *Houtekamer and Mitchell* (2001). The inflation factor is used to replace the forecast ensemble according to

$$\boldsymbol{\psi}_j = \rho(\boldsymbol{\psi}_j - \overline{\boldsymbol{\psi}}) + \overline{\boldsymbol{\psi}}, \tag{A.14}$$

with ρ slightly greater than one (typically 1.01). The purpose is to account for a slight under representation of variance due to the use of a small ensemble.

Bishop et al. (2001) used an implementation of the EnKF in an observation system simulation experiment. Ensemble predicted error statistics were used to determine the optimal configuration of future targeted observations. The application typically looked at a case where additional targeted measurements could be deployed over the next few days and the deployment could be optimized to minimize the forecast errors in a selected region. The methodology was named "ensemble transform Kalman filter" and it was further examined by *Majumdar et al.* (2001).

Reichle et al. (2002) give a nice discussion of the EnKF in relation to the optimal representer solution. They find good convergence of the EnKF towards the representer solution with the difference being caused by the Gaussian assumptions used in the EnKF at analysis steps. These are avoided in the representer method which solves for the maximum likelihood smoother estimate.

Anderson (2001) proposed a method denoted the "ensemble adjustment Kalman filter" where the analysis is computed without adding perturbations to the observations. A drawback may be the required inversion of the measurement error covariance when this is nondiagonal. This method becomes a variant of the square root algorithm used by *Bishop et al.* (2001).

Bertino et al. (2002) applied the EnKF and the RRSQRT filter with a model for the Odra estuary. The two methods were compared and used to assimilate real observations to assess the potential for operational forecasting in the lagoon. This is a relatively linear model and the EnKF and the RRSQRT filter provided similar results.

Eknes and Evensen (2002) examined the EnKF with a 1–D three component marine ecosystem model with focus on sensitivity to the properties of the assimilated measurements and the ensemble size. It was found that the EnKF could handle strong nonlinearities and instabilities which occur during the spring bloom.

Allen et al. (2002) takes the *Eknes and Evensen* (2002) work one step further by applying the method with a 1–D version of ERSEM for a site in the Mediterranean Sea. They showed that even with such a complex model it is possible to find an improved estimate by assimilating in situ data into the model.

Haugen and Evensen (2002) applied the EnKF to assimilate sea level anomalies and sea surface temperature data into a version of the Miami Isopycnic Coordinate Ocean Model (MICOM) by *Bleck et al.* (1992) for the Indian Ocean. The paper provided an analysis of regionally dependent covariance functions in the tropics and subtropics and also the multivariate impact of assimilating satellite observations.

A.6 Chronology of ensemble assimilation developments 259

Mitchell et al. (2002) examined the EnKF with a global atmospheric general circulation model with simulated data resembling realistic operational observations. They assimilated 80 000 observations daily. The system was examined with respect to required ensemble size, and the effect of localization (local analysis at a grid point using only nearby measurements). It was found that severe localization could lead to imbalance, but with large enough ratio of influence for the measurements, this was not a problem and no digital filtering was required. In the experiments they also included model errors and demonstrated the importance of this to avoid filter divergence. This work is a significant step forward and it shows promising results with respect to using the EnKF with atmospheric forecast models.

Whitaker and Hamill (2002) proposed another version of the EnKF where the perturbation of observations are avoided. The scheme was tested for small ensemble sizes (10–20 members) where it had a clear benefit on the results when compared to the EnKF which has larger sampling errors with such small ensemble sizes.

Nævdal et al. (2002) used the EnKF in a reservoir application to estimate model permeability. They showed that there could be a great benefit of using the EnKF to improve the model through parameter estimation, and that this could lead to improved predictions.

Brusdal et al. (2003) discussed an application similar to the Indian Ocean application by *Haugen et al.* (2002), but focussed on the North Atlantic. In addition, this paper included an extensive comparison of the theoretical background of the EnKF, EnKS and the SEEK filter, and also compared results from these methods.

Natvik and Evensen (2003a,b) presented the first realistic 3–D application of the EnKF with a marine ecosystem model. These papers proved the feasibility of assimilating SeaWiFS ocean colour data to control the evolution of a marine ecosystem model. In addition several diagnostic methods were introduced which can be used to examine the statistical and other properties of the ensemble.

Keppenne and Rienecker (2003) implemented a massively parallel version of the EnKF with the Poseidon isopycnic coordinate ocean model for the tropical Pacific. They demonstrated the assimilation of in situ observations and focussed on the parallelization of the model and analysis scheme for computers with distributed memory. They also showed that regionalization of background covariances has negligible impact on the quality of the analysis.

Bertino et al. (2003) used the EnKF with a simple ecosystem model and introduced a transformation of the biological variables, based on a Gaussian anamorphosis, to make the ensemble predictions more Gaussian. It was shown that this resulted in a more consistent analysis update. This approach seems promising for handling modest deviations from Gaussianity in the predicted ensemble.

Lisæter et al. (2003) presented the first application of the EnKF, or any advanced data assimilation system, with a coupled ice-ocean general circulation model, for assimilation of sea-ice concentration data. The results of the study were positive and it was concluded that the assimilation of the sea-ice concentrations had a positive impact on the evolution of the Arctic ocean and sea ice. The study lead to an improved understanding of seasonally dependent correlations between ice and ocean variables and it became clear that a simple OI based assimilation scheme would not be able to handle this case properly.

Evensen (2003) gave a review of the EnKF and introduced a new notation or formulation of the method in the *ensemble space*. The paper discussed issues related to the formulation of *red model noise*, the computation of an efficient *local analysis*, the assimilation of *non-synoptic* and *nonlinear* measurements. It further, reformulated the EnKS in the new ensemble notation and presented an ensemble optimal interpolation (EnOI) scheme. The use of the EnKF and EnKS for *parameter and bias estimation* was illustrated. Much focus was given to the practical implementation of the analysis scheme, and an efficient but approximate algorithm was presented which is correct for large ensemble sizes and small number of measurements. The algorithm was later shown to perform poorly for large number of measurements by *Kepert* (2004), and better algorithms where this approximation is avoided has later been proposed by *Evensen* (2004).

Zang and Malanotte-Rizzoli (2003) compared an implementation of a reduced rank extended Kalman filter with the EnKF. The model used was a quasi-geostrophic ocean model, which, dependent on a viscosity parameter exhibit various degrees of nonlinearity. It was found that the EnKF could handle both the strongly and weakly nonlinear cases, while the reduced rank extended Kalman filter only worked well with the nearly linear model.

Lorentzen et al. (2003) used the EnKF for parameter estimation in a two-phase flow model. It was shown that the tuning of parameters resulted in more consistent solutions.

Nævdal et al. (2003) continued the development of the EnKF for estimation of permeability in the whole reservoir, and again showed promising results.

Kivman (2003) used the EnKF for sequential parameter estimation in the nonlinear and stochastic Lorenz system. Results were compared with the sequential importance resampling filter (SIR) which was shown to perform better than the EnKF in this case. This can be expected since the SIR solves the full problem without any linearizations, but on the other hand it has a problem handling models with large state spaces, where huge ensemble sizes may be needed.

van Leeuwen (2003) used a sequential importance resampling filter (SIR) with the Korteweg-De Vries equation and compared results with the EnKF. It was found that the SIR could handle the nonlinearity better than the EnKF. In fact the linear analysis in the EnKF sometimes produced slightly negative values for some of the ensemble members, which causes the model to go unstable. This might be avoided by including some numerical corrections on the updated results. Still the general conclusion that the SIR will better

A.6 Chronology of ensemble assimilation developments

handle nonlinear or non-Gaussian distributions, is valid, but at the expense of integrating a much larger ensemble.

Tippett et al. (2003) gave a summary of the square root filters developed by *Bishop et al.* (2001), *Anderson* (2001) and *Whitaker and Hamill* (2002). See also the general discussion of ensemble methods in a "local least squares framework" given by *Anderson* (2003).

Snyder and Zhang (2003) used the EnKF to assimilate simulated Doppler radar observations of radial velocity in a nonhydrostatic, cloud scale model. The results suggested that the EnKF can handle the nonlinearities in the dynamics of moist convection.

Crow and Wood (2003) examined the EnKF for assimilation of remotely sensed observations of surface brightness temperature in a land surface moist model. Even though the distributions are characterized by skewness it was concluded that the EnKF resulted in improved estimates.

Anderson (2003) discussed different ensemble Kalman filters and in particular presented a local least squares framework for ensemble filtering, which lead to an efficient two step update procedure, consisting of the computation of the update increments followed by the ensemble member update. Relations to Bayesian estimation and some nonlinear filters were also discussed.

Kepert (2004) discussed the approximate analysis scheme from *Evensen* (2003) and the use of a low-rank representation of the measurement error covariance matrix. It was clearly pointed out that the analysis scheme from *Evensen* (2003) performed poorly with large number of measurements. Further, the use of a low-rank representation of the measurement error covariance matrix could lead to a loss of rank in the ensemble during the update step.

Evensen (2004) introduced a new square root implementation of the EnKF. It was shown that the square root schemes also need to include a randomization step, similar to the perturbation of measurements used in the traditional EnKF, to provide consistent results. In addition a sub-space pseudo inversion algorithm was introduced which significantly reduced the computational cost when many observations are used. It was also shown that the problems using a low rank measurement error covariance matrix, as pointed out by *Kepert* (2004), can be avoided by ensuring that the low rank measurement error covariance matrix is fully contained in the space defined by the measurement of the ensemble perturbations, i.e. the matrix $\boldsymbol{S}$.

Lawson and Hansen (2004) compared deterministic ensemble filters, i.e. square root filters without randomization, with stochastic ensemble filters like the traditional EnKF with measurement perturbations. They identified some problems with the deterministic filters which gave updated ensembles with rather poor properties. The reason for these results was explained in Chap. 13 where it was shown that a randomization of the ensemble perturbations need to be introduced.

Annan and Hargreaves (2004) discussed an application of the EnKF for parameter estimation in the Lorenz model. The estimation was performed in a climatological sense, to produce a model with the correct climatology.

Ott et al. (2004) provide an extended discussion on the use of the local analysis computation from *Evensen* (2003).

Hunt et al. (2004) have elaborated further on the use on non-synoptic measurements as was discussed by *Evensen* (2003), and point out that the EnKF can easily be used to consistently assimilate measurements taken at times differing from the actual analysis time. This is particularly useful when assimilating a stream of data with high frequency in time.

Zou and Ghanem (2004) discussed the use of the EnKF for multiscale data assimilation. This relates to the assimilation of measurements of processes with different scales.

Nohara and Tanaka (2004) used the EnKF to update the forecast ensemble in atmospheric ensemble predictions.

Dowell et al. (2004) extended the work by *Snyder and Zhang* (2003) to assimilate real observations from Doppler radar in a supercell storm event. They studied the impact of the choice of initial ensemble and localization in the EnKF, and in general obtained acceptable results.

Gu and Oliver (2004) examined the EnKF for combined parameter and state estimation in a standardized reservoir test case. They obtained promising results using a fairly small ensemble size but also pointed out several issues for further investigation.

Annan et al. (2005) performed another parameter estimation study, using the EnKF, now for an intermediate complexity atmospheric general circulation model, with the objective of tuning the model climatology. It is concluded that the EnKF provides a promising alternative to traditional Bayesian sampling methods used for these problems and it can handle the curse of dimensionality.

Nerger et al. (2005) compared the EnKF in its traditional implementation with the singular evolutive extended Kalman filter (SEEK) from *Pham et al.* (1998), and the more sophisticated singular evolutive interpolated Kalman filter (SEIK) by *Pham* (2001). It should be noted that the EnKF with improved sampling as discussed in Chap. 11 will have similar properties and computational cost as the SEIK filter. Further, with the new subspace inversion schemes the analysis in the EnKF will be more computationally efficient than the SEIK, and it also handles a non-diagonal measurement error covariance matrix.

Caya et al. (2005) have compared the EnKF with 4DVAR for radar data assimilation in an atmospheric general circulation model. Simulated data were used, and several aspects of the EnKF versus 4DVAR were discussed.

Hacker and Snyder (2005) used the EnKF for assimilation of surface observations in a 1D atmospheric planetary boundary layer model. Results showed that the simulated observations could be assimilated by the EnKF, and effectively constrained the evolution of the model.

Zhang et al. (2005) examined the possibilities of applying an ensemble Kalman filter with a global ocean circulation model used for ENSO forecasting. They found that the EnKF based predictions appeared to improve upon previous 3DVAR results.

Hamill and Whitaker (2005) studied how one could account for model errors related to unresolved scales in a dynamical model. Parameterizations such as covariance inflation and additive errors were examined.

Leeuwenburgh (2005) assimilated simulated along track radar altimeter data into an ocean general circulation model for the tropical Pacific, to examine how well the subsurface dynamics could be recovered from surface measurements. The assimilation of altimeter data using the EnKF showed a positive impact and it was concluded that it might lead to improved ENSO forecasts.

Houtekamer et al. (2005) has further discussed their implementation of the EnKF in a near operational setting and compared its performance with the 4DVAR implementation. They conclude that operationally interesting results can be obtained with the EnKF using an ensemble of moderate size and the current development will continue.

Moradkhani et al. (2005) discussed a dual state-parameter estimation problem in hydrological models using the EnKF. It is concluded that their method is a useful alternative to traditional parameter estimation methods.

Gao and Reynolds (2005) compared the EnKF with another method named randomized maximum likelihood. They used the same reservoir example as *Gu and Oliver* (2004), and pointed out certain similarities between the methods.

Liu and Oliver (2005a,b) examined the EnKF for facies estimation in a reservoir simulation model. This is a highly nonlinear problem where the reservoir consists of sand and shale classes of vastly different porosity and permeability. Thus, the pdf for the petro-physical parameters will be multimodal, and it is not clear how the EnKF can handle this. A method was used, where the facies distribution for each ensemble member is represented by two normal distributed Gaussian fields, using a method named truncated pluri-Gaussian simulation (*Lantuéjoul*, 2002).

Wen and Chen (2005) provided another discussion on the use of EnKF for estimation of the permeability field in a two dimensional reservoir simulation model, and they also examined the impact of ensemble size in their experiments.

Lorentzen et al. (2005) provided another example where the EnKF was applied with the model from *Gu and Oliver* (2004), and focused on the stability of the results with respect to the choice of initial ensemble.

Skjervheim et al. (2005) used the EnKF to assimilate 4D seismic data. It was shown that the EnKF could handle the large data sets and that a positive impact could be found despite the high noise level in the data.

Zafari and Reynolds (2005) used simple but highly nonlinear models to examine the validity of the linear update scheme used in the EnKF. They concluded that the EnKF has problems with multimodal distributions where the mean is not a good estimator, but on the other hand obtained reasonable results with a less nonlinear but more realistic reservoir model. They also showed that the rerun algorithm proposed by *Wen and Chen* (2005) is inconsistent and should not be used.

Torres et al. (2006) used the EnKF in application with the ERSEM ecosystem model. They used the EnKF on transformed variables as in *Bertino et al.* (2003) and found this to improve the results.

A.6.2 Other ensemble based filters

The EnKF can also be related to some other sequential filters such as the singular evolutive extended Kalman (SEEK) filter by *Pham et al.* (1998), *Brasseur et al.* (1999), *Carmillet et al.* (2001) (see also *Brusdal et al.*, 2003, *Nerger et al.*, 2005, for a comparison of the SEEK and the EnKF); the reduced rank square root (RRSQRT) filter by *Verlaan and Heemink* (2001); and error subspace statistical estimation (ESSE) by *Lermusiaux and Robinson* (1999a,b) and *Lermusiaux* (2001) which can be interpreted as an EnKF where the analysis is computed in the space spanned by the EOFs of the ensemble.

A.6.3 Ensemble smoothers

Some publications have focussed on the extension of the EnKF to a smoother. The first formulation was given by *van Leeuwen and Evensen* (1996) who introduced the ensemble smoother (ES). This method has later been examined in *Evensen* (1997) with the Lorenz attractor; applied with a QG model to find a steady mean flow by *van Leeuwen* (1999a) and for the time dependent problem in *van Leeuwen* (2001); and for fish stock assessment by *Grønnevik and Evensen* (2001). *Evensen and van Leeuwen* (2000) re-examined the smoother formulation and derived a new algorithm with better properties named the ensemble Kalman smoother (EnKS). This method has also been examined in *Grønnevik and Evensen* (2001) and *Brusdal et al.* (2003).

A.6.4 Ensemble methods for parameter estimation

There are now several publications discussing the potential of using the EnKF for parameter estimation. We refer to *Evensen* (2003) which outlines how the model state can be augmented with a set of poorly known parameters, and the joint model state and parameters are then updated simultaneously. Applications of the EnKF for parameter estimation include, *Nævdal et al.* (2002), *Lorentzen et al.* (2003), *Kivman* (2003), *Nævdal et al.* (2003), *Annan and Hargreaves* (2004), *Gu and Oliver* (2004), *Annan et al.* (2005), *Moradkhani et al.* (2005), *Lorentzen et al.* (2005), *Gao and Reynolds* (2005), *Wen and Chen* (2005), *Liu and Oliver* (2005a,b), *Skjervheim et al.* (2005), and *Zafari and Reynolds* (2005).

A.6.5 Nonlinear filters and smoothers

Another extension of the EnKF relates to the derivation of an efficient method for solving the nonlinear filtering problem, i.e. taking non-Gaussian contributions in the predicted error statistics into account when computing the analysis. These are discarded in the EnKF (see *Evensen and van Leeuwen*, 2000), and a fully nonlinear filter is expected to improve the results when used with nonlinear dynamical models with multimodal behaviour where the predicted error statistics is far from Gaussian. Implementations of nonlinear filters in the assimilation community have been based on either kernel approximations or particle interpretations, e.g. see *Miller et al.* (1999), *Anderson and Anderson* (1999), *Pham* (2001), *Miller and Ehret* (2002) and *van Leeuwen* (2003). See also the particle filter web page

http://www-sigproc.eng.cam.ac.uk/smc,

which contains a number of references to people and publications relevant to sequential Monte Carlo Methods and particle filtering. More research is needed before these can claimed to be practical for realistic high dimensional systems.

References

Allen, J. I., M. Eknes, and G. Evensen, An ensemble Kalman filter with a complex marine ecosystem model: Hindcasting phytoplankton in the Cretan Sea, *Annales Geophysicae*, *20*, 1–13, 2002.

Anderson, J. L., An ensemble adjustment Kalman filter for data assimilation, *Mon. Weather Rev.*, *129*, 2884–2903, 2001.

Anderson, J. L., A local least squares framework for ensemble filtering, *Mon. Weather Rev.*, *131*, 634–642, 2003.

Anderson, J. L., and S. L. Anderson, A Monte Carlo implementation of the nonlinear filtering problem to produce ensemble assimilations and forecasts, *Mon. Weather Rev.*, *127*, 2741–2758, 1999.

Annan, J. D., and J. C. Hargreaves, Efficient parameter estimation for a highly chaotic system, *Tellus*, *56A*, 520–526, 2004.

Annan, J. D., J. C. Hargreaves, N. R. Edwards, and R. Marsh, Parameter estimation in an intermediate complexity earth system model using an ensemble Kalman filter, *Ocean Modelling*, *8*, 135–154, 2005.

Azencott, R., *Simulated Annealing*, John Wiley, New York, 1992.

Barth, N., and C. Wunsch, Oceanographic experiment design by simulated annealing, *J. Phys. Oceanogr.*, *20*, 1249–1263, 1990.

Bennett, A. F., *Inverse Methods in Physical Oceanography*, Cambridge University Press, 1992.

Bennett, A. F., *Inverse Modeling of the Ocean and Atmosphere*, Cambridge University Press, 2002.

Bennett, A. F., and B. S. Chua, Open-ocean modeling as an inverse problem: The primitive equations, *Mon. Weather Rev.*, *122*, 1326–1336, 1994.

Bennett, A. F., and R. N. Miller, Weighting initial conditions in variational assimilation schemes, *Mon. Weather Rev.*, *119*, 1098–1102, 1990.

Bennett, A. F., L. M. Leslie, C. R. Hagelberg, and P. E. Powers, Tropical cyclone prediction using a barotropic model initialized by a generalized inverse method, *Mon. Weather Rev.*, *121*, 1714–1729, 1993.

Bennett, A. F., B. S. Chua, and L. M. Leslie, Generalized inversion of a global numerical weather prediction model, *Meteorol. Atmos. Phys.*, *60*, 165–178, 1996.

Bentsen, M., G. Evensen, H. Drange, and A. D. Jenkins, Coordinate transformation on a sphere using a conformal mapping, *Mon. Weather Rev.*, *127*, 2733–2740, 1999.

Bertino, L., G. Evensen, and H. Wackernagel, Combining geostatistics and Kalman filtering for data assimilation in an estuarine system, *Inverse Methods*, *18*, 1–23, 2002.

Bertino, L., G. Evensen, and H. Wackernagel, Sequential data assimilation techniques in oceanography, *International Statistical Review*, *71*, 223–241, 2003.

Bishop, C. H., B. J. Etherton, and S. J. Majumdar, Adaptive sampling with the ensemble transform Kalman filter. Part I: Theoretical aspects, *Mon. Weather Rev.*, *129*, 420–436, 2001.

Bleck, R., An oceanic general circulation model framed in hybrid isopycnic-cartesian coordinates, *Ocean Modelling*, *4*, 55–88, 2002.

Bleck, R., C. Rooth, D. Hu, and L. T. Smith, Salinity-driven thermohaline transients in a wind- and thermohaline-forced isopycnic coordinate model of the North Atlantic, *J. Phys. Oceanogr.*, *22*, 1486–1515, 1992.

Bohachevsky, I. O., M. E. Johnson, and M. L. Stein, Generalized simulated annealing for function optimization, *Technometrics*, *28*, 209–217, 1986.

Bouttier, F., A dynamical estimation of forecast error covariances in an assimilation system, *Mon. Weather Rev.*, *122*, 2376–2390, 1994.

Bowers, C. M., J. F. Price, R. A. Weller, and M. G. Briscoe, Data tabulations and analysis of diurnal sea surface temperature variability observed at LOTUS, *Tech. Rep. 5*, Woods Hole Oceanogr. Inst., Woods Hole, Mass., 1986.

Brasseur, P., J. Ballabrera, and J. Verron, Assimilation of altimetric data in the mid-latitude oceans using the SEEK filter with an eddy-resolving primitive equation model, *J. Marine. Sys.*, *22*, 269–294, 1999.

Brusdal, K., J. Brankart, G. Halberstadt, G. Evensen, P. Brasseur, P. J. van Leeuwen, E. Dombrowsky, and J. Verron, An evaluation of ensemble based assimilation methods with a layered OGCM, *J. Marine. Sys.*, *40-41*, 253–289, 2003.

Burgers, G., P. J. van Leeuwen, and G. Evensen, Analysis scheme in the ensemble Kalman filter, *Mon. Weather Rev.*, *126*, 1719–1724, 1998.

Carmillet, V., J.-M. Brankart, P. Brasseur, H. Drange, and G. Evensen, A singular evolutive extended Kalman filter to assimilate ocean color data in a coupled physical-biochemical model of the North Atlantic, *Ocean Modelling*, *3*, 167–192, 2001.

Caya, A., J. Sun, and C. Snyder, A comparison between the 4DVAR and the ensemble Kalman filter techniques for radar data assimilation, *Mon. Weather Rev.*, *133*, 3081–3094, 2005.

Chen, W., B. R. Bakshi, P. K. Goel, and S. Ungarala, Bayesian estimation via sequential Monte Carlo sampling: Unconstrained nonlinear dynamic systems, *Int. Eng. Chem. Res.*, *43*, 4012–4025, 2004.

Chilés, J.-P., *Geostatistics: Modeling Spatial Uncertainty*, Wiley Series in Probability and Statistics, John Wiley & Sons, 1999.

Chua, B. S., and A. F. Bennett, An inverse ocean modeling system, *Oceanogr. Meteor.*, *3*, 137–165, 2001.

Courtier, P., Dual formulation of variational assimilation, *Q. J. R. Meteorol. Soc.*, *123*, 2449–2461, 1997.

Courtier, P., and O. Talagrand, Variational assimilation of meteorological observations with the adjoint vorticity equation II: Numerical results, *Q. J. R. Meteorol. Soc.*, *113*, 1329–1347, 1987.

Courtier, P., J. N. Thepaut, and A. Hollingsworth, A strategy for operational implementation of 4DVAR, using an incremental approach, *Q. J. R. Meteorol. Soc.*, *120*, 1367–1387, 1994.

Crow, W. T., and E. F. Wood, The assimilation of remotely sensed soil brightness temperature imagery into a land surface model using ensemble Kalman filtering: A case study based on ESTAR measurements during SGP97, *Advances in Water Resources*, *26*, 137–149, 2003.

Derber, J., and A. Rosati, A global oceanic data assimilation system, *J. Phys. Oceanogr.*, *19*, 1333–1347, 1989.

Doucet, A., N. de Freitas, and N. Gordon (Eds.), *Sequential Monte Carlo Methods in Practice*, Statistics for Engineering and Information Science, Springer-Verlag New York, 2001.

Dowell, D. C., F. Zhang, L. J. Wicker, C. Snyder, and N. A. Crook, Wind and temperature retrievals in the 17 May 1981 Arcadia, Oklahoma, supercell: Ensemble Kalman filter experiments, *Mon. Weather Rev.*, *132*, 1982–2005, 2004.

Duane, S., A. D. Kennedy, B. J. Pendleton, and D. Roweth, Hybrid Monte Carlo, *Phys. Lett. B.*, *195*, 216–222, 1987.

Echevin, V., P. De Mey, and G. Evensen, Horizontal and vertical structure of the representer functions for sea surface measurements in a coastal circulation model, *J. Phys. Oceanogr.*, *30*, 2627–2635, 2000.

Egbert, G. D., A. F. Bennett, and M. G. G. Foreman, TOPEX/POSEIDON tides estimated using a global inverse model, *J. Geophys. Res.*, *99*, 24,821–24,852, 1994.

Eknes, M., and G. Evensen, Parameter estimation solving a weak constraint variational formulation for an Ekman model, *J. Geophys. Res.*, *102*, 12,479–12,491, 1997.

Eknes, M., and G. Evensen, An ensemble Kalman filter with a 1–D marine ecosystem model, *J. Marine. Sys.*, *36*, 75–100, 2002.

Evensen, G., Using the extended Kalman filter with a multilayer quasi-geostrophic ocean model, *J. Geophys. Res.*, *97*, 17,905–17,924, 1992.

Evensen, G., Sequential data assimilation with a nonlinear quasi-geostrophic model using Monte Carlo methods to forecast error statistics, *J. Geophys. Res.*, *99*, 10,143–10,162, 1994a.

Evensen, G., Inverse methods and data assimilation in nonlinear ocean models, *Physica D*, *77*, 108–129, 1994b.

Evensen, G., Advanced data assimilation for strongly nonlinear dynamics, *Mon. Weather Rev.*, *125*, 1342–1354, 1997.

Evensen, G., The ensemble Kalman filter: Theoretical formulation and practical implementation, *Ocean Dynamics*, *53*, 343–367, 2003.

Evensen, G., Sampling strategies and square root analysis schemes for the EnKF, *Ocean Dynamics*, *54*, 539–560, 2004.

Evensen, G., and N. Fario, A weak constraint variational inverse for the Lorenz equations using substitution methods, *J. Meteor. Soc. Japan*, *75(1B)*, 229–243, 1997.

Evensen, G., and P. J. van Leeuwen, Assimilation of Geosat altimeter data for the Agulhas current using the ensemble Kalman filter with a quasi-geostrophic model, *Mon. Weather Rev.*, *124*, 85–96, 1996.

Evensen, G., and P. J. van Leeuwen, An ensemble Kalman smoother for nonlinear dynamics, *Mon. Weather Rev.*, *128*, 1852–1867, 2000.

Evensen, G., D. Dee, and J. Schröter, Parameter estimation in dynamical models, in *Ocean Modeling and Parameterizations*, edited by E. P. Chassignet and J. Verron, pp. 373–398, Kluwer Academic Publishers. Printed in the Nederlands., 1998.

Gao, G., and A. C. Reynolds, Quantifying the uncertainty for the PUNQ-S3 problem in a Bayesian setting with the RML and EnKF, SPE reservoir simulation symposium (SPE 93324), 2005.

Gauthier, P., Chaos and quadri-dimensional data assimilation: A study based on the Lorenz model, *Tellus, Ser. A*, *44*, 2–17, 1992.

Gauthier, P., P. Courtier, and P. Moll, Assimilation of simulated wind lidar data with a Kalman filter, *Mon. Weather Rev.*, *121*, 1803–1820, 1993.

Gelb, A., *Applied Optimal Estimation*, MIT Press Cambridge, 1974.

Gradshteyn, I. S., and I. M. Ryzhik, *Table of Integrals, Series, and Products: Corrected and enlarged edition*, Academic Press, Inc., 1979.

Grønnevik, R., and G. Evensen, Application of ensemble based techniques in fish-stock assessment, *Sarsia*, *86*, 517–526, 2001.

Gu, Y., and D. S. Oliver, History matching of the PUNQ-S3 reservoir model using the ensemble Kalman filter, SPE Annual Technical Conference and Exhibition (SPE 89942), 2004.

Hacker, J. P., and C. Snyder, ensemble Kalman filter assimilation of fixed screen-height observations in a parameterized PBL, *Mon. Weather Rev.*, *133*, 3260–3275, 2005.

Hamill, T. M., and C. Snyder, A hybrid ensemble Kalman filter–3D variational analysis scheme, *Mon. Weather Rev.*, *128*, 2905–2919, 2000.

Hamill, T. M., and J. S. Whitaker, Accounting for the error due to unresolved scales in ensemble data assimilation: A comparison of different apporaches, *Mon. Weather Rev.*, *133*, 3132–3147, 2005.

Hamill, T. M., S. L. Mullen, C. Snyder, Z. Toth, and D. P. Baumhefner, Ensemble forecasting in the short to medium range: Report from a workshop, *Bull. Amer. Meteor. Soc.*, *81*, 2653–2664, 2000.

Hamill, T. M., J. S. Whitaker, and C. Snyder, Distance-dependent filtering of background error covariance estimates in an ensemble Kalman filter, *Mon. Weather Rev.*, *129*, 2776–2790, 2001.

Hansen, J. A., and L. A. Smith, Probabilistic noise reduction, *Tellus, Ser. A*, *53*, 585–598, 2001.

Haugen, V. E., and G. Evensen, Assimilation of SLA and SST data into an OGCM for the Indian ocean, *Ocean Dynamics*, *52*, 133–151, 2002.

Haugen, V. E., G. Evensen, and O. M. Johannessen, Indian ocean circulation: An integrated model and remote sensing study, *J. Geophys. Res.*, *107*, 11-1–11-23, 2002.

Heemink, A. W., M. Verlaan, and A. J. Segers, Variance reduced ensemble Kalman filtering, *Mon. Weather Rev.*, *129*, 1718–1728, 2001.

Houtekamer, P. L., and H. L. Mitchell, Data assimilation using an Ensemble Kalman Filter technique, *Mon. Weather Rev.*, *126*, 796–811, 1998.

Houtekamer, P. L., and H. L. Mitchell, Reply, *Mon. Weather Rev.*, *127*, 1378–1379, 1999.

Houtekamer, P. L., and H. L. Mitchell, A sequential ensemble Kalman filter for atmospheric data assimilation, *Mon. Weather Rev.*, *129*, 123–137, 2001.

Houtekamer, P. L., H. L. Mitchell, G. Pellerin, M. Buehner, M. Charron, L. Spacek, and B. Hansen, Atmospheric data assimilation with an ensemble Kalman filter: Results with real observations, *Mon. Weather Rev.*, *133*, 604–620, 2005.

Hunt, B., E. Kalnay, E. Kostelich, E. Ott, D. J. Patil, T. Sauer, I. Szunyogh, J. A. Yorke, and A. V. Zimin, Four dimensional ensemble kalman filtering, *Tellus, Ser. A*, *56A*, 273–277, 2004.

Jazwinski, A. H., *Stochastic Processes and Filtering Theory*, Academic Press, San Diego, Calif., 1970.

Kalman, R. E., A new approach to linear filter and prediction problems, *J. Basic. Eng.*, *82*, 35–45, 1960.

Kepert, J. D., On ensemble representation of the observation-error covariance in the ensemble Kalman filter, *Ocean Dynamics*, *6*, 561–569, 2004.

Keppenne, C. L., Data assimilation into a primitive-equation model with a parallel ensemble Kalman filter, *Mon. Weather Rev.*, *128*, 1971–1981, 2000.

Keppenne, C. L., and M. Rienecker, Assimilation of temperature into an isopycnal ocean general circulation model using a parallel ensemble Kalman filter, *J. Marine. Sys.*, *40-41*, 363–380, 2003.

Kirkpatrick, S., C. D. Gelatt, and M. P. Vecchi, Optimization by simulated annealing, *Science*, *220*, 671–680, 1983.

Kivman, G. A., Sequential parameter estimation for stochastic systems, *Nonlinear Processes in Geophysics*, *10*, 253–259, 2003.

Lantuéjoul, C., *Geostatistical Simulation: Models and Algorithms*, Springer-Verlag, 2002.

Lawson, W. G., and J. A. Hansen, Implications of stochastic and deterministic filters as ensemble-based data assimilation methods in varying regimes of error growth, *Mon. Weather Rev.*, *132*, 1966–1981, 2004.

Leeuwenburgh, O., Assimilation of along-track altimeter data in the Tropical Pacific region of a global OGCM ensemble, *Q. J. R. Meteorol. Soc.*, *131*, 2455–2472, 2005.

Leeuwenburgh, O., G. Evensen, and L. Bertino, The impact of ensemble filter definition on the assimilation of temperature profiles in the Tropical Pacific, *Q. J. R. Meteorol. Soc.*, 2006, to appear.

Lermusiaux, P. F. J., Evolving the subspace of the three-dimensional ocean variability: Massachusetts Bay, *J. Marine. Sys.*, *29*, 385–422, 2001.

Lermusiaux, P. F. J., and A. R. Robinson, Data assimilation via error subspace statistical estimation. Part I: Theory and schemes, *Mon. Weather Rev.*, *127*, 1385–1407, 1999a.

Lermusiaux, P. F. J., and A. R. Robinson, Data assimilation via error subspace statistical estimation. Part II: Middle Atlantic Bight shelfbreak front simulations and ESSE validation, *Mon. Weather Rev.*, *127*, 1408–1432, 1999b.

Lisæter, K. A., J. Rosanova, and G. Evensen, Assimilation of ice concentration in a coupled ice-ocean model, using the ensemble Kalman filter, *Ocean Dynamics*, *53*, 368–388, 2003.

Liu, N., and D. S. Oliver, Critical evaluation of the ensemble Kalman filter on history matching of geologic facies, SPE reservoir simulation symposium (SPE 92867), 2005a.

Liu, N., and D. S. Oliver, Ensemble Kalman filter for automatic history matching of geologic facies, *J. Petrolium Sci. and Eng.*, *47*, 147–161, 2005b.

Lorentzen, R. J., G. Nævdal, and A. C. V. M. Lage, Tuning of parameters in a two-phase flow model using an ensemble Kalman filter, *Int. Jour. of Multiphase Flow*, *29*, 1283–1309, 2003.

Lorentzen, R. J., G. Nævdal, B. Vallés, A. M. Berg, and A.-A. Grimstad, Analysis of the ensemble Kalman filter for estimation of permeability and porosity in reservoir models, SPE 96375, 2005.

Lorenz, E. N., Deterministic nonperiodic flow, *J. Atmos. Sci.*, *20*, 130–141, 1963.

Madsen, H., and R. Cañizares, Comparison of extended and ensemble Kalman filters for data assimilation in coastal area modelling, *Int. J. Numer. Meth. Fluids*, *31*, 961–981, 1999.

Majumdar, S. J., C. H. Bishop, B. J. Etherton, I. Szunyogh, and Z. Toth, Can an ensemble transform Kalman filter predict the reduction in forecast-error variance produced by targeted observations?, *Q. J. R. Meteorol. Soc.*, *127*, 2803–2820, 2001.

McIntosh, P. C., Oceanic data interpolation: Objective analysis and splines, *J. Geophys. Res.*, *95*, 13,529–13,541, 1990.

Metropolis, N., A. W. Rosenbluth, M. N. Rosenbluth, A. H. Teller, and E. Teller, Equation of state calculations by fast computing machines, *J. Chem. Phys.*, *21*, 1087–1092, 1953.

Miller, R. N., Perspectives on advanced data assimilation in strongly nonlinear systems, in *Data Assimilation: Tools for Modelling the Ocean in a Global Change Perspective*, edited by P. P. Brasseur and J. C. J. Nihoul, vol. I 19 of *NATO ASI*, pp. 195–216, Springer-Verlag Berlin Heidelberg, 1994.

Miller, R. N., and L. L. Ehret, Ensemble generation for models of multimodal systems, *Mon. Weather Rev.*, *130*, 2313–2333, 2002.

Miller, R. N., M. Ghil, and F. Gauthiez, Advanced data assimilation in strongly nonlinear dynamical systems, *J. Atmos. Sci.*, *51*, 1037–1056, 1994.

Miller, R. N., E. F. Carter, and S. T. Blue, Data assimilation into nonlinear stochastic models, *Tellus, Ser. A*, *51*, 167–194, 1999.

Mitchell, H. L., and P. L. Houtekamer, An adaptive Ensemble Kalman Filter, *Mon. Weather Rev.*, *128*, 416–433, 2000.

Mitchell, H. L., P. L. Houtekamer, and G. Pellerin, Ensemble size, and model-error representation in an Ensemble Kalman Filter, *Mon. Weather Rev.*, *130*, 2791–2808, 2002.

Moradkhani, H., S. Sorooshian, H. V. Gupta, and P. R. Houser, Dual state-parameter estimation of hydrological models using ensemble Kalman filter, *Advances in Water Resources*, *28*, 135–147, 2005.

Muccino, J. C., and A. F. Bennett, Generalized inversion of the Korteweg–de Vries equation, *Dyn. Atmos. Oceans*, *35*, 227–263, 2001.

Nævdal, G., T. Mannseth, and E. Vefring, Near well reservoir monitoring through ensemble Kalman filter, Proceeding of SPE/DOE Improved Oil recovery Symposium (SPE 84372), 2002.

Nævdal, G., L. M. Johnsen, S. I. Aanonsen, and E. Vefring, Reservoir monitoring and continuous model updating using the ensemble Kalman filter, SPE Annual Technical Conference and Exhibition (SPE 84372), 2003.

Natvik, L. J., and G. Evensen, Assimilation of ocean colour data into a biochemical model of the North Atlantic. Part 1. Data assimilation experiments, *J. Marine. Sys.*, *40-41*, 127–153, 2003a.

Natvik, L. J., and G. Evensen, Assimilation of ocean colour data into a biochemical model of the North Atlantic. Part 2. Statistical analysis, *J. Marine. Sys.*, *40-41*, 155–169, 2003b.

Natvik, L.-J., M. Eknes, and G. Evensen, A weak constraint inverse for a zero dimensional marine ecosystem model, *J. Marine. Sys.*, *28*, 19–44, 2001.

Neal, R. M., Bayesian training of backpropagation networks by the Hybrid Monte Carlo method, *Technical Report CRG-TR-92-1*, Department of Computer Science, University of Toronto, 1992.

Neal, R. M., Probabilistic inference using Markov chain Monte Carlo methods, *Technical Report CRG-TR-93-1*, Department of Computer Science, University of Toronto, 1993.

Nerger, L., W. Hiller, and J. Schröter, A comparison of error subspace Kalman filters, *Tellus*, *57A*, 715–735, 2005.

Nohara, D., and H. Tanaka, Development of prediction model using ensemble forecast assimilation in nonlinear dynamical system, *J. Meteor. Soc. Japan*, *82*, 167–178, 2004.

Ott, E., B. Hunt, I. Szunyogh, A. V. Zimin, E. Kostelich, M. Corazza, E. Kalnay, D. J. Patil, and J. A. Yorke, A local ensemble Kalman filter for atmospheric data assimilation, *Tellus, Ser. A*, *56A*, 415–428, 2004.

Park, J.-H., and A. Kaneko, Assimilation of coastal acoustic tomography data into a barotropic ocean model, *Geophysical Research Letters*, *27*, 3373–3376, 2000.

Pham, D. T., Stochastic methods for sequential data assimilation in strongly nonlinear systems, *Mon. Weather Rev.*, *129*, 1194–1207, 2001.

Pham, D. T., J. Verron, and M. C. Roubaud, A singular evolutive extended Kalman filter for data assimilation in oceanography, *J. Marine. Sys.*, *16*, 323–340, 1998.

Pires, C., R. Vautard, and O. Talagrand, On extending the limits of variational assimilation in nonlinear chaotic systems, *Tellus, Ser. A*, *48*, 96–121, 1996.

Reichle, R. H., D. B. McLaughlin, and D. Entekhabi, Hydrologic data assimilation with the ensemble Kalman filter, *Mon. Weather Rev.*, *130*, 103–114, 2002.

Robert, C. P., and G. Casella, *Monte Carlo Statistical Methods*, Springer Texts in Statistics, second ed., Springer, 2004.

Skjervheim, J.-A., G. Evensen, S. I. Aanonsen, B. O. Ruud, and T. A. Johansen, In corporating 4D seismic data in reservoir simulatuion models using ensemble Kalman filter, SPE 95789, 2005.

Skjervheim, J.-A., S. I. Aanonsen, and G. Evensen, Ensemble Kalman filter with time difference data, *Computational Geosciences*, 2006, submitted.

Snyder, C., and F. Zhang, Assimilation of simulated doppler radar observations with an ensemble Kalman filter, *Mon. Weather Rev.*, *131*, 1663–1677, 2003.

Stensrud, D. J., and J. Bao, Behaviors of variational and nudging assimilation techniques with a chaotic low-order model, *Mon. Weather Rev.*, *120*, 3016–3028, 1992.

Talagrand, O., and P. Courtier, Variational assimilation of meteorological observations with the adjoint vorticity equation. I: Theory, *Q. J. R. Meteorol. Soc.*, *113*, 1311–1328, 1987.

Tippett, M. K., J. L. Anderson, C. H. Bishop, T. M. Hamill, and J. S. Whitaker, Ensemble square-root filters, *Mon. Weather Rev.*, *131*, 1485–1490, 2003.

Torres, R., J. I. Allen, and F. G. Figueiras, Sequential data assimilation in an upwelling influenced estuary, *J. Marine. Sys.*, *60*, 317–329, 2006.

van Leeuwen, P. J., The time mean circulation in the Agulhas region determined with the ensemble smoother, *J. Geophys. Res.*, *104*, 1393–1404, 1999a.

van Leeuwen, P. J., Comment on "Data assimilation using an ensemble Kalman filter technique", *Mon. Weather Rev.*, *127*, 6, 1999b.

van Leeuwen, P. J., An ensemble smoother with error estimates, *Mon. Weather Rev.*, *129*, 709–728, 2001.

van Leeuwen, P. J., A variance-minimizing filter for large-scale applications, *Mon. Weather Rev.*, *131*, 2071–2084, 2003.

van Leeuwen, P. J., and G. Evensen, Data assimilation and inverse methods in terms of a probabilistic formulation, *Mon. Weather Rev.*, *124*, 2898–2913, 1996.

Verlaan, M., and A. W. Heemink, Nonlinearity in data assimilation applications: A practical method for analysis, *Mon. Weather Rev.*, *129*, 1578–1589, 2001.

Wackernagel, H., *Multivariate Geostatistics*, Springer-Verlag, 1998.

Wen, X.-H., and W. H. Chen, Real time reservoir model updating using the ensemble Kalman filter, SPE reservoir simulation symposium (SPE 92991), 2005.

Whitaker, J. S., and T. M. Hamill, Ensemble data assimilation without perturbed observations, *Mon. Weather Rev.*, *130*, 1913–1924, 2002.

Yu, L., and J. J. O'Brien, Variational estimation of the wind stress drag coefficient and the oceanic eddy viscosity profile, *J. Phys. Oceanogr.*, *21*, 709–719, 1991.

Yu, L., and J. J. O'Brien, On the initial condition in parameter estimation, *J. Phys. Oceanogr.*, *22*, 1361–1364, 1992.

Zafari, M., and A. Reynolds, Assessing the uncertainty in reservoir description and performance predictions with the ensemble Kalman filter, SPE 95750, 2005.

Zang, X., and P. Malanotte-Rizzoli, A comparison of assimilation results from the ensemble Kalman filter and a reduced-rank extended Kalman filter, *Nonlinear Processes in Geophysics*, *10*, 477–491, 2003.

Zhang, S., M. J. Harrison, A. T. Wittenberg, A. Rosati, J. L. Anderson, and V. Balaji, Initialization of an ENSO forecast system using a parallelized ensemble filter, *Mon. Weather Rev.*, *133*, 3176–3201, 2005.

Zou, Y., and R. Ghanem, A multiscale data assimilation with the ensemble Kalman filter, *Multiscale Model. Simul.*, *3*, 131–150, 2004.

Index

3DVAR, 256, 262
4DVAR, 38, 73, 75, 112, 257, 262, 263

adjoint
   method, 55, 73, 78, 112, 113, 117
   model, 38, 59, 69, 71, 78, 182
   operator, 72, 111, 113
   representer, 56, 57, 60, 183
   variable, 54, 110, 111, 182
augmented variable, 180, 251, 253, 254, 264
auto-correlation, 83, 84

Bayes' theorem, 6, 7, 15, 95, 97, 99, 101, 103, 141
best estimate, 2, 28, 29, 36, 39, 42, 44, 193, 256
best-guess estimate, 123
bias, 3, 11, 186, 189, 192, 260
Brownian motion, 40, 178

canonical equations, 80
cap-rock, 240
central limit theorem, 12
chaotic dynamics, 36, 72, 73, 75, 236, 256
closure scheme, 33, 36, 38, 43, 44
coloured noise, 106, 177, 184, 185
complex conjugate, 160
conjugate gradient, 68, 69
convolution, 64, 69, 77, 184
correlation, 11
cost function, 52, 75, 78, 96, 112, 139, 141–143, 145, 146, 154, 252

covariance, 9, 10
covariance inflation factor, 257, 263

de-correlation length, 29, 63, 158, 160–162, 175, 210, 224
detailed balance, 80
Dirac delta function, 16, 18, 51, 97, 141
discretization, 55, 76, 98, 104

eigenvalue decomposition, 163, 196, 212, 215
elliptic boundary value problem, 49
ensemble
   average, 37–39, 143, 176, 234, 253
   generation, 157, 160
   member, 144, 164, 165, 167, 199, 233, 243
   smoother, 99, 124, 256
   variance, 169, 177, 199
ensemble optimal interpolation, 235, 255, 260
estimate
   unbiased, 14
estimator
   biased, 12, 185
   unbiased, 2, 10, 13, 14
expected value, 8

faults, 240
Fokker-Planck equation, 40
Fourier
   coefficients, 159, 160
   space, 159, 160
   spectrum, 159, 160

transform, 84, 159, 160, 162
functional inverse, 17, 52, 59

gas-oil contact, 240
Gauss-Markov interpolation, 122, 137
Gaussian anamorphosis, 259
generalized inverse formulation, 47, 95, 101, 103, 107, 114, 117, 182
genetic algorithms, 78, 79, 92, 108, 117
geostatistics, 157
gradient descent, 55, 77, 78, 82, 85, 87, 92, 93, 111, 114, 115, 182, 253

Hamiltonian, 79
Hessian, 34, 76, 139
Hilbert space, 72
history matching, 239
hydro-carbons, 240

ill-posed inverse problem, 63
improved sampling, 164, 167, 195
influence functions, 19, 41, 56, 69, 134, 183
innovation, 21, 125, 127, 250, 253, 257
integration by parts, 54
Ito interpretation, 40

Jacobi matrix, 34

Kalman gain, 25, 42
Kolmogorov's equation, 40, 120, 155
Korteweg-De Vries equation, 116, 260
kriging, 137

lagged smoother, 136
Lagrangian function, 112
Lagrangian multipliers, 112
likelihood function, 7, 15, 97, 103, 141, 142
local analysis, 170, 218, 233, 249, 259, 260
Lorenz equations, 36, 72, 73, 75, 78, 81, 82, 92, 126, 131, 137, 256, 257, 260, 261, 264

Markov chain, 80
Markov Chain Monte Carlo, 9, 40
Markov process, 40, 98–101
matrix
  Hermitian, 215

normal, 215
orthogonal, 164, 165
random orthogonal, 165
maximum likelihood estimate, 1, 3, 15–17, 79, 107, 139, 142, 146, 186, 258
measurement functional, 16, 105, 140, 154, 181
  direct, 51, 96, 110
  nonlinear, 252
Metropolis algorithm, 79, 80, 139
minimum variance estimate, 16
modal trajectory, 107, 126, 193
Monte Carlo, 87, 119–121
  hybrid, 79–81, 92, 93
  method, 38, 79, 164
  sampling, 39, 157
Monte Carlo methods, 39, 44, 203
Moore-Penrose inverse, 209
multivariate analysis, 252, 256, 258

Newton method, 162
normal distribution, 6

over-determined system, 47, 97

parameter estimation, 95, 107, 111–113, 116, 154, 193, 239, 264
particle filter, 121, 257, 265
pdf, 5
  bimodal, 154
  conditional, 6
  Gaussian, 122, 127, 146
  joint, 6, 7, 97, 98, 119, 120, 126, 127, 146, 154, 193
  marginal, 6, 146, 154, 193
  multimodal, 155
  non-Gaussian, 43, 122
  posterior, 6, 7, 100, 107, 146, 154
  prior, 7, 104, 107, 119–122, 137
  unimodal, 154
permeability, 239
pluri-Gaussian simulation, 263
porosity, 239
probability distribution function, 5
pseudo inverse, 208, 209

QR decomposition, 200
quasi-random sampling, 164

regularization, 53, 77, 84, 103, 116
reproducing kernel, 22, 67
reservoir
    oil, 239
    pressure, 240
    simulation model, 239

sample mean, 9
sampling errors, 43, 163, 195
sand, 239
saturation
    gas, 240
    oil, 240
    water, 240
Schur product, 257
sequential processing of measurements, 99, 101, 131, 137
shale, 239
simulated annealing, 80–82, 92
singular value decomposition, 164, 197, 219

stochastic model, 39, 119, 120, 126, 185
Stratonovich interpretation, 40
substitution methods, 76, 78, 93

tangent linear operator, 34, 38, 44, 71, 72, 111, 113
Taylor expansion, 33
transmissibility, 242

unobservable field, 22, 23

variance, 8, 10
variational derivative, 17, 23, 53, 182
variogram, 157
    exponential, 157
    Gaussian, 158
    spherical, 158

water-oil contact, 240
white noise, 73, 75, 175, 185, 186

Printing: Krips bv, Meppel
Binding: Stürtz, Würzburg